SPACE COLONIES

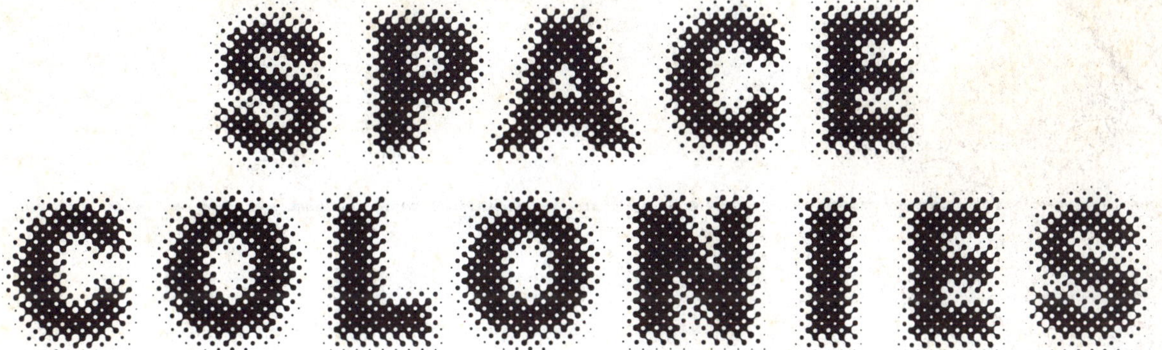

Edited by Stewart Brand

a Penguin Book

Edward S. Curtis

CONTENTS

Vision

4 A CoEvolution Book
5-7 The sky starts at your feet
8-11 **THE HIGH FRONTIER** *by Gerard O'Neill*
12-21 **TESTIMONY OF GERARD K. O'NEILL:** Space Colonization and Energy Supply to Earth
22-30 **"IS THE SURFACE OF A PLANET REALLY THE RIGHT PLACE FOR AN EXPANDING TECHNOLOGICAL CIVILIZATION?"** *Interviewing Gerard O'Neill*
30-32 3-D Universe ■ The High Frontier ■ L-5 News ■ National Space Institute

Debate

33-73 **COMMENTS ON O'NEILL'S SPACE COLONIES**
34 *Ken Kesey* ■ *Russell Schweickart* ■ *Lewis Mumford*
35 *Lynn Margulis* ■ *Richard Raymond* ■ The real backers
36-37 *Dean Fleming* ■ *Wendell Berry*
38-39 *Wilson Clark* ■ Space for peace ■ *E. F. Schumacher* ■ *Stephanie Mills* ■ *Steve Durkee* ■ Extra energy
40-41 *Steve Baer* ■ The SF vote ■ *Dennis Meadows* ■ *Hazel Henderson*
42 Allure ■ Daughter ■ *Carl Sagan* ■ Rather
43 *Paul and Anne Ehrlich* ■ Pioneer ■ Down the road
44-45 *William Irwin Thompson* ■ The rest of it ■ *George Wald* ■ *David Brower*
46-49 *Michael Phillips* ■ *Ant Farm* ■ *John Todd*
50-51 *Brother David Steindl-Rast* ■ Info isn't heavy ■ Planet parenthood ■ *Richard Brautigan* ■ Juvenile space
52 *Peter Warshall* ■ Space virus or child ■ Arab Space Program
53 *Anne Norcia* ■ *Marc LeBrun* ■ Metallurgical paradise ■ Hot Gaia ■ *Erik Eckholm*
54-55 *Garrett Hardin* ■ *R. Buckminster Fuller* ■ Seed
56-60 *Paolo Soleri*
60-61 *Julia Brand* ■ Better foolishness ■ *David Shetzline* ■ Partial atmospheric pressure
62-68 *John Holt* ■ Technical debate
69 Investment space ■ Space war ■ *Gurney Norman* ■ *Gary Snyder* ■ Sufficient error ■ Brazil's space colony
70-71 *Gerard O'Neill*
72-73 *Stewart Brand*

Penguin Books Ltd, Harmondsworth, Middlesex, England
Penguin Books, 625 Madison Ave, New York, NY 10022, USA
Penguin Books Australia Ltd, Ringwood, Victoria, Australia
Penguin Books Canada Ltd, 2801 John St, Markham, Ont, Canada L3R 1B4
Penguin Books (N.Z.) Ltd, 182-190 Wairau Rd, Auckland 10, New Zealand

Copyright © POINT, 1977. All rights reserved. ISBN 0 14 00. 4805 7
Printed in the U.S.A. by Waller Press, San Francisco, California
First printing, September 1977 — 22,000
Second printing, October 1977 — 15,000

Cover *Don Davis' painting of a "Model III" cylindrical Space Colony, a portion of which appears on the front cover, has inspired more belief and roused more ire than any other artifact associated with Space Colonies so far. The man-made idyll is too man-made, too idyllic, or too ecologically unlikely — say the ired. It's a general representation of the natural scale of life attainable in large rotating environment — say the inspired. Either way, it makes people jump.*

The front cover design is by David Wills. The back cover is by the Green Man Cooperative, Taos, New Mexico, S. N. Durkee, prop. It was used as front cover for the Spring 1976 CQ.

— SB (Stewart Brand)

74-81 **SPACE COLONIES SHOULD KEEP AWAY FROM THE GOVERNMENT FOR A WHILE** *Astronaut Russell Schweickart on the phone*

82-85 **THE DEBATE SHARPENS** *Wendell Berry angry*

86-88 "Spaceship Earth" comes home to roost ■ Driving or driven to space

89 Evolution of Space industrialization ■ Satellite surveillance ■ Other voices

90-91 Life spreads before the sun ■ If the Sun Dies

92-93 **ECOLOGICAL CONSIDERATIONS FOR SPACE COLONIES**

94-95 Space agriculture and Space cops ■ Space agriculture retort

96-97 The smile of Timothy Leary ■ Space just means bigger weapons ■ 8 am of the world ■ Limits to Growth — wronger than ever ■ Grandiose failure

98-103 **JACQUES COUSTEAU AT NASA HEADQUARTERS**

104-109 **THE SPACE COLONIES IDEA 1969-1977** *by Eric Drexler*

Space

110-113 **WHO'S EARTH** *by Russell Schweickart*

114-115 **"WHERE!?"** *by Carl Sagan* ■ Steel asteroids ■ A House in Space

116-119 **THERE AIN'T NO GRACEFUL WAY** *Astronaut Russell Schweickart talking to Peter Warshall*

120-127 **CONTROVERSY IS RIFE ON MARS** *interviewing Carl Sagan and Lynn Margulis*

128 Education for Space Work ■ The Directory of Aerospace Education

129 Colonies in Space ■ Space Age Review ■ Earth/Space News

130 Space Settlements ■ Space Manufacturing Facilities ■ Spaceflight & JBIS

131-132 Concise Atlas of the Universe ■ Sky and Telescope

133 **OLBERS' PARADOX** *by Lawrence Ferlinghetti*

134-137 **SOLAR SAILING** *by Eric Drexler* ■ Heliogyro solar sail

138-145 **OPERATING ON THE EDGE** *Russell Schweickart talking*

146-149 **JERRY BROWN — FROM LIMITS ON EARTH TO POSSIBILITIES IN SPACE**

150-154 **INSTEAD OF FRICTIONLESS ELEPHANTS** *talking with Gerard O'Neill*

155 The long view ■ Thanks

156-158 **INDEX**

159 Business

160 Announcement of **The CoEvolution Quarterly**

Editor Stewart Brand [SB]

Office Manager Andrea Sharp

Assistant Editor Anne Herbert

Production Manager Susan Goodrick

Typesetting Evelyn Eldridge, Andrew Main,
Joe Bacon, Mary Fran McCluskey

Cover and logo design David Wills

Paste-up David Wills, Susan Goodrick,
Don Ryan, Candice Kollar,
Diana Fairbanks, Andrew Main,
Carol Kramer

Index Sara Betty Berenson

Illustrations Don Ryan, Tom Parker,
G. Russ Youngreen, Jay Kinney,
Anne Norcia, Larry Todd,
Arthur Okamura, Carol Kramer,
Philip Garner

Proofreading Pam Cokeley, Michael Lewis

Printing Waller Press, San Francisco
Chong Lee, San Francisco
(photo prints)

A CoEvolution Book

Most of this book is Used Information. It is reprinted from various issues of **The CoEvolution Quarterly**, *a California-based peculiar magazine.*

You can look at that news two ways. If you operate by the Bread Model of Information, it's terrible news. You've been gypped — stale information.

On the other hand if you view information as something fundamentally different from bread, there's the possibility of good news. Having lived longer, the information here may be wiser, more co-evolved with the world. It may be more refined, having cycled complexly through the minds and responses of 40,000 **CQ** *readers. And it's been through two editorial distillations; the less-than-wonderful and out-of-date may have been extracted.*

Also, being linked up with a magazine, the book does not stop here. The Space Debate continues in the **CQ**, *and you're invited to participate with comments, item suggestions and reviews, and articles. Send them to* **The CoEvolution Quarterly, Box 428, Sausalito, California 94965.** *We pay $10 - $200 for anything we publish, including letters.*

If you are a regular **CQ** *subscriber you'll be glad to know that about 20% — some 32 pages — of this book is brand new (1977) material which hasn't appeared in the magazine and won't. This includes most of the book reviews, the Governor Jerry Brown material,*

two Eric Drexler articles, and the two most recent interviews with Space Colony designer Gerard O'Neill and astronaut Russell Schweickart.

The book is organized into three sections — Vision, Debate, and Space. The Vision is Gerard O'Neill's domain — progressing from broad propaganda to technical details to anecdotes. The Debate section is probably the most unique to this book since no one else has published the highly intelligent attacks that have been stung into life by the Space Colony idea. And the third section — Space — is natural history, accounts from people such as astronaut Schweickart and space scientist Sagan and politician Brown. Finishing, if you please, on a note of reality.

By the way, "co-evolution" is a term of recent coinage, co-conceived by biologists Paul Ehrlich and Peter Raven to explain something terribly obvious but not before formally recognized about living organisms. They spend most of their adaptive effort getting along with other life which is likewise busily competing, cooperating, and avoiding at them. Life co-evolves with life.

That includes us. So as you study your work, your yard, your watershed, your bio-community and human community, your weather, your access to tools, your night sky, and your prospects in Space, be aware that they are studying you.

— SB

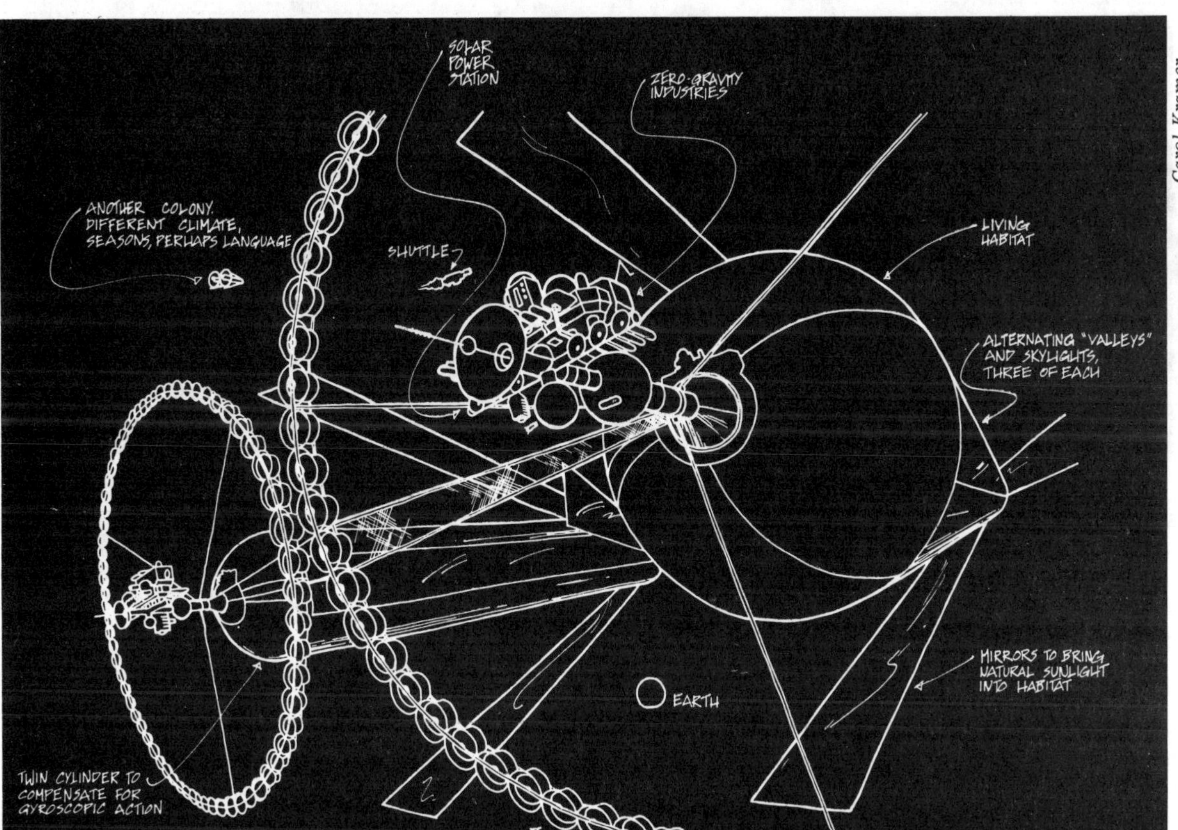

The sky starts at your feet

This book is about how to take Space personally.

Gerard O'Neill's vision of Space Colonies has turned the universe inside out for people. Instead of seeing the space program as a "boondoggle for scientists" (Herman Kahn), suddenly they can see Space as a path, or at least a metaphor, for their own liberation. And those who are critics of high technology — who abound in this book — can leverage their arguments from Space industrialization as the quintessence of what they are fighting. What's new is that people are extrapolating from the future and outside instead of just from the past and inside.

The subject is FREE SPACE. That's the technical term for everyplace outside the Earth's atmosphere. It's a political term — and increasingly, as exploration and argument proceed into orbit, a political reality.

Speaking of terms, the use of the term "Space Colony" has been expressly forbidden by the U.S. State Department because of anti-colonial feelings around the world. So NASA has shrugged and adopted "Space Settlements" — unpoetic terminology since the last thing you do in Space is settle. We're sticking to "Space Colonies". It's more accurate; this time there's a difference in that no Space natives are being colonized; and the term reminds us of things that went badly and went well in previous colonizations.

If we're lucky we may enact a parallel with what happened in Europe when America was being colonized. Intellectual ferment — new lands meant new possibilities; new possibilities meant new ideas. If you can try things you think up things to try.

O'Neill's scheme invites you to give your imagination a Space Colony (see sketch above) of one million inhabitants, each of whom has five acres of "land". Believe that it is readily possible — maybe inevitable — by 2000 AD. Have you any thoughts about how to organize its economy, politics, weather, land use, education, culture? Any thoughts about how to organize your life to get there?

O'Neill invites you to imagine an inside-out planet, cylindrical, with at the end caps mountain ranges which have the interesting property that as you climb higher your weight decreases. Near the top (center of rotation) at .1 g (1/10 Earth gravity) you can don wings and take flight. Or take a long slow

> **The sky starts at your feet. Think how brave you are to walk around.**
> — *Anne Herbert*

dive into a swimming pool. Or watch someone else's slow-motion splash. At the foot of the mountain you might have a round river allowing you to canoe downstream several miles past the other two "valleys" and back to your home.

The details of the design and of the speculation change constantly, but O'Neill invites anyone to challenge the overall scale, engineering, budget, and schedule of the project. If you can find the fatal flaw you could bring this nonsense to a stop. Or, failing that, participate in the design and imagination of how Space Colonies might, in fact, work.

One thing that impresses me about the Space environment is that, hostile as it is to us pulpy organisms, it is wholly benign for electronic and mechanical machinery, much better for them than this corrosive, weighty Earth's surface. An engineering friend of mine, Michael Callahan, used to speculate that the machines have been longing for years to get into Space. They're using us to get there and when they've

Ocean, ocean, you'll get me yet.
— Ken Kesey

succeeded they'll throw us away. Or, maybe they'll give us something wonderful we don't even know we need. In whatever philosophical and technical configuration they are, we shall be obliged to rely upon machines to make Space habitable for us.

What got me interested in Space Colonies a few years ago was a chance remark by a grade-school teacher. She said that most of her kids expected to live in Space. All their lives they'd been seeing "Star Trek" and American and Russian Space activities and drew the obvious conclusions.

Suddenly I felt out-of-it. A generation that grew up _with_ Space, I realized, was going to lead to another generation growing up in Space. Where did that leave me?

For these kids there's been a change in scope. They can hold the oceans of the world comfortably in their minds, like large lakes. Space is the ocean now.

According to a Navy man (a Commander J. Henry Glazer) I was listening to last night in Governor Brown's office in Sacramento, it is _naval_ experience that is going to best inform the navigation, construction, command regimes, and life-on-board of _spaceships_ — once we get past the brief airplane-like period of the space shuttle. Astronaut Rusty Schweickart swallowed his Air Force loyalty and agreed.

And for those who long for the harshest freedoms, or who believe with Buckminster Fuller that a culture's creativity requires an Outlaw Area, Free Space becomes what the oceans have ceased to be — Outlaw Area too big and dilute for national control.

What's in it for Earth, then? Well, say the most dogmatic Space Colony proponents, you could solve,

Wrong attitude

Larry Todd

in order, the Energy Crisis, the Food Crisis, the Arms Race, the Population Problem, and maybe even the Climatic Shift.

Liberals and environmentalists hoot in derision. Then a year later some of them are back for a second look. And some of those, accurately perceiving possible benefits and possible frightful hazards of Space colonization, begin to participate in the debate and design.

Whenever the universe turns inside-out, as happens from time to time in any civilization, you get a lot of disruption and confusion, but you also can get a fresh angle on old problems, public and personal.

As a nomad once told me, "Think for a while about cows and fences and grazing. It's not just in your mind — the grass IS greener on the other side of the fence."

— SB

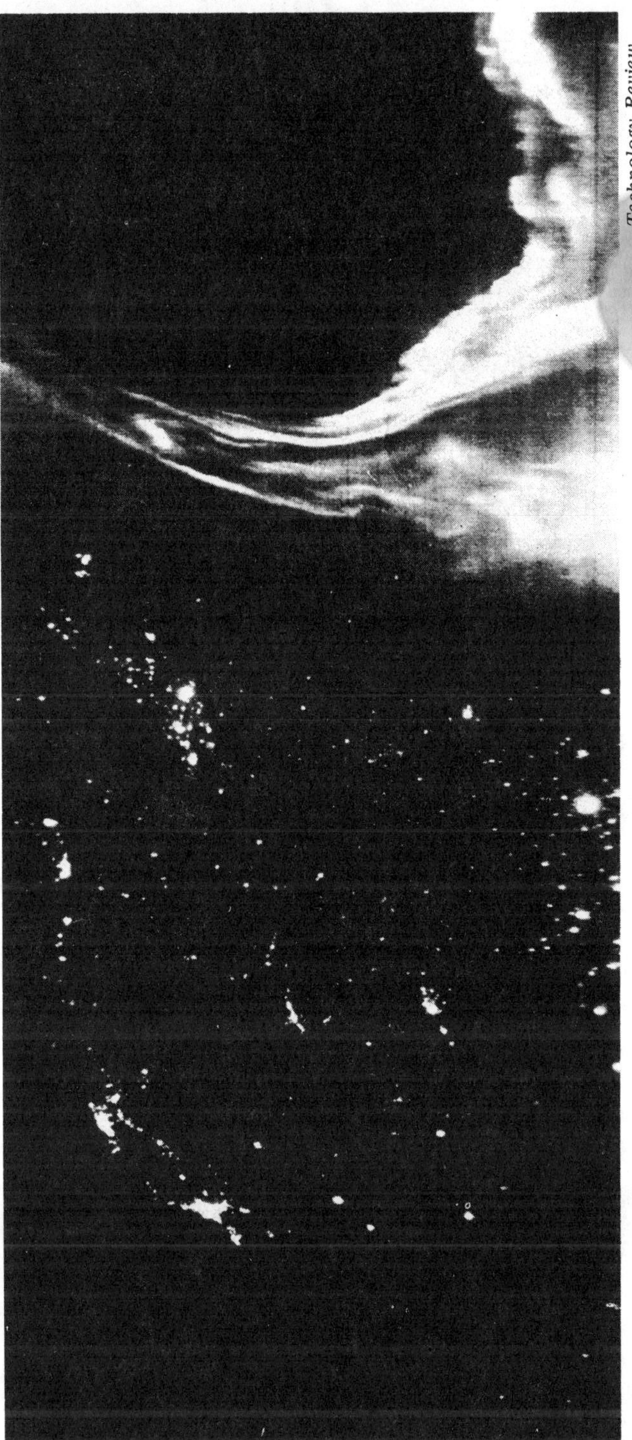

Technology Review

"*Northern lights*" *over North American photographed from an Air Force satellite January 5, 1973. The human population centers show up clearly in this infrared photo — the two bright areas at lower left are Los Angeles and the San Francisco Bay Area, the bright area at upper right is St. Paul-Minneapolis.*

Nothing looks the same from Space. If you live in a satellite, the Earth is something that goes on in your sky.

The High Frontier

BY GERARD O'NEILL

During the past decade a number of premises about the basic problems of the world have become very widely accepted. The more important of these accepted ideas are:

    1) That for the forseeable future every significant human activity must be confined to the surface of the earth.

    2) That the material and energy resources of the human race are just those of our planet.

    3) That any realistic solutions to our problems of food, population, energy and materials must be based on a kind of zero-sum game, in which no resources can be obtained by one nation or group without being taken from another.

Given those premises, logic has driven most observers to the conclusion that long-term peace and stability can only be reached by some kind of systematic global arrangement, with tight constraints to insure the sharing, equable or otherwise, of the limited resources available. I find it personally shocking that many such observers, even those who profess to a deep concern for humankind, accept with equanimity the need for massive starvation, war or disease as necessary precursors to the achievement of such a systematic global arrangement.

In what follows, I will deliberately depart from my usual style. I will <u>not</u> hedge all my statements with cautious limits and buttress them with footnotes, as I would before a scientific audience or as I certainly shall when I testify before a Congressional subcommittee a few weeks from now. Rather, I will be assertive in style, so as to make clear by its shock-value how fundamentally different one new concept is.

If the studies which we have carried out at Princeton University continue to survive technical review, then I must tell you that in my opinion the three basic premises on which most discussions of the future have been based are simply wrong. The human race stands now on the threshold of a new frontier, whose richness surpasses a thousand fold that of the new western world of five hundred years ago.

That frontier can be exploited for all of humanity, and its ultimate extent is a land area many thousands of times that of the entire Earth. As little as ten years ago we lacked the technical capability to exploit that frontier. Now we have that capability, and if we have the willpower to use it we can not only benefit all humankind, but also spare our threatened planet and permit its recovery from the ravages of the industrial revolution.

These statements may sound like empty rhetoric. In the next few minutes I would like to sketch for you how they can be proven to be true. It is not necessary to have a technical background to appreciate these facts. Indeed, one of the most

Dr. Gerard O'Neill, 50, is a high-energy physicist best known in his field for originating the colliding-beam storage ring, which has been adopted in nuclear accelerators throughout the world. Since 1974 he has become better known to the general public as the designer and promoter of very large scale Space Colonies.

He is a professor at Princeton (though this year, 1977, he's teaching at MIT), a former Navy non-com, and a holder of the International Diamond Badge for soaring (about 1% of glider pilots have one).

This talk was given at the World Future Society convocation in Washington D.C. in Spring, 1975. It was perhaps the least well-attended of the hundred presentations at the conference. Futurists were more interested in problems than solutions that year. O'Neill's remarks that day converted me from mild interest in the Space Colonies to obsession.

—SB

Figure 3. Possible interior design of a first, small-size space community. It could be large enough to provide comfortable apartments, shops, parks, small rivers and lush vegetation.

surprising aspects of the new opportunities is that they do not require new technology for their realization.

The high frontier which I will describe is space, but not in the sense of the Apollo program, a massive effort whose main lasting results were scientific. Nor is it space in the sense of the communications and observation satellites, useful as they are. Least of all is it space in the sense of science-fiction, in which harsh planetary surfaces were tamed by space-suited daredevils. Rather, it is a frontier of new lands, located only a few days travel time away from the Earth, and built from materials and energy available in space.

These are the facts which force a revolution in our thinking:

1) Solar energy: as everyone knows, the Sun is a virtually inexhaustible source of clean energy. It is difficult to use on Earth as more than a small supplement to other sources, though, for two reasons:

a) Unreliability: though solar energy is available full time in space, on Earth it is cut off by nighttime, by seasonal variation in the day-length, and by clouds.

b) Low average intensity: the cost of any solar power installation is the amortization cost of the equipment, because the source is free. The amount of solar energy which flows unused, in a year, through each square meter of free space is ten times as much as falls on an equal area in even the most cloud-free portions of Arizona or New Mexico. A given solar-energy installation in space, therefore, is potentially able to operate at a tenth the cost at which it could operate on Earth.

2) Materials: If we build new lands in space, starting from the Earth, we are the "gravitationally disadvantaged." We are at the bottom of a gravitational well 4000 miles deep, from which materials can only be lifted into space at great cost. Our technique must exploit the fact that the Moon has a gravitational well only 1/20 as deep, and as we now know from the Apollo samples is a rich source of metals, glass, oxygen, and soil. In the long run, we can use the fact that the asteroids are also a source of materials: the three largest asteroids alone contain

9

enough materials for the construction of new lands with a total area many thousands of times as large as that of the Earth.

Briefly I will describe for you first the long-term then the immediate possibilities on this new frontier. As I do, remember that everything I describe is well within the limits of present-day, conventional materials, and of present technology. If we were to start now, with determination and drive, in my opinion the first space community could be in place, with its productive capacity benefiting the Earth, before 1990. Other people who have made such estimates put the date about a decade later, but the surprising fact is that the agreement is that close. The reason is that the job requires straightforward engineering, not any basically new science — nothing as new and advanced as hydrogen fusion power, for example. To estimate conservatively what sort of habitat might be practical in the long term, perhaps 40 to 100 years from now, I'll sketch a space colony which is about as big as may be practical using present-day materials.

Recently the NASA/Ames laboratory completed a painting of the exterior of such a colony: note industry sites. A colony would be big enough to model some of the most desirable areas of the Earth. A portion of the island of Bermuda, or a section of the California coast like Carmel could be easily fit within one of the "valleys" of a Model III colony.

The date of realization of colonies of <u>that size</u> does not depend on materials or engineering — those we have already. Rather, it depends on a balance between productivity, a rising living standard and the economies possible with automation. Under the space-colony conditions of virtually unlimited energy and materials resources, a continually rising real income for all colonists is possible — a continuation rather than the arrest of the industrial revolution. Reasonable estimates of three per-cent per year for the real income rise, 8% for interest costs and 10% per year for automation advances put the crossover date (the date when <u>large</u> colonies become economically feasible) about 40 to 50 years from now — well within the lifetimes of most of the people who are now alive.

Colony construction is a bootstrap process, in which one starts small, the first small colony builds the next larger, and so on. The first space community might house about 10,000 people. My coworkers and I sometimes call it Model 1, or Island One. We think it could be built in 15 to 25 years, at a cost per year probably not much higher than that of Project Apollo, and not more than one or two tenths of the annual cost of Project Independence. To be realized, of course, its immediate economic return must be far higher, because of interest amortization

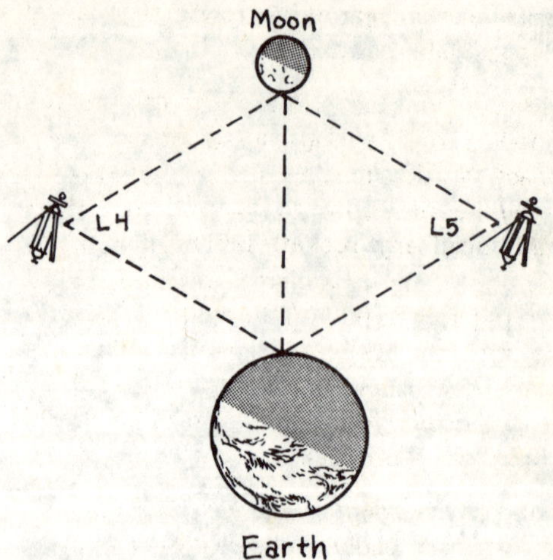

Figure 1. Location of the Lagrange points L4 and L5. Each is on the orbit of the Moon and is the third point of an equilateral triangle, the Earth and Moon being the other two points. Space communities could be located on stable orbits about either L4 or L5.

with discounted economics.

Island One will need careful design, with spot shielding for cosmic-ray protection. It may run in an enriched oxygen atmosphere. At the L5 Earth-Moon Lagrange libration point, though small, Island One could be a far more attractive environment for living than most of the world's population now experiences.

To obtain from an investment not much larger than that of Project Apollo a return many thousands of times as great, we need a trick: the trick is to exploit the low gravity and vacuum environment of the Moon, to obtain from the lunar surface 98% of the material needed for Island One. For the entire construction of Island One the excavation left on the moon will be only 7 X 200 X 200 yards.

The machine to transport the lunar material is called a mass driver; it exists only on paper, but it can be designed and built with complete assurance of success because it requires no high-strength materials, no high accelerations or temperatures, and its principles are fully understood. Running only 25% of the time, the mass driver could lift 500,000 tons of material to L5 in the 6 year construction time of Island One. An identical machine, located in space, could be a very effective reaction motor for the shifting of heavy payloads, in the 100,000 ton range.

Once Island One is completed, it can become a manufacturing facility with a unique advantage, <u>for those products</u> whose end-use is in high orbit above the Earth. The economic payoff from Island One can be estimated by noting that a typical industry on Earth yields about 20 tons of finished products per man-year of labor, with a dollar value averaging a dollar per pound.

The lift cost from Earth to L5 or to geosynchronous orbit now averages a thousand dollars per pound. The space shuttle and simplified derivatives of it can bring that cost down to 300-400 $/pound. Very advanced large rockets, taking many years and tens of billions of dollars to develop, might reduce that cost to 50-100 $/pound. Taking for conservatism the lowest of those rates, the productivity of Island One would be worth 20 billion dollars per year, in addition to the intrinsic value of goods it produces.

That might be academic were it not that there exists a product, badly needed, whose end-use is in high orbit: the product is satellite solar power. For several years design groups at Boeing Aircraft and at the Arthur D. Little Co. have studied the concept of locating large solar power stations in geosynchronous orbit, where sunlight is available 99% of the time, to beam down microwave power to the earth for conversion to ordinary AC or DC power. Oddly enough, the microwave link problem appears solved in principle: 56% efficiency has been demonstrated, and the goal is only 63% to 70%: the stumbling block to realization has just been the lift costs.

By using Island One as a manufacturing facility, it appears that the barrier can be broken and economical solar power achieved. The key point is that the use of lunar materials eliminates the need for incurring the high-lift costs from the Earth, and so appears capable of giving solar electric power on Earth at rates initially competitive, and eventually much lower than coal-fired or nuclear power plants. This is of prime importance for world peace, because the energy source is inexhaustible, and these power stations can be built for any nation that needs them.

If exploited, this approach would also eliminate the need for developing breeder reactors, whose pluto-nium production would always be subject to theft and use by terrorists.

It should also be emphasized that the provision of unlimited low-cost energy to the developing nations will probably be the most effective contribution we could make to solving the world's food problem, because the cost of chemicals for high-yield agriculture is almost entirely the cost of energy for their production.

Throughout this discussion you may have noticed that I have been careful to avoid prophecy. I am pointing out a realistic possibility: the drivers to make it happen are there: immediate jobs, of the high-technology kind which economic studies have shown generate wealth throughout an economy; the direct payback of a much-needed, marketable product. Evidently deeper human needs are also driving this development, because of all the correspondence I receive on this subject only 1% is in opposition to it, and only another 1% is in any way irrational or unintelligent. Yet I do not say it will happen, or when. I hope it will, and soon.

While carefully avoiding prophecy, in closing I would like to show the potential power of the space-colonization approach by showing what could happen if the high frontier were to be explored and exploited by all of humankind: if the construction of new lands in space were to take place on the fastest possible time-scale.

If that were to happen, by about 2018 emigration to better land, better living conditions, better job opportunities and greater freedom of choice and opportunity in small scale, eventually independent communities could even become a viable option for more people than the population increase rate. That time is less than forty-five years away.

If the new option is taken, it would be naive to assume that its benefits will be initially shared equably among all of humankind. The world has never worked that way, and since people do not change there is no reason to suppose that it will work that way in this case. But the resources of space are so great that even nations which achieve only after a long delay the ability to use them will still find an abundance remaining. There are in my opinion at least five or six nations or groups of nations which possess the technical and economic ability to carry out the construction of Island One, the first real beachhead in space, on their own. Some one of these nations or groups may get there first, or — in my opinion preferably — they may do so together. It seems to me that we have in this case the opportunity for a cooperative international program which could have a real stabilizing effect on world tensions; and, knowing that the resources of space are so great, we who may be among those first to exploit them can well afford to provide for our less advantaged fellow humans the initial boost that will permit their exploiting these new resources for themselves. Suddenly given a new world market of several hundred billion dollars per year, the first group of nations to build space manufacturing facilities could well afford to divert some fraction of the new profits to providing low-cost energy to nations poor in mineral resources, and to assisting underdeveloped nations by providing them with initial space colonies of their own.

If we use our intelligence and our concern for our fellow human beings in this way, we can, without any sacrifice on our own part, make the next decades a time not of despair but of fulfilled hope, of excite-ment, and of new opportunity. ■

Space Colonization and
Energy Supply to the Earth

TESTIMONY

OF

DR. GERARD K. O'NEILL

**BEFORE THE
SUB-COMMITTEE ON
SPACE SCIENCE AND APPLICATIONS
OF THE
COMMITTEE ON SCIENCE AND TECHNOLOGY
UNITED STATES HOUSE OF REPRESENTATIVES
JULY 23, 1975**

INTRODUCTION

Within the past year a new possibility for the direction and motivation of our thrust into space has reached the stage of public discussion. It is called space colonization, or the development of space manufacturing facilities. Our present American leadership in space technology gives us a unique opportunity to play a central role in that new development, if we act with decision and speed.

The central ideas of space colonization are:

1) To establish a highly-industrialized, self-maintaining human community in free space, at a location along the orbit of the moon called L5 (*Figure 1*), where free solar energy is available full time.

2) To construct that community on a short time scale, without depending on rocket engines any more advanced than those of the space shuttle.

3) To reduce the costs greatly by obtaining nearly all of the construction materials from the surface of the moon.

4) At the space community, to process lunar surface raw materials into metals, ceramics, glass and oxygen for the construction of both additional communities and of products such as satellite solar power stations. The power stations would be relocated in synchronous orbit about the earth, to supply the earth with electrical energy by low-density microwave beams.

5) Throughout the program, to rely only on those technologies which are available at the time, while recognizing and supporting the development of more advanced technologies if their benefits are clear.

THE SPACE COLONY CONCEPT

Although it has precursors in the works of many authors, the modern idea of space colonies originated from several questions, posed six years ago as an academic exercise:

1) Is it possible, within the limits of 1970's technology, using only the ordinary construction materials with which we are already familiar, to build communities in free space rather than on a planetary surface like the earth, the moon, or Mars?

2) Can these communities be large enough, and sufficiently earth-like, to be attractive to live in; small worlds of their own rather than simply space stations?

3) Would such colonies have unique advantages from an economic viewpoint, so that they could justify the costs of their construction and contribute in a productive way to the total human community?

4) If such colonies were built, would their further development be such as to relieve the earth of further exploitation by the industrial revolution, and to open up a new frontier to challenge the best and highest aspirations of the human race?

Surprisingly, six years of continued research has confirmed, in even more increasing detail, that the answer to all four of these questions is a strong "yes."

GEOMETRIES

The largest colonies now foreseeable would probably be formed as cylinders, alternating areas of glass and interior land areas. From those land areas a resident would see a reflected image of the ordinary disc of the sun in the sky, and the sun's image would move across the sky from dawn to dusk as it does on Earth. Within civil engineering limits no greater than those under which our terrestrial bridges and buildings are built, the land area of one cylinder could be as large as 100 square miles. Even a colony of smaller dimensions could be quite attractive.

Rotation of the cylinder would produce earth-normal gravity inside *(Figure 3)*, and the atmosphere enclosed could have the oxygen content of air at sea-level on earth. The residents would be able to choose and control their climate and seasons.

Agriculture for a space community would be carried out in external cylinders or rings, with atmospheres, temperatures, humidity and day-length chosen to match exactly the needs of each type of crop being grown. Because sunshine in free space is available 24 hours per day for 12 months of the year, and because care would be taken not to introduce into the agricultural cylinders the insect pests which have evolved over millennia to attack our crops, agriculture in space could be efficient and predictable, free of the extremes of crop-failure and glut which the terrestrial environment forces on our farmers.

INDUSTRY

Non-polluting light industry would probably be carried on within the cylindrical living-habitat, convenient to homes and shops. Heavy industry, though, could benefit from the convenience of zero gravity. Through an avenue on the axis of the cylinder, workers in heavy industry could easily reach external, non-rotating factories *(Figure 4 — back cover)*, where zero gravity and breathable atmospheres would permit the easy assembly, without cranes, lift-trucks or other handling equipment, of very large, massive products. These products could be the components of new colonies, radio and optical telescopes, large ships for the further human exploration of the solar system, and power plants to supply energy for the earth.

12

Gerard O'Neill (left) and model of Space Colony before the sub-committee.

LIMITS OF GROWTH

In the early years of this research, before the question of implementation was seriously addressed, it seemed wise to check whether an expansion into space would soon encounter "growth limits" of the kind which humankind is now reaching on earth, and which have been vividly described for us by Professor Jay Forrester of Massachusetts Institute of Technology, in studies supported by the Club of Rome.

If the space colonization program is begun, its technical and economic imperatives seem likely to drive it rather quickly toward the exploitation of asteroidal rather than lunar materials. Long before the results of mining activity on the moon became visible from the earth, the colony program would be obtaining its materials from the asteroids. Given that source, the "limits of growth" are absurdly high: the total quantity of materials within only a few known large asteroids is quite enough to permit building space-colonies with a total land area more than ten thousand times that of the earth.

ENERGY WITHOUT GUILT

The efficiencies of a space community, regarded as an island of a technological human civilization, stem from the abundance and full-time dependability of free solar energy in that environment, and from the possibility of controlling the effective gravity, over a wide range from zero to more than earth-normal, by rotation. In contrast, industrial operations on earth are shackled by a strong gravity which can never be "turned off"; those on the moon would be similarly limited, although the limit would be lower.

In a space colony, the basic human activities of living and recreation, of agriculture, and of industry could all be separated and non-interfering, each with its optimal gravity, temperature, climate, sunlight and atmosphere, but could be located conveniently near to each other. Energy for agriculture would be used directly in the form of sunlight, interrupted at will by large, very low-mass aluminum shades located in zero gravity in space near the farming areas. The day-length and seasonal cycle would therefore be controllable independently for each crop.

Process heat for industry would be obtained with similar economy; in space, temperatures of up to several thousand degrees would be obtainable at low cost, simply by the use of low-mass aluminum-foil mirrors to concentrate the ever-present sunlight. In space, a passive aluminum mirror with a mass of less than a ton and a dimension of about 100 meters, could collect and concentrate, in the course of a year, an amount of solar energy which on earth would cost over a million dollars at standard electricity busbar rates.

Electrical energy for a space community could be obtained at low cost, within the limits of right-now technology, by a system consisting of a concentrating mirror, a boiler, a conventional turbogenerator and a radiator, discarding waste heat to the cold of outer space (*Figure 5*). It appears that in the environment of a space community residents could enjoy a per capita usage of energy many times larger even than what is now common in the United States, but could do so with none of the guilt which is now connected with the depletion of an exhaustible resource.

THE BOOTSTRAP METHOD

Until recently, it had been assumed that the only practical way to locate or assemble an object in a high orbit was to build it or its components on earth, and then to lift it out of the earth's gravity, through the atmosphere, by rockets. One might fairly call this the "brute force" method. In space colonization, we would like to use a far more economical alternative, a kind of "end run" instead of a power play through the middle. It is outlined in *Figure 6*.

Here on the surface of the earth we are at a very low point in the gravitational map of the solar system. In energy terms, we are at the bottom of a gravitational well which is 4,000 miles deep. This is reflected in the fact that we must accelerate a spacecraft to a speed of more than 25,000 miles per hour before it can escape the earth's gravity and go as far as lunar orbit. In a sense, we are the "gravitationally disadvantaged."

We are fortunate that we have another source of materials, which lies at a much shallower point in the gravitational map of the solar system. The energy required to bring materials from the moon to free space is only 1/20 as much as from the earth. Further, the moon has no atmosphere: a disadvantage if we wanted to live there, but a great advantage if we want to obtain from the moon materials at low cost. On the moon we could assemble a launching device for the acceleration to escape velocity of lunar surface raw materials. Such a machine does not require high-strength or high-temperature materials, and the methods for building it are well understood. One design of that kind is called a mass-driver (*Figure 7*): it would be a linear electric motor, forming a thin line several miles long, which would accelerate small 10-pound vehicles we call buckets. At lunar escape speed the bucket would release its payload, and would then return on a side track for reuse. Only the payload would leave the mass-driver, so nothing expensive would be thrown away. The mass-driver would be an efficient machine, driven by a solar-powered or nuclear electric plant, and our calculations show that in six years of time it could launch to escape distance from 300 to 1000 times its own mass. A collector at escape distance from the moon would accumulate materials, and there, with the full solar energy of free space, they would be processed to form the metals, glass and soil of the first space community.

With the help of that economy measure, the mass lifted from the earth need only be a few percent of the mass of the colony itself. We would have to bring the components of the mass-driver and of a lunar outpost (*Figure 8*), components of a construction station in lunar orbit for the processing and assembly of materials, and those elements, mainly carbon, nitrogen and hydrogen, which are rare on the moon. By so avoiding the need for prior development of advanced high-capacity lift vehicles, we could also carry out the construction of the first colony on a fast time scale, possibly beginning as early as 1980-82 when the space shuttle will come into operation. For the lifting of freight to low orbit, we would need one new vehicle, of a type which the aerospace experts call a "dumb booster": a freight rocket based on the same type of engines already developed for the shuttle. For operations in space above low orbit a chemical tug would be sufficient. My recommendation would therefore be strongly supportive of a recently-initiated NASA study of the design of a shuttle-derived heavy-lift vehicle, and of a chemical tug whose segments could be lifted to orbit by the shuttle.

In this approach, we would establish a productive beachhead in space as early as possible, and as the resulting traffic increased would let its revenue assist in paying for the further development of more advanced launch vehicles.

LUNAR MATERIALS

At the time of the Apollo project we did not think of the moon as a resource base. The moon landings, originally motivated by national pride and a sense of adventure, became scientific expeditions and as such returned a high payoff in knowledge.

Now, though, it is time to cash in on Apollo. It was impossible to plan in a rational way a program of space colonization until the Apollo lunar samples were returned for analysis. From those samples we now have the analyses of the lunar soil and rock. *Table 1* summarizes representative data from soils at the Apollo II landing site:

TABLE 1

UNSELECTED APOLLO II SOIL SAMPLE

Oxygen	40%	Titanium	5.9%
Silicon	19.2%	Aluminum	5.6%
Iron	14.3%	Magnesium	4.5%
Calcium	8.0%		

This unselected sample is more than 30% metals by weight.

The baseline mass-driver would be capable of transferring from the moon from 1/2 million to 2 million tons of such materials within a six-year period: that is, from 28,000 to over 100,000 tons of aluminum, 70,000 to 280,000 tons of iron, and corresponding amounts of the other lunar materials. Strangely, though the lunar surface is devoid of life, its most abundant element is the one which we need in every breath we take: oxygen. That oxygen, transported to free space and unlocked from its binding metals by solar energy, would be usable not only for an atmosphere but to fuel rocket engines, reducing by 85% the requirement for fuel carried from the earth.

The lunar surface materials are poor in carbon, nitrogen and hydrogen; in the early years of space colonization these elements would have to be brought from earth. They would be reused, not thrown away. For every ton of hydrogen brought from earth, nine tons of water could be made at the colony site, the remaining eight tons being oxygen from the processing of lunar oxides.

The removal of half a million tons of material from the surface of the moon sounds like a large-scale mining operation, but it is not. The excavation left on the moon would be only 5 yards deep, and 200 yards long and wide: not even enough to keep one small bulldozer occupied for a five-year period.

A few years after the first space community is built we can expect that transport of asteroidal materials to L5 will

Brayton Cycle Schematic

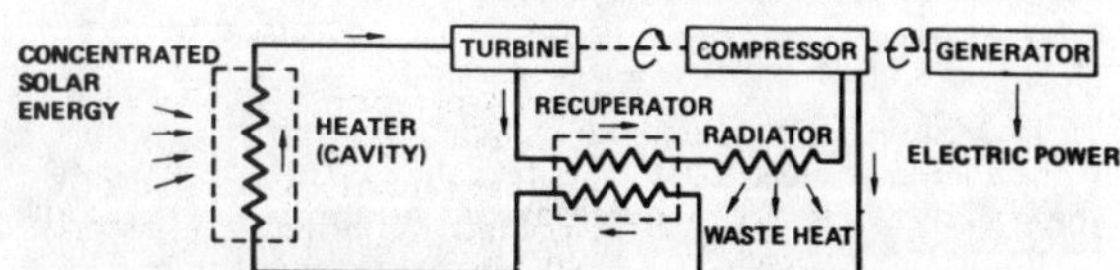

5. Schematic of a closed-cycle turbogenerator using helium as a working gas. A large fossil-fuel power plant using this kind of turbine is now being installed for commercial power generation at Oberhausen, in West Germany.

become practical. No great technical advance is required for that transition; the energy-interval between the asteroids and L5 is only about as great as between the earth and L5. Once the asteroidal resources are tapped, we should have not only metals, glass and ceramics, but also carbon, nitrogen and hydrogen. These three elements, scarce on the moon, are believed to be abundant in the type of asteroid known as carbonaceous chondritic. Therefore I add my support to those who for several years have been recommending an unmanned rendezvous-probe mission to a selected asteroid. Such a mission has already been studied in detail by NASA, and is well within present technical feasibility. If conducted in the late 1970's or early 1980's, with the aim of assaying a carbonaceous chondritic asteroid for its C, N, H content, such a mission would serve the same function that oil well prospecting now serves on earth: the finding and proving of necessary resources for subsequent practical use.

ISLAND ONE

The first space community will be economically productive only if talented, hard-working people choose to live in it, either permanently or for periods of several years. It must therefore be much more than a space-station; it must be as earth-like as possible, rich in green growing plants, animals, birds, and the other desirable features of attractive regions on earth.

Within the materials limits of ordinary civil engineering practice, and within an overall mass budget of 1/2 million tons (about the same as the mass of a super-tanker), several designs for this first "Island in Space" have evolved. One such geometry is shown in *Figure 3*; I am indebted to Field Enterprises, Inc. for permission to show this figure, which is from the 1976 edition of **Science Year**.

All of the geometries we have studied are pressure vessels, spherical, cylindrical or toroidal, containing atmospheres and rotating slowly to provide a gravity as strong as that of the earth. With gravity, good long-term health can be maintained; the colonists should experience none of the bone-calcium loss suffered by the Skylab astronauts in their zero-gravity, non-rotating environment.

Physiology experiments in rotating rooms on earth indicate that humans can acclimatize to quite high rotation rates, some to as much as one rotation every six seconds. A fraction of the space-community population will, though, "commute" daily between the rotating earth-gravity environment and zero or low-gravity work areas. We must therefore hold the rotation rate to a rather low value, to avoid inner-ear disturbances. It is quite possible that our lack of information is forcing us toward unnecessary conservatism on this point. It would be quite useful to carry out long-term physiology experiments during the space-shuttle program, to examine rotation effects in the space environment. On earth our simulation of these effects can never be more than approximate.

Conservatism on this requirement has, though, led us quite recently to a new and possibly more attractive alternative design (*Figures 10, 11*). It allows for natural sunshine, a hillside terraced environment, considerable bodies of water for swimming and boating, and an overall population density characteristic of some quite attractive modern communities in the U.S. and in southern France.

It is startling to realize that even the first-model space-community could have a population of 10,000 people, and that its circumference could be more than one mile. From the valley area, where as in *Figure 3* streams could flow, a ten-minute walk could bring a resident up the hill to a region of much-reduced gravity, where human-powered flight would be easy, sports and ballet could take on a new dimension, and weight would almost disappear. It seems almost a certainty that at such a level a person with a serious heart condition could live far longer than on earth, and that low gravity could greatly ease many of the health problems of advancing age. In *Figures 10* and *11*, the outer ring is a toroidal volume used for agriculture. It too would rotate to provide earth-gravity, but more slowly; its rotation would compensate for the gyroscopic action of the main living habitat, and permit the axis of the habitat always to point toward the sun.

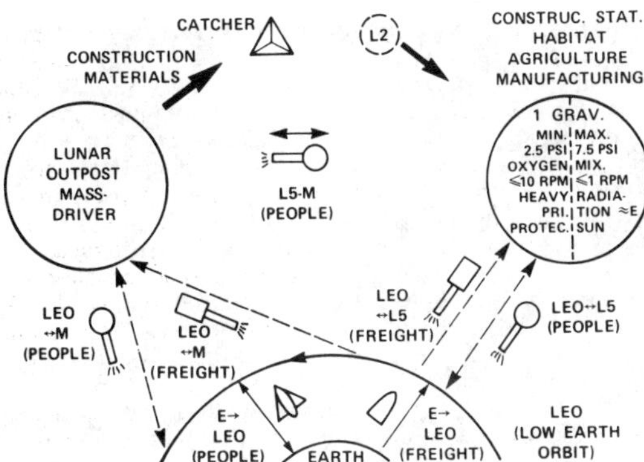

6. Schematic of transportation flow for space colonization. With the space-shuttle and simpler vehicles easily derived from it at low development cost, a lunar mining outpost and an L5 construction-station would be set up. A large fraction (over 95%) of all materials for colony-construction and later manufacturing would then be obtained from the lunar surface by an automated, high-efficiency launch system.

Just beyond the hemispherical ends, a few minutes from the residential areas, there could be large assembly areas, with low or zero gravity. In one design now being studied these areas would be cylindrical, rotating once every 70 seconds, and would provide 1-1/2% of earth-gravity. There, a ton of mass would weigh only 30 pounds, but tools and equipment would stay put when set "down." Workers commuting to those areas would experience rotation-rate changes of no more than one rpm.

COST DRIVERS IN SPACE-COLONY CONSTRUCTION

During the past six months, independent cost estimates for the construction of Island One have been made by the NASA Marshall Space Flight Center. These are not at the stage of an official report, but excellent cooperation and communication between Princeton and NASA/MSFC has allowed identification of some important cost-drivers in the construction of a first colony. These are:

1) Frequency and efficiency of crew-rotation between the earth and L5, and between the earth and the moon, during the construction period.

2) Extent of resupply needed during construction: This item can vary over a wide range, depending on the atmospheric composition needed at the construction station, and whether food is brought in water-loaded or dry form.

3) Atmospheric composition: The structural mass of Island One is proportional to the internal atmospheric pressure, but independent of the strength of the artificial gravity produced by rotation. Nitrogen constitutes 79% of our atmosphere on earth, but we do not use it in breathing: to provide an earth-normal amount of nitrogen would cost us two ways in space-colony construction, because structure masses would have to be increased to contain the increased pressure, and because nitrogen would have to be imported from the earth. A final choice of atmospheric mix would be based on a more complete understanding of fire-protection.

Parenthetically, the tragic Apollo fire of 1967 is not a valid guide in making this choice. It occurred in a confined capsule, with no water supply available, and in an atmosphere of nearly pure oxygen at almost 15 pounds per square inch of pressure — nearly five times earth-normal. A space colony would operate at 1/5 to 1/6 of that oxygen pressure, in a very large environment, with abundant water available everywhere.

A modest program of experiments on earth could add greatly to knowledge on this point, and might save a great deal of money. Lacking such experiments, present designs are conservative, based on carrying a substantial pressure of nitrogen.

COSTS AND PAYOFFS

A range of costs for large-scale engineering projects is listed in *Table 2*, for scale:

TABLE 2
APPROXIMATE COSTS OF ENGINEERING PROJECTS, IN 1975 DOLLARS

a) Panama Canal	2 Billion Dollars
b) Space Shuttle Development	5-8 Billion Dollars
c) Alaska Pipeline	6 Billion Dollars
d) Advanced Lift Vehicle Development	8-25 Billion Dollars
e) Apollo	39 Billion Dollars
f) Super Shuttle Development	45 Billion Dollars
g) Manned Mission to Mars	100 Billion Dollars
h) Project Independence	600-2000 Billion Dollars

(The re- or devaluation of the dollar forward or backward to 1975 makes each of the numbers in Table 2 uncertain by at least 25%.)

The Apollo project provided trips to the moon for a total of twelve men, at a cost of about 3 billion dollars per man. In space colonization we are considering, for Island One, a thousand times as many people for a long duration rather than for only a few days. With the cost savings outlined earlier, it appears that we can accomplish this thousand-fold increase at a cost of at most a few times that of the Apollo project.

It does not appear worthwhile to make a new, detailed cost estimate at this time for the establishment of Island One. Design details are changing as additional people join the studies, new optimizations and new solutions to technical problems are being found, and the actual cost of construction will clearly depend not only on that work in progress, but on the details of project management.

Rather, I will summarize in *Table 3* estimates made up to this time, characterizing the approach used for each estimate.

TABLE 3
PRELIMINARY ESTIMATES OF COST FOR L5 PROJECT (ESTABLISHMENT OF ISLAND ONE) IN 1975 DOLLARS

a) Physics Today, September 1974 (G.K. O'Neill) **33 Billion Dollars (0.85A)**	Spartan. No crew rotation; oxygen atmosphere; little resupply. Power plants on moon and L5 at 10 Kg/Kw.
b) Internal unpublished report, NASA/MSFC, Jan. 1975 (E. Austin, et al.) as modified April 1975. **200 Billion Dollars (5.1 A)**	Luxurious. Includes chemical and nuclear tugs, super shuttle development, orbital bases, oxygen/nitrogen mix, extensive crew rotation, resupply at 10 lbs./man-day, power plants at 100 Kg/Kw.
c) NASA/MSFC re-estimate April 1975 (E. Austin) as reported to meeting at NASA Headquarters (J. Yardley, J. Disher, R. Freitag and others) **140 Billion Dollars (3.6A)**	High. Unnecessary lift systems removed, but still includes oxygen/nitrogen mix, crew rotation, resupply at 10 lbs./man-day, power plants at 100 Kg/Kw.

(Note: The unit "A" is the cost of Project Apollo in 1975 dollars.)

Detailed conversations with NASA personnel involved in cost estimation indicates a desire on their part, natural enough, to include in the estimates a contingency factor for problem areas not yet identified. The higher estimates listed above appear to include such contingency factors. Within the uncertainties characteristic of the early phase of any project, a figure of 100 billion dollars with limits of 50 billion dollars either way may be as close an estimate as can be made at this time; that is, 5% to 15% of Project Independence, or 2.5 times the cost of Project Apollo.

[more →]

8. *Artist's view of a lunar mining outpost and mass-driver, powered by solar energy. All the materials for construction of the first 10,000 person space community could be obtained from an excavation 5 yards deep by 200 yards long and wide.*   Copyright © 1975 Field Enterprise Educational Corporation.

The payoffs from the existence of Island One can be estimated in several ways. One, crude but reasonable, is to assign to the material output of Island One's industries an added value, per pound of finished products, equal to the lift cost of bringing similar products from the earth. For shuttle-derived heavy lift vehicles, and productivities typical of heavy industry on earth, that added value is in the range of 40-160 billion dollars/year; equal, that is, in one year to the whole cost of construction of the first colony. That added value exists only for those finished products whose end use is in high orbit (geosynchronous, L5 or beyond). One such product, of prime importance at this time, is satellite solar power stations.

ENERGY FOR THE EARTH

Both the oil-consuming nations and the underdeveloped third world are vulnerable to the threat of supply cutoff from the Middle East. The only permanent escape from that threat lies in developing an inexhaustible energy source with a cost so low that the source can eventually be used to produce synthetic fuels economically.

The intensive development of nuclear energy does not seem to be an adequate solution: nuclear power is moderately expensive (15 mils/KWH) and its use encounters considerable public resistance. Nuclear proliferation and radioactive waste disposal are real problems.

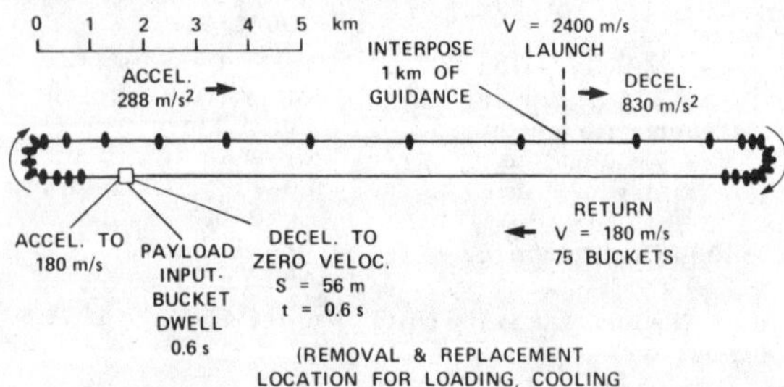

7. *Schematic of an electromagnetic mass-driver. Small "Buckets" supported magnetically would each be accelerated to lunar escape velocity. Over a one-kilometer drift space the direction and speed of the bucket would be sensed and adjusted by additional magnetic coils. The bucket would then release its payload, and return to pick up another. The payload would climb out of the moon's gravity, arriving at a low speed for collection and processing.*

Fossil fuels are scarcer now, and intensive strip-mining for coal will almost inevitably further damage the environment. Solar energy on the earth is an unreliable source, suitable for daytime peak loads in the American southwest, but not clearly competitive in most applications.

Solar energy converted to electricity in space, beamed to earth by microwaves, and reconverted here to ordinary electricity, is being studied with increasing seriousness (*Figures 12, 13*). Already an overall transmission efficiency of 54% has been demonstrated in tests. Delay in realization of satellite solar power stations (SSPS) is mainly due to the problem of lift costs: even for the lightest power plants which seem attainable, and for the lowest lift costs which a very advanced (non-shuttle-derived) launch vehicle could achieve, the economics of the SSPS seem to be only marginal.

Our studies indicate that the construction of SSPS units at the space colony, from lunar material processed at L5, should be economically quite competitive even from the start. The energy interval between L5 and geosynchronous orbit is small, so SSPS units built at L5 could be relocated rather quickly and easily in operational orbits, to supply energy for the earth.

Construction of solar power plants at L5 would overcome four basic objections that have been leveled at the ground-launched SSPS concepts:

1) That they can demonstrate economic feasibility only if a whole series of goals can be reached, each within close limits.

2) That since those achievements could at best only be reached by pushing the state of the art very hard, there is no room for dramatic reductions of energy cost with further development.

3) Ground-launch methods depend critically on the achievement of very low lift costs to geosynchronous orbit. This would require development costs of some tens of billions of dollars, and the technology involved is not well enough understood that success would be certain.

4) In ground-launched SSPS concepts the entire weight of the power plant has to be carted up through the atmosphere. The quantities involved (up to half a million tons per year, if the SSPS program is to be of substantial benefit) are high enough that environmentalist objections, particularly regarding the ozone layer of the atmosphere, might be strong enough to hamper the program seriously, as has happened in the case of nuclear power.

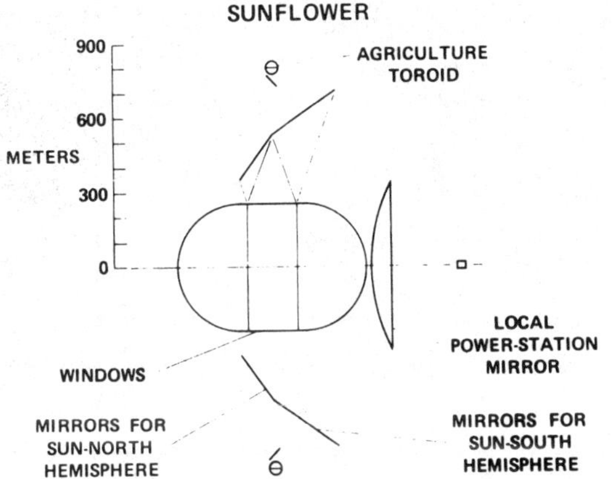

10. A new design, developed by the 1975 NASA/Ames-Stanford University Summer Study on Space Colonization, for an initial space community. Its petal-shaped mirrors, its tracking of the sun, its reliance on solar energy, and its property of being a warm habitat for life in the cold of space all suggest the name "Sunflower." It could house a 10,000 person work force in a comfortable earth-like environment.

With construction at L5, the technologies of power plant development and of rocketry need not be strained. No advanced rocket vehicles are needed, and power plant technology of the present day (*Figure 5*) is sufficient. This contrast is evident in *Table 4*, in which the critical parameters of SSPS design and construction are compared for two earth-launched systems and for one built at a space community. In every case the target figure required for SSPS construction at L5 is more conservative than for either of the earth-launched systems, generally by a large factor.

TABLE 4
SATELLITE SOLAR POWER STATION
DESIGN PARAMETERS
(required for economic viability)

	Earth-launched turbogenerator (Boeing Aircraft Study)	Earth-launched photovoltaic (A.D. Little Co.)	L5 built turbogenerator (this report)
Power plant mass per unit power	5 Kg/Kw	0.8 Kg/Kw	10-15 Kg/Kw
Component lift cost from earth	$77/Kg	$220/Kg	($940/Kg)
Efficiency of transmission	70%	$65%	55-63%
Interest rate	8%		10%
Busbar power cost (initial)	25 mils		15 mils

In *Table 4*, the lift cost from earth is not of great importance in the L5 construction case, because only a small amount of mass from the earth would be required in building an SSPS at L5. The figure listed is, though, the same one used for cost estimates of the construction of the space-colony itself.

The economics of SSPS construction at L5 requires a fresh viewpoint: in that construction almost no materials or energy from the earth would be required. The colony itself, once established, would be self-sustaining, and its residents would be paid mainly in goods and services produced by the colony.

In the summary which follows, the economic input to the combined colony/SSPS program is taken as the total development and construction cost of the first colony, the cost of lifting the materials needed from the earth for subsequent colonies and for non-colony-built SSPS components, a payment in dollars on earth of $10,000/person-year to every colonist, representing that portion of salaries convertible to goods and services on earth (for subsequent use on visits or, if desired, on retirement) and a carrying charge of 10% interest on the total investment (outstanding principal) in every year of the program.

The economic output (yield) from the program is taken as the revenue from power at busbar rates, initially 15 mils/Kwh. The SSPS plants are assumed to be in base-load service, at 95% utilization. To support that assumption, busbar rates are reduced at four-year intervals, to 10 mils/Kwh.

This should be regarded as only the first approximation to an accurate economic analysis. It is equivalent to discounted economics with a 10% discount rate. Knowledge of the input parameters is not yet precise enough to justify analysis in greater detail.

We have examined several cases, in each of which the first space-colony is used as a production site for construction of additional colonies as well as for solar power plants. This "regenerative" effect is essential: a real solution to national and international energy problems can only be achieved by the production of many, not just a token few, satellite power stations. For a high production rate the total number of space colonies must be increased, so that a total work force of 100,000 - 200,000 people in space can be maintained.

Figures 14 - 16 present the results of the analyses. In all cases, it was assumed that the construction of the first colony would require six years of effort, and that thereafter each colony could replicate itself in two years. This tripling of production rate represents devoting 4000 people of a 10,000-person colony to new-community construction (vs. 2000 people available at the construction site during the building of Island One) and in addition, an assumed learning-curve efficiency increase by a modest factor of 1.5.

The remainder of the work force, 6000 persons, was assumed to be committed to SSPS construction, and to produce two SSPS units per year. The productivity implied, 13-25 tons/person-year, is similar to that of heavy industry on earth. (The use of photovoltaic cells, if their progress makes them competitive, is not ruled out. Silicon, their principle constituent is abundant in the lunar raw material.)

[more ➔]

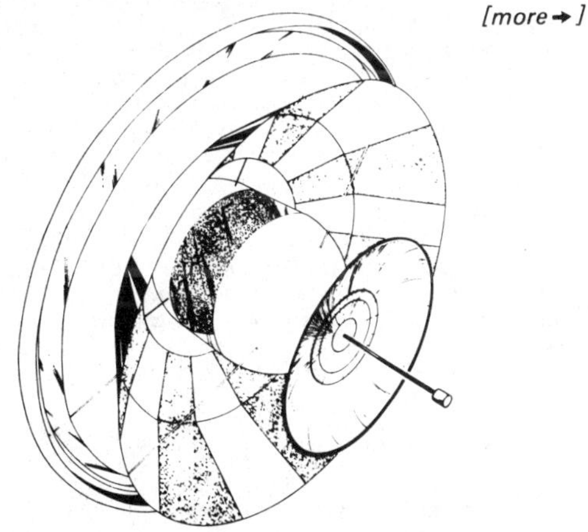

11. An angle view of "Sunflower." The interior circumference of the habitat could be over one mile. Agriculture could be carried on in an external, counter-rotating toroid, shown here as an outer ring.

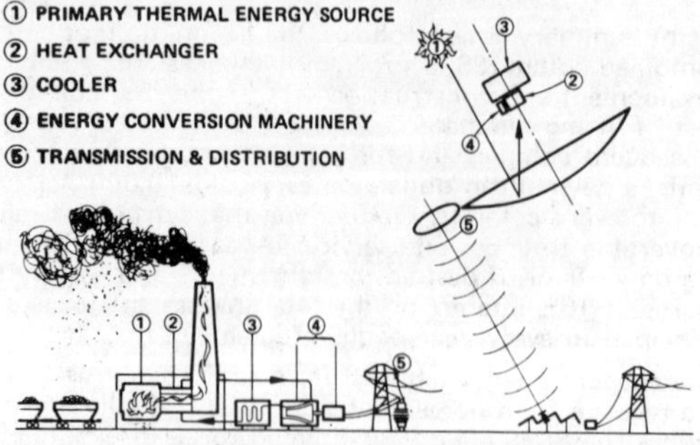

12. Schematic of a satellite solar power system. The power plant shown, being studied by the Boeing Corporation, uses turbogenerator machinery. An alternative, based on photovoltaic solar cells, is under study by the A.D. Little Co., Raytheon, Grumman Aircraft and the Jet Propulsion Laboratory.

The question of productivity and the effects of automation within the weather-free, zero-gravity environment of a space community's assembly region deserves intensive study; so far it has been possible only to verify that the estimates given are consistent with earthbound experience. I anticipate that the residents of the early space communities will be nearly all employed in production, support services being automated as far as possible.

In *Figure 14*, a time-line is developed based on making an early start, with the shuttle and a shuttle-derived freight vehicle. A medium-to-high estimate (96 Billion Dollars) of the cost of Island One is assumed, and an additional 82 Billion Dollars for the transport of carbon, nitrogen, hydrogen and colonists to the later colonies is added. New-colony construction is halted after the 16th colony, due to market saturation.

By the 13th year of this program (the year 1995, given a starting date of 1982 for major construction activity, implying intensive design beginning by 1976) the L5-built SSPS plants could fill the entire market for new generator capacity in the U.S. Given the rapid growth of the manufacturing capacity and the possibility of busbar power cost reductions, true "energy independence" for the nations taking part in the L5 project could occur before the year 2000, with a shift to production of synthetic fuels. In the words of one exuberant young economist at the NASA/Ames-Stanford University 1975 Summer Study, "We can put the Middle East out of business!" In my own view, I would far prefer to see a cooperative multinational program formed, based on participation by all interested nations. If the L5 project continues to look feasible, it would be in the interest not only of energy-consuming industrial nations, but of the OPEC nations to take part in it, because if these numbers are correct, the market value of Middle Eastern oil could drop irreversibly before the end of this century.

A cost-benefit analysis of the *Figure-14* case has been made, and yields a benefit/cost ratio of 2.7. A favorable benefit/cost ratio also results from a variety of different input assumptions, with assumed total program costs up to 280 Billion Dollars. The favorable result is sharply sensitive to only two parameters: speed and interest rates. An interest-rate reduction to 8% approximately doubles the benefit/cost ratio; an increase to 13% reduces it to near 1.0. A stretch-out of the program would be disastrous as regards both energy benefits and the benefit/cost ratio.

Figure 15 indicates how rich a source of wealth the space-colony program could become. By year 11 (1993 on the fastest-possible time-scale) the energy flowing to the power grids on earth from L5-built SSPS units could exceed the peak flow rate of the Alaska pipeline. By year 17 the total energy so provided could exceed the total estimated capacity of the entire Alaska North Slope oil-field.

Figure 16 shows the effect of delay (as for example to develop advanced lift vehicles prior to space-community construction). The benefit/cost ratio would not be greatly improved, and total program costs would be reduced only by a factor of two, even if vehicle development costs and later operating costs would be delayed by the full 7-year development time of the new vehicles. This does not, therefore, seem to be a wise route to take, but requires further study.

THE U.S. AS ENERGY EXPORTER

The underdeveloped third-world nations are now trying to industrialize, in order to increase their living standards and economic security. If the example of the industrialized world is valid, their success in that attempt may be a powerful element in reducing the runaway population growth rates which now threaten their progress and, in the long run, political stability.

Because of widespread concern over decreasing energy and materials supplies, we are now viewed by many as exploiters of scarce resources. This has been a significant factor in hostility toward the U.S. and toward other industrial nations. With a program of power plant construction at L5 we could return, at little cost in energy and materials from the earth, to our traditional role as a generous donor of wealth to those in need. In this case the wealth we could provide would be in the form of energy to third-world nations, and ultimately of "beachhead" colonies for their own progress. The L5 project would give us the opportunity to act with generosity, yet with little cost to our own national resources.

RESPONSE FROM GOVERNMENT AND THE PUBLIC

It is a tribute to some remarkably perceptive men within NASA and the NSF that, despite their unfamiliarity a year ago with the modern concept of space colonization, they have now encouraged its development and have even begun to support it with a small amount of funding (approximately $40,000 in 1975).

For a person with a technical education, it is logical to assume, given a new concept, that "If I haven't heard of it before, it must be as far off as the 21st Century." Usually that attitude is justified. Space colonization, though, is a curious exception. It is a technical concept realizable without any new breakthroughs in materials technology or technical understanding. We are unfamiliar with it only because, until the Apollo samples were returned, no one could have put together all the necessary components of a space-colony program in the form of a complete system with defensible numbers.

In contrast to that situation, we have examples of development programs which <u>do</u> require breakthroughs in the understanding of new physical phenomena, but which have become accepted parts of our research effort simply because we have been hearing about them for a long time. One

Turbomachinery Power Converter

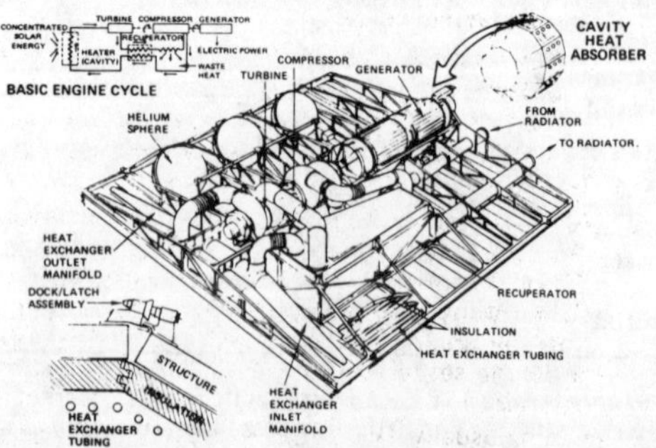

13. Details of power-satellite turbomachinery. Total mass of the satellite depends strongly on the peak operating temperature of the system; for an earth-launched satellite, for which lift-costs are critical, the motivation toward high operating temperatures is strong.

classic example is hydrogen fusion power. It has been discussed in public for thirty years, and has been worked on in research for more than twenty years. In effect, it has become institutionalized. Although no responsible advocate of fusion power will commit himself as to when fusion power will become economically competitive, the idea has been around for so long that its eventual success is accepted as inevitable by most people. (My own view is that fusion power research should continue to be supported, on what I would regard as the off-chance that it might someday be competitive with L5-built satellite power stations.)

Space colonization, and the construction of satellite power stations at L5, requires no such breakthrough in the understanding of a new physical regime. It is mainly civil engineering on a large scale, in a well-understood, highly predictable environment. It does not even require the development of a new rocket engine. Some, fortunately a substantial number, of responsible administrators in NASA have been quick to grasp this distinction, and to see the potentialities of space colonization for the agency and for the public. For others, though, it has been almost an embarrassment, because the assignment of space colonization to its proper place in time-sequence (that is, <u>now</u>) implies that all previous planning has omitted an important option. In the case of NASA, proper recognition of the space colony concept is further impeded by the orders previously given to the agency, and never rescinded: to plan on constant or decreasing funding levels, to bring up no surprises, and as far as possible to become invisible.

The evidence of the past year indicates that in terms of public response space colonization may become a phenomenon at least as powerful as the environmental movement. Since the first small, informal conference on that topic, in May 1974, a rapidly increasing number of articles about it have appeared, in many newspapers and magazines, and all have been quite favorable. Several are still in press at this time. Radio and television coverage has also increased rapidly.

Popular response in letters to Princeton has been strong. Of these letters, more than 99% are favorable. Also, encouragingly, less than 1% of all mail is in any way irrational. Many of the correspondents offered volunteer help, and are actively working at the present time in support of the space-colonization concept. The letters express the following reasons why this concept, in contrast to all other space options now extant, is receiving such broad support:

    1) It is a right-now possibility. It could be realized within the immediate future.

    2) In contrast to the elitisim of the Apollo project or of a manned mission to Mars, it offers the possibility of direct personal participation by large numbers of ordinary people. Many of the correspondents, from hard-hat construction workers to highly-educated professional people, see themselves as prospective colonists.

    3) In contrast to such technical options as the supersonic transport, nuclear power or the strip-mining of coal, it is seen as offering the possibility of satisfying real needs while preserving rather than further burdening the environment.

    4) It is seen as opening a new frontier, challenging the best that is in us in terms of technical ability, personal motivation and the desire for human freedom. Many correspondents refer to space colonization by analogy to the discovery of the New World or to the settlement a century ago of the American frontier.

One letter, unusually well-expressed but otherwise not atypical, concludes:

"I would greatly appreciate being informed of your own personal assessment of what can and should develop out of your space colonization ideas. If they do in fact have the social and human potential that they appear to me to have, any unnecessary delay in their realization would seem to me to be unthinkably irresponsible."

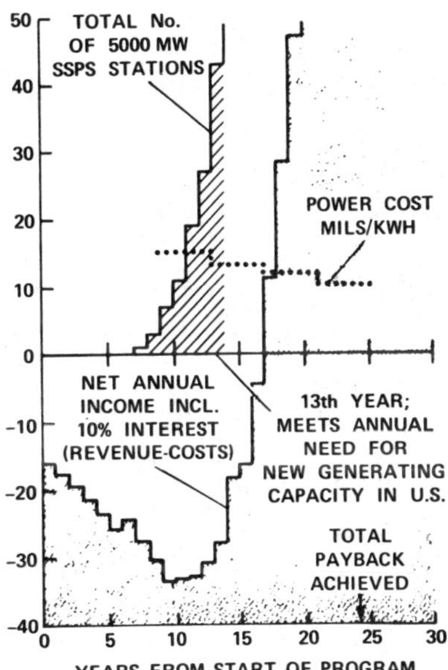

14. *Costs and benefits of a space-colonization program, based on conservatively high estimates for costs (178 Billion Dollars over 14 years) plus interest rates (10%). Cost estimates assume only shuttle-derived lift vehicles. A six-year construction time for the first community, and a two-year replication time thereafter, are assumed, with a productivity of two satellite power stations per colony per year after initial startup. By the 13th year, power plant capacity so produced would meet U.S. needs. In this scenario the benefit/cost ratio, including all construction, development and interest charges, would exceed 2.7. Power costs, initially 15 mils/Kwh, would be reduced in steps to achieve market penetration. Revenue would be obtained only from the sale of energy, not of power stations.*

CURRENT RESEARCH

During 1975 the major events in space colonization have been the Princeton University Conference (co-sponsored by NASA, the NSF, Princeton University and the American Institute of Aeronautics and Astronautics; Cf. ref. 5 when available), and the NASA/Ames-Stanford University Summer Study on Space Colonization (ref. 6 when available).

Writing at the mid-point of the Summer Study, the principle results so far can be listed as:

    1) Verification that shuttle-derived lift vehicles would be adequate for the establishment of Island One.

    2) Verification that agricultural-yield figures used in ref. 2 were conservative by approximately a factor 2.

    3) New, tighter requirements on allowable rotation rates.

    4) Verification that productivity figures so far in use are in the right general range.

    5) More detailed analysis of discounted economics, verifying a high benefit/cost ratio.

    6) New, more detailed results in the areas of colony geometry, materials processing, and mass-driver payload guidance.

In the period since May 1974, when this concept first came to public attention, research on it has progressed at what I would describe as the fastest possible rate. In the year beginning in September 1975 this progress will slow unless some extraordinary mechanism is found to provide funding for in-depth studies to be carried out by the government agencies and the private sector. A level of 0.5 - 1.0 Million Dollars is probably adequate; to provide more at this time would probably result in some waste and inefficiency.

APPENDIX

(This section is keyed to the titled headings of the main text, and is intended for the reader with technical training, who may wish to check independently some of the most important numbers or statements.)

INTRODUCTION

L5: An orbit about L5, stable in the four-body problem of the sun, earth, moon and colony, has been shown by Kamel and earlier authors. Cf. references in PTA (ref. 2). Occultation of the sun in that orbit is rare and brief. L4 is equally usable.

High-orbit products: The possibility of returning material products to the earth's surface from L5 is not considered in this document.

THE SPACE-COLONY CONCEPT

Authors: Tsiolkowsky in Russia, Bernal in England, and Cole in the U.S.A. all wrote books which bear on the concept of space colonies. Clarke, Stroud and others have also considered portions of the problem.

GEOMETRIES

The image of the sun's disc would rotate about its center, but the disc is so nearly circular that this rotation would not be detectable by the naked eye.

Civil engineering limits: A standard safety factor of 1.67 is used, as in the building industry on earth. (Corresponding factors are 1.5 for commercial aircraft, and as low as 1.2 for military aircraft.) For aluminum/silicon alloy, cold-drawn, with an ultimate strength of 60,000 psi, the yield point is 50,000 psi and the working stress is here taken as 30,000 psi. For hot-formed aluminum, 20,000 psi is used. The same safety factor is used for iron and titanium. Diameters up to four miles are assumed, with total atmospheric pressure of 5 psi minimum. See PTA for formulas. (Mass table in PTA for model 1 has a non-propagating error: for 20,000 tons aluminum read 80,000 tons metals.)

INDUSTRY

Axis of rotating habitat contains avenue-passage and passes through a hollow bearing. Bearing forces are small, typically one ten-millionth of colony weight in one gravity.

LIMITS OF GROWTH

M.I.T. Studies: Cf. references in PTA. Asteroidal materials: Total volume of proven asteroids is estimated as 1/2500 of volume of the earth (Cf. Allan, Astrophysical Constants). Economic imperative is construction of a new colony adjacent to an asteroid, so that economic productivity can be achieved without prior moving of materials. Relocation of a colony to L5 from the asteroidal region would require about 30 years at an expenditure of 7% of total colony mass.

ENERGY WITHOUT GUILT

The energy intensity (insolation) in space is 1.4 Kw/m^2, or 1.23×10^8 KWH/year for a 100 meter square. This would cost $1.8 \times 10^6 at a busbar rate of 15 mils. The lower figure used in the text allows for reflection losses. Mirror assumed is .001 inches aluminum, with a factor three multiplier for support frames.

For an initial community of 10,000 persons, an electrical power plant of 100 megawatts is assumed (10 Kw/person). For the USA in 1975, average usage of electrical energy is at the rate of about 2 Kw/person, and peak capacity is equivalent to 2.5 Kw/person.

THE BOOTSTRAP METHOD

The velocity intervals from low earth orbit to lunar parking orbit (LPO), to L5 or to geosynchronous orbit (GSO) are all approximately equal, in the range 11.1 - 11.4 Km/sec for minimum-energy two-impulse burns. Escape velocity from the moon is 2.4 Km/sec. With kinetic energy = $1/2$ mv^2, escape from the earth therefore requires 21.4 times as much energy as from the moon. Spiral orbits (low thrust) require more energy.

The mass driver: A description and table of parameters for this machine is listed in PTA. Further study results will be available in references 4 and 6.

Magnetic fields are held below 10,000 gauss, and accelerations to less than 29 gravities. The nominal repetition rate is 1 Hz, for payloads of 9 Kg each. The peak transfer rate is therefore 780 metric tons per day. The range of a factor 4 quoted in the text allows for turnoff during the lunar night, and for reliability down to 50%.

Guidance is by magnetic trimming during a one-kilometer inertial drift-space, roll/pitch/yaw and position sensing being done by laser interferometry before payload release.

In PTA an estimate of 10,000 tons for lift-needs from earth to L5 was given, and 3,000 tons for transfer from the earth to the moon, based on a "Spartan" approach: oxygen atmosphere, construction work force stay time until completion of the first community, and food supply in dehydrated form. Another extreme was given by NASA/MSFC, based on a nitrogen-mix atmosphere, extensive atmospheric make-up from earth, frequent crew rotation and food resupply in wet form. It was about a factor three higher (unpublished internal report, no number). The extremes are therefore 2% - 6% of an estimated 500,000 ton total mass.

In current discussions of vehicle-systems, a distinction is drawn between lift vehicles made of building-blocks each of which is already under development for the space shuttle (e.g., SRB's, SSME's, avionics) and lift vehicles requiring extensive new development. For the space-colonization program only the former are required. Several papers in ref. 5 (Tischler, Davis, Salkeld) cover this topic.

Construction station: PTA estimate was 1000 tons. A more detailed estimate (G. Driggers, ref. 5) gives 2500 tons.

LUNAR MATERIALS

The source for Table 1 is ref. 7. Samples from other Apollo landing sites have generally greater amounts of aluminum and smaller amounts of iron. The lunar surface rocks often have higher metal content, but are neglected here.

The structural aluminum considered for use in colony-building is an alloy of aluminum and silicon, the most plentiful of lunar elements after oxygen.

The fuel estimate made is based on the usual 6:1 oxygen/hydrogen mixture (fuel-rich) commonly used for LOX-hydrogen rocket engines.

D. Criswell (ref. 5) has calculated the yields of carbon, nitrogen and hydrogen which could be obtained by sifting lunar soils for the fine-grained material, and then heating that material. The rare light elements are concentrated in the finer grains, and can be extracted by that process. In this report no advantage is taken of that option.

Asteroidal materials: As noted earlier, the velocity interval from the earth to L5 is about 11.4 Km/sec. A selection of ten asteroids, whose orbital elements are well-known, was checked. It was found that in all cases the total velocity interval required for transfer to L5 was close to 10 Km/sec. Correction to match the orbital plane with that of the earth was an important term.

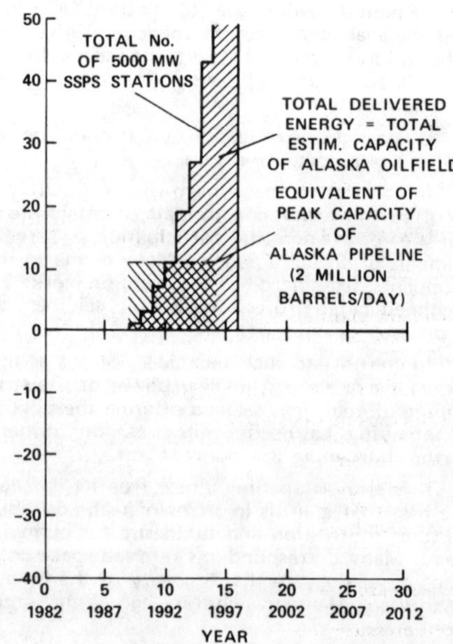

15. Effect of an early decision to drive toward space colonization at the earliest possible date. By year 11 from the start of the program the energy flow to the earth from satellite power stations built at L5 could exceed the peak capacity of the Alaskan pipeline. By year 16, the total energy supplied to date could exceed the total estimated oil reserves of the Alaska North Slope.

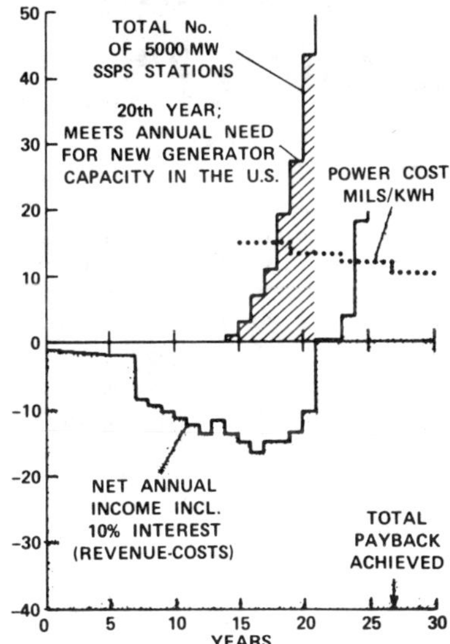

16. Effect of delay in startup, to wait for advanced lift vehicles. The most optimistic estimates for low development and operation costs for such vehicles are assumed. Total program cost is reduced only by a factor 2, for a factor 12 reduction in lift costs. Benefits, including energy independence, are delayed by 7 years.

ISLAND ONE

The design of Figures 10 and 11 has a habitat-interior diameter of 540 meters and a circumference of 1.05 miles. Total interior non-window surface area is over 900,000 m^2, about half of which is at 70% or more of earth gravity. The counter-rotating toroidal agriculture ring provides 400,000 to one million meters^2 for photosynthetic crop-growing, plus additional covered areas for processing and storage.

In order that the entire colony maintain its axis always pointed toward the sun, yet not require thrusters, the total rotational angular momentum must be zero. In the "Sunflower" design this is accomplished by devoting about 20% of the total mass to the agricultural ring.

The low-gravity work areas described are nominally 40 meters in diameter (412 ft. circumference or floor width) and can be of any desired length. Six of them, each 200 meters long, would provide approximately three times the total high-bay assembly area of the General Electric Large Turbine Division plant at Schenectady, New York, where a large fraction of the turbogenerator capacity of the USA is built.

COST-DRIVERS IN SPACE-COLONY CONSTRUCTION

Atmospheric composition: A typical design for a hemisphere diameter of 540 meters has the following contributions to total internal pressure:

Aluminum weight	3.2 cm	0.13 psi
Soil or structures weight	30 cm	1.08 psi
Atmosphere		7.50 psi
Total		8.71 psi

In this typical case the atmospheric pressure accounts for 86% of the total structural requirement. With a full 14.7 psi of atmospheric pressure the figure would be 92%.

COSTS AND PAYOFFS

In Table 2, items (a) and (c) are from the Exxon Corporation (Smithsonian Magazine, April 1975, p. 117).

Item (e) assumes a cost of 23 Billion Dollars as of 1967 and an average of 7% inflation since that year.

Item (f) is based on an unpublished NASA/MSFC Study Document, "Space Colonization by the Year 2000 - An Assessment."

Item (g) is from J.N. Wilford, New York Times, July 13, 1975, quoting Vance Brand, U.S. Astronaut.

Value added by location in high orbit: A fully employed population, a productivity of 20 tons/person-year, and lift costs in the range $100 - $400 per pound are assumed.

Busbar power costs: Present figures average 15 mils/Kwh for nuclear power, 17 mils/Kwh for fossil-fuel power. Peak-shaving power earns revenue at a much higher rate, but the energy generated by peak-shaving generators is a small fraction of the total.

Solar energy arriving on the land area of the continental U.S. averages about 1/10 of the amount which intercepts equal area in free space. For base-load power, the capital cost of the system must provide for a December/January day length, storage for extended bad weather, and a high demand.

Fifty-four percent efficiency has been demonstrated in 1975 by a JPL group, in cooperation with Raytheon (also Cf. ref. 8).

Microwave power transmission has its own environmental problems, but they appear to be less serious than those of nuclear or fossil-fuel power (Cf. refs. 8 and 9).

The velocity interval from L5 to geosynchronous (spiral orbit transfer) is 1.1 Km/sec and is in full sunshine. Transfer could be by a mass-driver, powered by the SSPS itself and used as a reaction engine. The reaction mass could be the wastes (for example liquid oxygen) from the industrial processing at L5. A transfer time of one month or less appears feasible.

Vehicle development costs: for an advanced (non-shuttle-derived) heavy lift vehicle, estimates of development cost from within the aerospace industry vary from 5 Billion Dollars to 25 Billion Dollars; of attainable launch costs to geosynchronous, from $77/Kg to $400/Kg.

The costs of SSPS construction at L5 (input for Figures 14 - 16) include lift costs for microwave transmitter magnets and initially for computers and controls, as well as items listed in the text.

Alaskan oil field comparison: 1 barrel of oil has an energy content of 5.24×10^9 joules (ref. 10). The peak capacity of the Alaska pipeline will be 2×10^6 barrels/day (ref. 11). For a high conversion efficiency of 48%, the pipeline will then supply 1.83×10^{18} joules annually. This is a rate of 5.8×10^{10} watts, or 58,000 megawatts, equivalent to less than 12 5,000 megawatt SSPS units.

The estimated total reservoir of oil in the Alaskan North Slope (the pipeline source) is 10^{10} barrels (ref. 11), or 2.1×10^{19} joules at 48% conversion efficiency. This is 134 SSPS-years, a total reached in the first nine years with the growth rates assumed for Figures 14 - 16. For comparison, the total proven reserve of oil in the Middle East is 33.8×10^{10} barrels (ref. 12).

CURRENT RESEARCH

One area requiring verification is semi-closed-cycle ecology. Many small islands have effective ecosystems more limited than that of the first colony, but verification is still required. Fortunately, total closure is unnecessary: "economic closure," the achievement of a closure level adequate to reduce to tolerable levels the lift costs for seeds, etc. from the earth, will be sufficient. Isolation and heat-sterilization can halt any runaway biological subsystem. ∎

REFERENCES

1. Nature, August 23, 1974.
2. Physics Today (referred to as PTA), September 1974.
3. Space Colonization and Energy Needs on Earth. G.K. O'Neill, Science.
4. Proceedings, 1974 Princeton Conference on Space Colonization. [See p. 130]
5. Proceedings, 1975 Princeton University Conference on Space Manufacturing Facilities. [See p. 130]
6. The Colonization of Space: Report, 1975 NASA/Ames-Stanford University Summer Study. [See p. 130]
7. Mason and Marsden, "The Lunar Rocks."
8. Solar Power via Satellite: Testimony of Dr. Peter E. Glaser, A.D. Little Inc., before the Committee on Aeronautical and Space Sciences, U.S. Senate, October 31, 1973.
9. Derivation of a Total Satellite Energy System, G.R. Woodcock and D.L. Gregory, AIAA Paper 75-640, 4/24/75.
10. J.C. Fisher, Physics Today, December 1973, p. 42.
11. Smithsonian Magazine, April 1975, p. 117 (Exxon Corporation).
12. L'Express, No. 1219, 18-24 November 1974.

"Is the surface of a planet really the right place for an expanding technological civilization?"

INTERVIEWING GERARD O'NEILL

Stewart Brand: What is the point of origin for you on space colonies?

Gerard O'Neill: My interest in space as something for people to be in, rather than simply to look at, goes back a long way. But the particular thing that started the space colonies concept, really, was a course that I taught at Princeton in 1969. It was the big, standard Freshman physics course, with about 320 students in it. "Physics 103."

I chose to do double-load teaching that whole year — my idea was that I would be the lecturer in that course, and I would also take a class section so I'd see the course from the top and the bottom at the same time. There were quite a few things that I wanted to do to improve the course. One of the things I felt most concerned about was that this was the peak time of disenchantment with anything in science and engineering. The students who were good at science, and particularly the students who were good at engineering, felt very defensive about it, because all of their friends and their roommates were saying that they weren't doing anything relevant. And I felt that, despite the bad times, improvements in the human condition could be reached by using science and engineering in the right ways, as opposed to the wrong ways.

So I thought it would be worthwhile, particularly for those few students who were so far ahead of everyone else that the ordinary coursework couldn't challenge them, to invite them to come to an extra seminar where I would try to find examples of problems to look at which could be of interest in their lifetimes, and which would be challenging on a large scale, and potentially very beneficial to the rest of humanity. And now having given you a non-answer to your question, I'll give you a complete non-answer, because I have to say that out of somewhere, and I don't know where, it occurred to me that the first reasonable question to ask was: "Is the surface of a planet really the right place for an expanding technological civilization?" And, of course, once you ask the right question, the right answer follows almost automatically. That's simply a question of working out the numbers.

SB: Now, this is how many students in the seminar?

O'Neill: Oh, not more than 6 or 8, I suppose. We met once a week for several weeks.

SB: You asked that question and what happened then?

O'Neill: Well, the students were able mainly to do library research, going and looking up in Encyclopedia Britannica how big the land area of the world was and things of that kind. I had to supply a good fraction of the calculations although they were able to do some of them, but within the time of the seminar I did encourage them to do calculations of how big could a rotating pressure vessel in space, to hold an atmosphere and provide a gravity, be made. That answer came back pretty quickly. It already started being interesting because it was several miles in diameter.

The reason is that if you're using the electromagnetic interaction — that is the ordinary interaction that holds solid matter together, instead of the very weak gravitational interaction (which is holding on to an atmosphere the difficult way, the way a planet does) — then you've got an enormous factor in your favor. What does that mean in terms of how much land area you can build with a rather small amount of material? The first answers that we came out with indicated that we were talking about more than a thousand times the land area of the Earth as the potential room for expansion. So those two numbers, the question of the largest structure size using ordinary materials and the question of what the limits to growth were, were enough to get me interested in the problem.

Another item that came in was energy, because it seemed pretty clear that solar energy was the obvious way to go. And also that if you were building large things, in the long run it would be better to do it in zero-gravity than in a planetary gravity.

SB: So, as I understand it, the question was asked, and the implied answer was, "No, the planet's surface is not the right place." And the implied next question is, "Where, then?" And then the answer was inside-out planets.

O'Neill: Well, the classical science fiction idea of colonization is always you go off and you find another planetary surface, like the moon or Mars. . . . That misadventure we sidetracked very quickly because first of all there just isn't that much area involved, and second, most of those other planetary surfaces are fairly unpleasant in terms of where they're located. They're the wrong distance from the sun, and they've got the wrong rotation times, and the wrong gravities usually. Besides all that, there was the fact that it didn't make sense once you could get out into the space beyond a planet to give up the fulltime solar energy that you could get if you just stayed there.

The sort of analogy that I like to use nowadays is to say that, "Here we are at the bottom of a hole which is 4,000 miles deep. We're a little bit like an animal who lives down at the bottom of a hole. And one day he climbs up to the top of the hole, and he gets out, and here's all the green grass and the flowers and the sunshine coming down. And he goes around and it's all very lovely, and then he finds another hole, and he crawls down to the bottom of that hole. And if we go off and try to get serious about colonizing other planetary surfaces, we're really doing just that. It's kind of atavistic but there really isn't any other excuse for it.

SB: I want to track a little more on the sequence of events for you and for the students discovering all this. You asked the question, . . . Were you already being electrified by all of this as it got started or did that come later?

O'Neill: I was already. I started getting interested quite
quickly. There was the question of how to make it as
earth-like as possible, because certainly right from the
beginning, my feeling certainly was, that I had no desire
to go the route of just inventing a big spaceship or something
that would be a space station. That had no interest for me
at all. This was to be something that was to be potentially
beneficial for a lot of people. It had to look an awful lot
like the Earth. So, one of the questions I worried about
quite a bit was how to provide earth-normal gravity, and a
normal atmosphere, and a normal appearance of the sun as
well. It was during that first few weeks I think that I came
up with this simple geometry of the 3-fold symmetry of the
mirrors and the alternating land and window areas, which
so far still looks fairly reasonable in terms of the constraints
that you put on it.

The fundamental thing, but one that didn't occur to me
until quite a long time later — because of course my work
on this was very occasional, a few minutes every few weeks
kind of thing — was the question of how you cope with the
angular momentum. The spinning cylinders are there, and
it's sort of not very elegant to go throwing away reaction
mass to try to process that angular momentum. But it wasn't
until sometime later, I have to admit, that it occurred to me
that the easy way was to make two of these things and to
hitch them together.

    SB: And they rotate in opposite directions?

O'Neill: Right. It should have occurred pretty quickly,
because it's an awful simple-minded idea, but it didn't.

    Mike Phillips: Does it occur in any natural form?

O'Neill: Well, again, if you ask the question in the right way,
it should occur to you very well. I mean the wrong way to
ask the question, and the way that I asked it first, was "How
do you apply forces to a gyroscope to make it precess?" The
right way to ask the question is "How can you have a
rotating object that doesn't have any angular momentum?"
and then you get the answer immediately. That's easy to
say after the fact.

    SB: Was there a point or a series of points where
    the practicality started to overwhelm you?

O'Neill: I think I began to realize, really, within the first
month or two that this was, in my opinion at least, something
very important. And that I somehow had to get it out into
the open and get it discussed. But then, as you probably
know, it was a long and very frustrating period before anything really came out on it. I talked about it a little bit to
some of my friends. I used to talk about it to my children.
I'd take them on walks in the woods, and speculate about
what life in a space colony would be like. In fact Tasha came
to one or two of the seminars at that time, because we had
met only within weeks of that same time.

    SB: Were you electrified, Tasha? Or was this just
    something he was doing?

Tasha O'Neill: Well, at first it was beyond me, anyway. I
had just come from Europe and I didn't understand anything
anyway. And I didn't understand anything about physics.
And then, you know, time went on, and it took hold of me
too. I understand far more by just listening over and over and
over . . .

O'Neill: You'd be surprised to hear her explain angular
momentum now.

    SB: Let me pursue this, because I am always
    interested in how ideas take root in a person and
    then in society. So here you are now with what
    must be a gradually increasing obsession and not
    many listeners. Did you take a strategic approach
    in any way? Or just keep talking and let things
    find you?

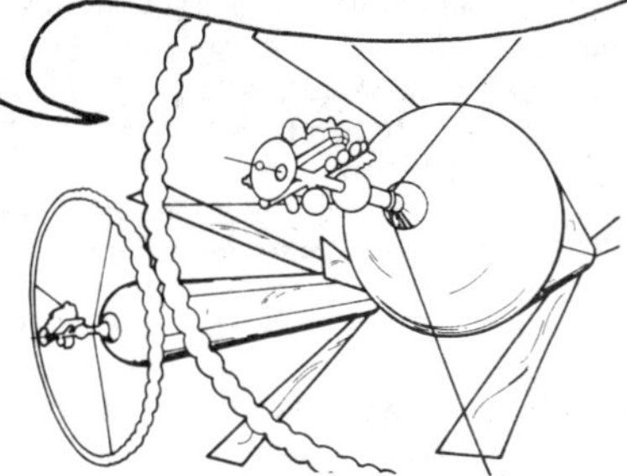

O'Neill: It was awfully irregular and unplanned. I spoke
about these things with a few friends and at one party, in
the home of some friends. There happened to be someone
who was associated in an indirect way with one of the
magazines and that lady got very interested and excited about
this and suggested that I write up an article for the magazine.
And that seemed to me to be a worthwhile idea so through
her, the magazine editor and I got in touch, and at his
suggestion I went ahead and wrote an article.

    SB: What magazine? What editor? What lady?

O'Neill: I think that was **Atlantic**.

    SB: That's the East coast phenomenon, Michael.

O'Neill: What's the East coast phenomenon?

    SB: That publishing occurs on the East coast. If
    you were in Nebraska, you'd still be there.

O'Neill: Well, the reaction was — the editor was quite
interested and he wrote back and asked an enormous number
of questions, and then I expanded on the article, answered all
the questions, and sent it back again. Again, long months
went by and finally he rejected it and said, "It's not that I'm
not interested, it's just that I have this feeling that I've asked
ten questions and you've given me ten answers and now that
suggests a hundred more questions, and it's not going to con-
verge somehow. Better that I not do anything about it."

    SB: That's the difference between our magazines.
    Atlantic converges; we don't.

O'Neill: So then it went through a period of — let's see,
that was 1970, and it was not until four more years that I was
able to get an article published.

    SB: All this time you're just muttering to friends?

O'Neill: No, I kept on. I rewrote the article again and . . .

Tasha: Collected rejection slips.

O'Neill: Yes, collected rejection slips. I wrote a letter to
Scientific American. I didn't send in the manuscript, but
they had published 2 articles of mine previous to that, so . . .

    SB: On what?

O'Neill: Subjects of physics. There was an article on spark
chambers, which was a technique of high-energy physics
which was in vogue in the 1960's, and an article on particle
storage rings, which happened to be something that I'd
started. So I wrote an historical article about that in the
60's. And since I had written for that magazine before, why
I felt that it might make sense to try them again. So I wrote
them a letter suggesting an article on this topic, and just in a
couple of paragraphs outlined what had been done. And I
got a very stuffy letter of rejection back. I think it was two
lines, and sort of immediately dismissed the whole thing.
They didn't even want to look at a manuscript.

So then I rewrote it again and that time I tried **Science**. And
that rejection was curious and more complicated. In all of
these cases, of course, the time to accumulate rejection slips
was very long. You know, many months would go by, in

each case. In the case of *Science*, they sent me the reviewers' comments. Both of the reviewers had recommended rejecting the article. One of them had gone into shock, really, there's no other way to say it. He had just said, "No one is thinking in this direction, and therefore it's got to be wrong." And the other one started to think about 2 or 3 possible objections but he didn't let his own mind carry him far enough logically to see the answers to the objections. Now that was a curious case. One of the reviewers, I'd never met. I don't even know to this day who it was. The other one, oddly enough, was my host at the first lecture that I ever gave on this subject. It was an odd coincidence, at Hampshire College in late 1972. A friend of mine said, "Look you're having all this trouble getting this idea out and under discussion, take it to the people. Give some talks at universities. And if you can't get the ideas out to the public any other way, take it to the students, they're a very good test of whether something is worthwhile or not."

> SB: This is what Buckminster Fuller did a while back when no one was buying. He went to students, and students bought, and then everybody else bought.

O'Neill: Well this fellow said, "Look, come and give the talk at our college" — this was Hampshire. And so I went up there in late 1972, and the man who was the dean of the engineering college, the college of science, was the host. The talk itself was a lot of help to me because it was an 8 o'clock in the evening lecture, and I talked for less than an hour, and then questions went on for more than an hour after that, and the students were very excited about it. Finally after something like 2 or 2½ hours had gone by, the host got up and said two things. "First," he said, "I want to say that when I first heard this idea I thought it was complete nonsense, and now I want to tell you that I've changed my mind. The second thing is that the speaker has an 8:30 class tomorrow and a 4-hour drive to get to it and we really have got to give him a chance to get to sleep. So let's just let those few people who want to ask questions, continue asking questions." And what happened was that about half the audience left, and the rest proceeded to take chairs closer to the front, and the whole discussion went on for about another hour.

When I left with my host to go back to his place to sleep the few hours that remained, he started asking me some very elliptical questions that I couldn't understand at first. He said, "You had all of these rejection slips now for all of these years. Does this personally get to you? Is this something that is personally threatening to you in some way?" And I laughed and said, "No, it doesn't. First, I really think that what I'm doing is worthwhile and, second, this is something completely beside my normal work, and I don't have any trouble getting my ordinary work accepted. I don't have to worry about my job and so on. So that I don't feel personally threatened by it at all. I'm sorry that people are so slow to catch on, but that's my only reaction." "Well, it was important to me to ask that," he said, "because I want to tell you that I was one of the reviewers who turned down your article for *Science*." And he said at that time that he would write to the editors of Science and tell that he had changed his mind.

> SB: Very nice. Did that then have some effect on *Science*?

O'Neill: I never got any direct reply from them on that. There was another man later on who got quite interested in that work who wrote them and I got a very stuffy short note from the editor, saying that in view of this suggestion that they would reconsider if I wished to resubmit the manuscript. By that time, *Physics Today* had already agreed to publish the article, and I wanted to be loyal to the first

people who had gone out on a limb and said that they would be willing to publish this work.

> SB: That was the first major publication . . .

O'Neill: That was the ONLY publication. The article was accepted about the beginning of 1974 and published in September of 1974. So there was a period of approximately four years during which I was trying to get it published and couldn't.

> SB: As this was going along, and before you talked with students at Hampshire were there friends of the idea that were sort of helping you stay afloat about it? Or were you pretty much all alone in your enthusiasm?

O'Neill: Well, there were personal friends whom I talked to, and who got quite interested and enthusiastic about it, but it was a sort of word of mouth thing among comparatively few people. And then I did start giving other lectures in colleges. — The second lecture was at Princeton. And I gave talks at quite a few colleges on the West coast in 1973. But that already represented a passage of about 3 years of time.

> SB: Is this with graphics and slides and things like that? Or straight. . .

O'Neill: I had some fairly primitive graphics.

> SB: During this time, how much were you actually working on the idea itself, refining it?

O'Neill: Very little. It would be a question of if I were in the course of a working trip someplace and had to spend a day or a night in some city where I didn't know anybody, I might work in my hotel room or I might spend a Saturday on it, or the middle of the night, or something like that. That was all the time I had.

> SB: So the **Physics Today** article, which was basically the same original article, slanted for . . .

O'Neill: I think it was about the 6th or 7th draft by then, but the ideas were basically the same. It had been improved and a whole lot of detail had been added, and of course as you say it was for a different audience. The original article had been written for a popular audience.

> SB: Then what was the response to the magazine article?

O'Neill: It was very strong. But already between the time that the article was accepted and when it actually appeared, you see, there was the May 1974 Princeton Conference, and that really was when the concept got known to a fair number of people.

> SB: Because of Sullivan in **The New York Times**?

O'Neill: Yes. And then the articles that followed from that.

> SB: Michael where did you come into the picture? And how?

> Mike Phillips: Gerry, you wanted to talk to somebody at POINT Foundation.

O'Neill: Yes. I wanted to have a little conference, and try to get some people in to talk about this and see whether there was anything fundamentally wrong in the ideas. And I thought, Well, look, we could probably put on the conference with no budget at all, in some kind of way. But darn it all, it ought to be possible to get a little bit of money to do a few things that would make it better. So why not waste the few hours to become educated in the question of how you go about getting money to do something which has not been done before in our society. And so I started calling foundations. And I very rapidly found that no matter what it says in the title of any foundation, and no matter what people may say about a foundation — that it's always looking for new ideas and things like that — it's really not true. The only kind of new ideas that foundations are looking for are those that follow directly along the lines of things which are already in the mainstream. And I got personal interest from a number of foundation executives, and so on, but not a penny of money from any of them.

SB: This is how many foundations you're talking about?

O'Neill: Oh, I must have tried 8 or 10, something like that. And then someone suggested the POINT Foundation.

Phillips: Probably David Hunter.

O'Neill: I don't know who it was, Mike, it could have been. Stern Foundation sounds familiar. Then I got in touch with Richard Austin, which was not easy, by the way, because it came by way of the Portola Institute. He had moved and now he was in San Francisco, and there was a sequence of following telephone numbers, and so on. Finally I talked to him, and he was his own absolutely unchangeable self and, you know, very genuine and very open right from the start. So that was how we set up that appointment with you and Richard and me.

SB: You were out here anyway or flew out for that?

O'Neill: No, I was out here anyway in connection with my high-energy physics work. There was no budget, so I couldn't go any other way. So, Mike instantly understood what I was talking about, and I gather you were almost out of money.

Phillips: It was getting down there, it was almost near the end [mumble] . . .

O'Neill: So you came up with the famous $600 dollars and that was what funded the first conference.

Phillips: I just got that figure out of the air.

SB: So the **Whole Earth Catalog** is responsible for the colonization of space.

Phillips: I didn't give it to him without restrictions though. You see, I thought they ought to go through Princeton. That Princeton had to accept it. The grant to him wouldn't be nearly as . . .

O'Neill: I think it had an excellent effect. In fact let's trace that effect. You see, it had never occurred to me to even talk to reporters about this. Our idea was, we'll invite a few friends, and we'll have a little conference. And because that was handled as an official university grant, a certain amount of boiler-plate went along with it. Part of that boiler-plate was that a statement of that grant came across the desk of the University Publicity Office. They automatically sent out a university publicity release on it, and that was the reason why reporters came down to the May conference, and articles started getting written.

SB: What reporters did you get?

O'Neill: Local ones, and Walter Sullivan came down from the **Times**.

SB: Do you happen to know what drew him down?

O'Neill: I think he's a good conscientious reporter and science writer, and there must have been something in it that interested him, but beyond that I don't know. Incidentally, I think he may be opposed to this idea now. The latest that I've heard is that he is personally somewhat negative on it, though he continues to report the work accurately and fairly.

SB: So, this was May 1974. The conference included who?

O'Neill: We had, let's see, Gary Feinberg, from Columbia, and Eric Hannah, who was a graduate student at Princeton at that time, and Eric Drexler, who was an undergraduate from MIT, who had found me rather than the other way

around, incidentally. He's a very interesting guy. And we had Joe Allen, a young NASA astronaut, who presumably will fly some of the shuttle missions. Joe had learned about my work quite indirectly. He had talked to someone who had been to a lecture that I'd given on the West coast. I think Cal-Tech, or something like that. So Joe was interested and he came up to the conference. And then there were two people from NASA headquarters, and one of them was in the advanced launch vehicle division of NASA, Manned Space Flight office. He was very helpful because he gave us some initial estimates of launch costs. Those are still within a factor of 1½ the same numbers that are being kicked around now, so we had already the basic answers from him at that time.

SB: It was just one day? Was it papers, or people chit-chatting, or what?

O'Neill: Well, it was really two days because we had people come in on a half-day on Thursday. Perhaps 10 - 15 of us, just sat quietly around some tables and tried to review some of these ideas and get in hand how these papers would mesh, how these informal talks would mesh on the following day. Freeman Dyson came on that day, and Freeman got very interested and he stayed for the whole conference in fact. Dave Anderson, a student from Columbia; George Hazelrigg from Princeton in the Aerospace Engineering Dept. So the first day was just a few people sitting informally around a table, and then the second day was open.

SB: So there were a reporter or two there, and Sullivan did what with it in the **Times**?

O'Neill: He wrote, I think, a very good article, which the editors of the **Times** chose to put on the front page. He called a few days later. He said his first reaction was to be surprised, Gosh, that they had put it on the front page, and then his second reaction was, Well, why not? So, then, everybody sort of picked up on it from there, and there was an article on it in **Time**, and lots of interview requests. The BBC was on the phone within a day or two, and CBC and various New York stations and so on, and stations out on the West coast.

SB: And you were doing the interviews?

O'Neill: I was doing the interviews. That was still in a time which I had the time to be able to do the interviews that got asked for.

SB: So there was a media flash for how long?

O'Neill: Well, it really never stopped. The Associated Press did an article in the summer of '74, quite a good article. Howard Benedict was the reporter, and he took care to check his facts very carefully. It was quite a good job. The **Los Angeles Times** asked me to do an op-ed piece, which I did for them, and they had it illustrated reasonably well. And since they're part of the L.A. Times - Washington Post syndicate, that got picked up by papers in many different places in the country. And then the **Physics Today** article came out in September. That prompted a large amount of response which was more from technically educated people. And also that prompted some of the counter-blasts. There were two that I would call carefully worked out, in the sense of someone at least sitting down and trying to work out some numbers. And it took me a lot of time, because I then had to sit down and answer these things in detail.

> Phillips: Were they creative? Did they result in new forms for . . .

O'Neill: It's a very good question. There was nothing that came out of them that was creative in the sense of suggesting a new possibility that had not been floating around earlier, or a new solution to a problem that hadn't been floating around earlier. But I think they were very helpful in forcing me to go into detail and justify on a numerical basis things which I hadn't taken the time to calculate, because I just had had a sort of hunch that the numbers would work out all right. I think that I certainly learned a lot in terms of additional insights into, say, how the economics of the whole thing might go than I would have without those criticisms. So it was probably quite worthwhile. Then, let's see, there was the New Scientist article — that was in late 1974.

> SB: You said then there was a second Princeton conference. When, this May?

O'Neill: That's right. We've been trying to measure in some way the exponentiation time for the whole thing, and at the second Princeton conference I just made a hasty calculation that during the last year the exponentiation time was something like 3 - 4 months.

> SB: "Exponentiation time" being what?

O'Neill: Time in which some level of interest, activity and so on is growing by a factor of e.

> SB: "Factor of e" being what?

O'Neill: 2.71 . . . Pocket calculators would automatically think in those terms too. Physicists are just used to it. So just for fun at that time, off the cuff, I tried to think of the ratio of the funding level of the 1975 Princeton conference which as far as I'm concerned will be the last one it will be necessary to hold of this general introductory types, to the funding level of the first one, which you paid for, or YOU [readers] paid for, depending on how you want to measure it. And the ratio was something like 14 or 15, and so working out the exponentiation time, it was just about four months. I'll admit that's a silly way to try to measure something.

> Phillips: The NASA Ames study is a hundred thousand dollars. May at Princeton was $12,000.

O'Neill: The big crunch will come now, because we're at the point where you have to sit down and do some serious studies. Bunches of people getting together and waving and shouting is not going to push the thing along. At this point you really have to do your homework and have serious, specific research which runs for a year or more, and on a much more fine-scaled set of topics. To do that right would take something between $500,000 and $800,000, I would say, spent over the next year. And I cannot imagine any way within the present set up of the government that that money could come out. Because the 2-year funding cycles that you normally have just don't permit that.

> SB: Summarize briefly what came out of the second Princeton conference, and what you expect to come out of the summer at NASA Ames.

O'Neill: Well, the NASA Ames study, I'm not sure what will come out of that, because that really is not a directed project study in the same way that the previous work has been. First of all it's funded out of the American Society of Engineering Education through a NASA grant, a continuing thing. It's really intended to be an educational process for young faculty members, not for aerospace professionals by any means. They came in for the summer with the intention of going through an exercise in systems design, and many of them arrived on the scene knowing next to nothing about the whole subject. So it's mainly an educational process. The challenge is to obtain a serious design study from a formal setup which is non-directive and quite different in outline. So far the Ames study is doing quite well, given its limitations.

> SB: In a sense this is like your original seminar.

O'Neill: Maybe so. The Princeton conference got us quite a bit farther, because that was a directed kind of thing where we had about 25 or 30 invited speakers, each of whom prepared a paper on a given topic. I think we can now make some fairly definite statements: one is that as far as the quantity of material that does have to be brought up from the earth — that is, what you need to give you the stepping-stone, to establish this sort of first beachhead in space — the costs, numbers and sizes of the vehicles involved are well within what people can do in the 1980's timeframe. They don't require anything more advanced than the space shuttle and the sort of vehicles that you could easily develop from it. There's nothing super advanced involved, no nuclear-powered rockets or anything like that.

The second solid thing that we could say is a consequence, not just of the Princeton conference but also of a lot of very thoughtful letters and calculations I've had from various people whom I would never have known about had it not been for all the publicity that's come to this business. The agricultural numbers that I used in Physics Today now seem to be very much on the conservative side. You could probably do quite a lot better than that in terms of yields per square meter to support people. But that's okay. We are farther along on the question of design of the mass-driver, which is really crucial to the whole thing. It's the electromagnetic machine for bringing lunar surface material from the surface of the moon up to the colony site. That's extremely important because practically all the material for the first colony, and everything for the products that it produces has got to come from the moon. There I would say we're somewhat further along now than we were a year ago, we seem to be qualitatively okay, and even quantitatively in terms of the basic details, but a lot of calculations still have to be done. And we are somewhat further along in terms of guidance methods and things of that kind. That's the sort of thing where nothing less than a serious, quiet study where four or five people sit down full-time on this subject for several months or a year would do the job.

And we had an interesting paper on Space Law. Apparently the building of the first space community would fit within all of the international treaties if you stick to, as I recall, three conditions. First thing, it's got to be non-military. The second, that if anything interesting, new research, comes out of it, like information about the surface composition of the moon or something of that kind, that it does have to be made available through the United Nations to anybody who wants the information. And the last is that, at least in some nominal form, the community has got to be under the jurisdiction of the nation or group of nations which establishes it. You cannot, at least deliberately, send people out to be absolutely on their own.

> SB: So there's a funny point now between the Princeton conference level and the next serious study level. What's your expectation as to how it will actually proceed over the next couple of years.

O'Neill: Well, we know that our support from NASA (which is small, but enormous by the standards of what we had just a year ago) is going to be continued and increased somewhat for this next year.

SB: Will you be administrator of that?

O'Neill: It's such a tiny thing, there isn't really much administration involved. Yes I'm the, whatever it is, the — I forget the title I'm supposed to have. It's the usual thing; you have to be responsible for these things.

SB: Are you getting any kind of on-going cadre of people tending to track along with the project?

O'Neill: Oh, very much so.

SB: There's how many people now?

O'Neill: Again it's a question of the level of involvement. There are now at least 15 or 20 people who are spending some significant fraction of their time on a volunteer basis, working on this, unpaid entirely. Some of them are spending quite a substantial amount of time that way. There are a very much larger number of people who have written offering to help if there's some way that they can within their limitations; they've got full-time jobs to hold down. So far I've felt that I ought to put out the newsletter myself.

SB: The newsletter, what's that?

O'Neill: It doesn't cost anything, you just have to write to my address in Princeton, and the secretary should send the newsletter. There's one coming out now. I just finished writing it. Prof. Gerard K. O'Neill, Physics Dept., Box 708, Princeton, NJ 08540.

SB: You're still being basically a high-energy physicist?

O'Neill: Oh, yes. There's no real government commitment, for example, to push this program hard at this point. I've spent many years building up my little group in experimental high-energy physics and I'm certainly not about to tear it apart in an unstable moment like this.

SB: What would it take for you to become a full-time space colonizer?

O'Neill: Well, if the president came to me and said, "Here is X billion dollars, we're going to go ahead with the thing and we want you to be involved with it." That would sure fetch me.

SB: Suppose NASA said, "Here's 5 years of personal salary to administer the growth of this program"?

O'Neill: That's not enough. I have a deep suspicion of governments, and really — although I'm not politically active — I know enough about politics to be very suspicious of it; I think I would have to see a really substantial committed kind of program going. I don't mean at the spending level of billions of dollars, but I'd like to see something where there's a very solid commitment to continue in the same sense as there was in the Apollo program.

SB: Of the level of Kennedy saying, 'We're going to be on the Moon in this decade." For some politician to make this go, he's going to have to say "by the year so-and-so." What year is that?

O'Neill: Arthur Kantrowitz, the president of AFCO-Everett was out visiting us a few days ago. He happens to be quite enthusiastic about this work, and he says that his answer for things of that kind is to say, "You'll have the result ten years after you've stopped laughing," which is I think, a pretty good answer. The most responsible answer I could give is to say that if I really had the responsibility for getting it done by a certain time and the authority to do it

in what I would consider the right way, then I would be willing to make a very strong commitment that it could be done in 15 years from time-zero. Whatever that time-zero is.

SB: This is Model One with an extra-territorial population of what?

O'Neill: Yes, Model One. Roughly 10,000 people. If you look at the growth rates that you could get from that first one, then you'd probably be talking about a quarter of a million people by the year 2,000. Because you'll be going up very fast after you get the first beachhead.

SB: And your graph I saw in Washington suggested a net population decrease on the planet's surface by . . .

O'Neill: I think the turn-over there is about 2018. Now that was based — first of all, I don't make it as a prediction — it was indicated as a technical possibility, and it was based on a time-zero of essentially now, which is certainly unrealistic politically, and a completion date of 13 years, so that would put Model One in place by 1988. Maybe you could even do it from now, technically, but it's probably more reasonable to say 13 - 20 years from a time of decision.

SB: Do you think there's no way to get the toothpaste back in the tube at this point. . . . that the idea is inevitable?

O'Neill: There are other possibilities. Civilization could tear itself apart with energy shortages, population pressures, and running out of materials. Everything could become much more militaristic, and the whole world might get to be more of an armed camp. Things of this kind might just not be done because no nation would dare to divert that much money away from military efforts. Or without war, it could be that the world will become poor, to the point where it can't afford to try things like this.

Of course, if neither of those possibilities occurs, then I do think there is some sort of inevitability about it. With that, of course, you can't associate a time-scale. It could be a long time.

SB: Who resists the idea in any large way? If anyone.

O'Neill: Well there was a while when I thought that elderly and famous professors of physics were the greatest opponents . . . In fact of all the mail I've gotten only about 1% has been in opposition to it.

SB: And what's your short roster of planetary problems that will be solved by this particular technique? Energy, population . . .

O'Neill: Well, yes, but by phrasing the question in that way it's difficult for me to answer except with a prediction or promise, and that's something that no decent scientist likes to make. I think it's very wrong to assume that something like this is going to promise happiness to all people, because people manage to make themselves unhappy in almost any circumstances.

SB: Well, now I'm going back to '69 when you and your students were on the defensive about relevance.

O'Neill: "What's the relevance?" That's a fairer way to ask the question. From the economic viewpoint, which is perhaps the narrowest, there are starting to be products which are needed, and whose end use is in very high orbits above the earth, like geo-synchronous orbit or even father out, escape distance or beyond. And for that class of products, a space community has a very strong advantage. We tried to run the numbers for several different kinds of products of that kind: processing lunar materials, satellites, solar power stations, very large radio telescopes, things like Project Cyclops, something that would listen for extra-terrestrial civilizations. We concluded that you could do Project Cyclops for 1/10 the price if you built it at a space colony, rather than on the earth. And things like large space-research vessels for going out to the outer parts of the solar system with a large research team, hundreds of people — very much in the spirit of Darwin's voyages. Exploring the planets in detail is something where you'd go out with a ship of several hundred people, that would be a self-contained community. It would be able to run for several years, and would go into orbit around one of the outer planets and send down small vessels to take samples and do surface explorations. You'd be doing your data analysis in real time, and any startlingly important results you would be beaming back to the vicinity of the Earth immediately. The detailed tapes you'd carry back at the end of the voyage.

One of the ideal industries for the community at L-5 is ship-building, because if you're going to build a big ship that's going to be in space, it's pretty absurd to build it at the bottom of the hole that's 4,000 miles deep and then try to haul the pieces up. As far as the satellite power stations are concerned, again we're going through the numbers, but it looks as if it can be done that way very much more cheaply than it could from the surface of the earth, again, because you're using lunar materials which can be obtained at low cost. Having cheap energy may do more toward feeding the population of Asia than almost anything else we can provide.

One product that can be built, once you have a first colony there, is more colonies. You can construct a community which is self-contained in terms of all of its basics, of relatively small size. You're not depending on a food distribution system that's a thousand miles long, for example, as we do here on the Earth. Or an energy distribution system that's 7,000 miles long, as we do here on the Earth. Suppose that all of those essentials were obtainable over a distance of only ten miles, and by a population which was as small as ten thousand or a hundred thousand. I would think that in the long run, the tendency toward community diversity, the diversity of governments, diversity of the ways people choose to live, the kinds of architecture that they choose to have, and so on, would be enormous. Which is, I think, in exact contrast to the way that things are going on the surface of the Earth at the present time. And I believe that if someone were to look back on this whole business from the vantage point of say a hundred years further in time, probably the economic factors, which loom so large to us, will seem then to be relatively unimportant, because they won't be able to appreciate from their presumably much wealthier vantage point, what our problems were like. But the question of diversity and of the opening up of new possibilities and new frontiers, both of the body and of the mind, I suspect will come to be regarded as the most important contribution that these ideas have made.

Phillips: I wanted to ask about the conference. What papers, or what people were the most exciting, or the most interesting?

O'Neill: That depends an awful lot on your point of view, of course. Naturally, I was most concerned over those speeches which bore directly on the question of the validity of the calculations that we'd been making up to the present time. So from my point of view, two of the most important papers of the conference were those by Hugh Davis of Johnson Space Center, and Del Tishler who was formerly at NASA headquarters. And those were very straight-forward launch vehicle papers, which might not be so exciting to someone who is interested in sociological questions for example.

Phillips: What about the materials talks?

O'Neill: Materials? There was a good talk by David Criswell, from the Lunar Science Institute in Houston, and it looks as if from the Apollo samples of lunar rock and the additional information that Dave was able to give us, we're in quite good shape for metals, glass, oxygen. We already knew a year ago that we don't have good sources of carbon, nitrogen, and hydrogen on the surface of the Moon. Dave went through an exercise which seems to indicate that we could, if we want to, get enough hydrogen out of the lunar surface by preselecting fine-grain material before we send it heating and processing a greater quantity of material than the amount out to L-5.

SB: It is my understanding that the lunar material is not what anyone would call ore in any Earth sense.

O'Neill: It's not so bad. There are lots of places on the earth where 1% ore is regarded as relatively good these days. And there are large areas of the lunar surface where just the ordinary dirt that you pick up out of the ground is as much as 10% aluminum, and around 30% in total metals.

SB: How does that compare to bauxite?

O'Neill: Not as good. Bauxite's a richer ore than that. But that's becoming a scarce resource. The Bureau of Mines is already conducting studies, some of which are up to the pilot plant stage, for the processing of some ores which turn out to be identical to the ones we would have to work with on the moon.

SB: Does working in vacuum in 1/6th gravity give you any advantages in working the ore?

O'Neill: Well the vacuum environment of the moon is vital for the question of low shipping costs. The mass-driver, the electromagnetic machine, can only work efficiently in vacuum. So that point is critical. The low gravity of the moon is also necessary for low transport costs.

Phillips: Vacuum processing might have some influence on ore.

O'Neill: Well, you have your choice, you see. You can do your processing either in vacuum or in an oxygen atmosphere if you want to, because the soils that you have on the moon are 40% oxygen by weight. In the long run you can get everything from the asteroid belt.

Phillips: Just because people haven't become emotionally attached to asteroids yet, and because they have become emotionally attached to the Moon?

O'Neill: Well if you look at the economics, it's certainly going to be much better to build new colonies by going out with the construction equipment to the vicinity of an asteroid, taking the material right there, for which you don't have to pay anything in terms of a velocity interval to get it away, build the colony, and then move in the people, because their mass is very small, so the cost of bringing them is very small. And you've got the colony already working and productive and beginning to pay back its construction costs, and then you can do whatever you like after that. Incidentally, the asteroid belt can very easily give the colonists Earth-normal sunlight. That's not a problem at that distance.

They have their choice. They can roughly over a period of one generation at a relatively low energy cost, work their way back into the vicinity of the Earth. Or, they could go the other way if they feel like it. You can imagine that during one generation, within 30 years or so, it would be technically quite possible for a big colony to move itself into even a polar orbit relative to the ecliptic. They could move themselves into an orbit so far removed from the rest of the human race that there'd be no interaction at all, except for communication of course, — if they chose to communicate. There might just be some totally absorbing colonies, that would be listening to what was going on from the outside but not saying anything at all.

> Phillips: What are the things that you want worked on right now? If you were to pick the 10 subjects.

O'Neill: That's an awful good question, Mike. I would love to see some really good work on chemical processing of the ore coming out from the moon. How to get the good quality glass, good quality metals, and any non-organic fibers you want, anything of that kind.

> Phillips: OK that's one. If there were 10 people you could set to work in teams . . .

O'Neill: Yes. I'd like to see some really good work on small-scale, closed-cycle agriculture. Some of that sort of stuff is already going on in some places. The question is to vary some things like oxygen concentrations, CO_2 concentrations, and so on. There are ways in which you could model this environment more thoroughly than it has been done up to now.

I think that somebody should be doing something which is really hard — it hasn't been done before and yet it's very accessible — and that is to analyze a number of the existing industries on the surface of the Earth in terms of their productivity, but measured in ways that we've not measured them before. Productivity in the sense of what is the total mass of the products that the industry turns out in the course of a year, divided by the total mass of the installed construction equipment of the factory itself. You can see why this is extremely important for us. It's never mattered to anybody before.

There's the question also of taking an industry, and saying what is the productivity of that industry measured in terms of the output products per person in the industry. But now possibly dividing the people into two classes, in a way that hasn't been done before. In an electronic age there need not be a distance of 50 feet between a design loft where someone is designing the next generation of products and the construction area where people are building this generation. There is no reason why, in this day and age, those things couldn't be a quarter of a million miles apart. And therefore you want to analyze an industry, seeing not just what is the total productivity in tons per man-year, but in tons per man of a kind that has to be associated with the production machinery, as opposed to all of the people who are selling, designing, administrating, and so on. You can imagine an industry which has one foot in each camp. You can imagine it with many of those activities carried on on the surface of the Earth, and a very good wide-band computer link-up between that area and the space community, and the actual construction going on in the space community.

> SB: Let's see, as you get into lunar distances, you're getting what kind of lag in communication?

O'Neill: Enough to be annoying for voice. It's like a second or two. In fact it's more than that. For a round trip time it would be almost 3 seconds.

> Phillips: You got a lot of things you could get started right now. In fact, by simply listing and describing them, you can get people to do them.

> SB: That's the kind of list I would be particularly interested in, because presumably some of our readers will go "Aha, that's me," and take off on it.

O'Neill: That would be very good. We need to have a more thorough and careful job done of looking at albedos and questions of thermal radiation from a space community, making sure there's no problem about holding the temperatures about right. We've been assured by the people down at JPL, who've sent all the spacecraft out, close to the sun and all the rest, that our particular problem is well within the range that they know how to handle. But it would be nice to look at that more thoroughly.

Let's see what other basic ones we've got. There's another class of problems which may take a while to get answered. It's the sort of thing that NASA would logically do, but I don't think they're prepared to go in this direction yet. The curious fact is that if you start designing one of these large communities, in terms of just the economics of how much in the way of materials you have to put in per person, it costs you very little to make earth-normal gravity. It turns out to be very easy. But to make earth-normal atmospheric pressure, costs you very badly, for two reasons. One is that 80% of our atmosphere is nitrogen, which we don't use at all. People can breathe pure oxygen atmospheres perfectly well. The Apollo astronauts were breathing pure oxygen atmospheres for days at a time. I've done it for hours at a time.

> SB: What about plants? Are they happy with that?

O'Neill: They couldn't care less. They take their nitrogen from the roots; they don't take it from the air. The main reason for going to anything but a pure oxygen atmosphere is just fire protection, apparently. And the situation that NASA has never had any reason to investigate, is one in which you have a pure oxygen atmosphere but have only 2 or 3 pounds per square inch pressure. That's about the same as the partial pressure of oxygen on the Earth, within our range of altitudes. In large volumes, as opposed to the confined volumes of the existing spacecraft, and with an abundant water source present. Now if you'll look at fire protection problems in that particular environment, I'm reasonably confident that you're going to find that fire protection is not that difficult, but that situation has not existed up to the present time, and no research has been done on it. All it would take really is a large vacuum tank, — where you can get rid of the nitrogen and have 3 psi of oxygen. To try to light some matches and see what happens when you've got a reasonable size space and that pressure and lots of water on hand to put out a fire.

> Phillips: What is the coriolus effect on a small colony? What are the interesting things that would occur?

O'Neill: Well, we've been told that somewhere between two rotations per minute and three rpm, the range for Model One, that only a very small fraction of the population would have any problems with initial seasickness. Pop-flies in baseball on Model One would be curious affairs. In your rotating coordinate system, you would see them assume a curve which they didn't really have.

> SB: Someone I talked to yesterday, said you wouldn't feel it walking around along a cylindrical axis, or along a cylindrical circumference, but if you changed elevation, you would feel it. If you jump.

O'Neill: I don't know if that's true. Maybe. Not at the lower rotation rates. Apparently the people who have been put in centrifuges on the earth have been able to acclimatize to spin rates as high as 10 per minute. Going around once every 6 seconds. But that's something which you're going to get with practice say over a 10-day period.

> SB: If "in wildness is the preservation of the world," then in what is the preservation of the space colony?

O'Neill: Making it wild, I think. The long-term plan, really dream, that I would have is a situation in which, in 50, 100, 150 years, it would be so cheap to replicate large communities that you would be building quite large ones, many, many square miles in land area for each one, and they would be very thinly populated. And so the natural development it seems to me, is toward a situation where you have a great many wild species involved, and as wild an environment as you choose to make. I would imagine one on which there is a lot of forest and park area and wild areas, and a relatively small amount which is manicured and put into the form that people like to have for their dwellings.

> SB: Now you're restating your question, whether a planet's surface is the best place for a wilderness?

O'Neill: Maybe so. But this situation that I was just describing, this possibility if pursued, is one that could occur both on the Earth and in the communities of course. Because the existence of the space communities as a place to which many people might choose to move would also be perhaps the only realistic non-violent way in which the Earth's population might really decrease.

> SB: I'm trying to imagine the trapped feeling that one might have. Travel between communities would be relatively easy. Travel to the Earth's surface and back would be relatively hard. Is that correct?

O'Neill: It would be interesting to compare it in terms of real income. Passage between the colonies and the Earth probably corresponds to passenger travel back and forth between Europe and the United States in, say, the 1700's. It's the kind of thing that Benjamin Franklin did to go and negotiate treaties in France. It was not the sort of thing that the ordinary guy was able to accomplish.

The cost of going back and forth to the Earth — I made some rough estimates on what that might be with the technology of let's say 20 - 30 years from now, still nothing far out like nuclear power or anything like that — and came out to about $3,000 per person for a round trip. Among the colonies it should be very easy, very cheap. From one community to another, even 5,000 miles away would probably be as little as $100 or something like that. A few dollars in energy costs is enough to launch a vehicle over that kind of distance.

> SB: Another question on the life of the idea. Who are your predecessors?

O'Neill: I think the one that is really the most relevant is Konstantin Tsiolkowsky — the Russian who did the early work on liquid rockets. I had been looking for his works, and I had made various attempts in libraries without ever connecting with the right ones. Finally just this spring, a friend gave me a couple of Tsiolkowsky's books. I'm still trying to get copies of my own and haven't been able to yet. But the book which is most relevent is called **Beyond the Planet Earth**. It was written more than 50 years ago, almost 75 years ago. It's a novel, and so of course he was able to duck lots of issues, but he had the essential ideas, I think, very much better than most people who came after him. In particular, he imagined his first voyagers out from earth, soon after liftoff got out of the eclipse area and started building greenhouses. That was the whole point. They had been out there for weeks when it occurred to one of them to say, "Maybe it would be nice to go over and have a look at the moon." So as an afterthought, an aside, they went over and visited the moon. In the course of the visit they found gold and diamonds, easy enough in a novel, and then they did the next sensible, logical thing, they went out and started exploring the asteroids as material sources. There were some people much, much later, like at least 25, more like 50 years later, that talked a little bit in those terms, but not even they, I think, saw the whole as clearly as Tsiolkowsky. ∎

3-D Universe

. . . astronomy has shown us that the stars are very far away, and they surround the earth in 3 dimensions, except I continue to experience them as "points of light in the sky." They might as well be glued to an acrylic sphere 20 miles away.

I've had occasional 3-d flashes, but only when stoned, and only when I wasn't trying to have a 3-d flash. One evening, stoned, I looked at the zodiacal light, and rather than being "a cone of light in the sky," it became a mass of dust and gas encircling the sun. Its size frightened me so much that a circuit breaker in my mind blew, and the zodiacal light returned to normal within a second. Even so, I was so shook up that I had to go back inside the house, to the comfort of a human-sized space.

Have any of your readers experienced the Milky Way in 3-d? Did they survive? Did it affect them permanently?

My wife says she can have 3-d flashes at will, when the mood is right. However, in asking around, I gather that such experiences are rare. This points out a dilemma which is common to all science, not just astronomy: little emphasis has been placed on how scientific knowledge can affect our <u>experience</u> of the universe. Instead, the knowledge is packed away in some abstract corner, isn't systematically integrated into our lives, and we remain, in many important respects, medieval peasants, only now we're smart.

The integration process completes a cycle: observe the universe, abstract knowledge from the observations, integrate the knowledge into our experience, observe the universe again . . . only now it's a WHOLE NEW UNIVERSE. This is an exciting point they never told me at the university, since they concentrated exclusively on the first half of the cycle.

But I imagine you Californians have been transforming the universe for some time now. When are you going to do an article on it? Do any of your readers have 3-d flashes? Do they have any ideas about how to have them at will? And how about experiencing fire as an oxidation process?

Gordon Solberg, Rimfire Ranch
Radium Springs, New Mexico

The High Frontier

The one book you must have if you're interested in Space Colonies is this one by Gerard O'Neill. His scheme has aroused so much rabid support and rabid opposition that O'Neill's gentle voice and responsible perspective has a critical balancing influence. I have seen environmentalists who at first blush loathed the idea of Space Colonies come away from O'Neill's book impressed and interested.

Having one individual most strongly identified with a grand vision such as Space Colonies is the healthiest way to proceed, I'm convinced. And I'm glad Gerry has the job on this one.

— SB

The High Frontier: Human Colonies in Space

Gerard O'Neill
1976; 288pp.

$2.25 postpaid
from:
Bantam Books, Inc.
414 E. Golf Road
Des Plaines, IL 60016
or Whole Earth

When I have considered the effect of our discovering, one day, signals from a more advanced civilization (note that it would be, with almost 100 percent certainty, millennia more advanced than we are because of our own position at the threshold of communication) it has seemed to me overwhelmingly probable that the first effect of the discovery, as soon as the excitement and the novelty have worn off a little, would be to kill our science and our art. What purpose to study the natural sciences? We already know that they are universal, so if a civilization now radioing to us is 50,000 years ahead in its knowledge, why continue to study and search for scientific truth on our own? Gone then the possibility of new discovery, or surprise, and above all of pride and accomplishment; it seems to me horribly likely that as scientists we would become simply television addicts, contributing nothing of our own pain and work and effort to new discovery.

In the arts, music and literature, the case may be somewhat more unclear; yet on earth the almost invariable consequence of contact between a primitive civilization and one more advanced is the stagnation of the arts in the former. Only in the form of a "tourist trade" does art survive, in most cases.

If this sequence of effects is of more than local significance, as I think it is, it will be quite obvious to any civilization more advanced than our own. I would then add one more assumption: that the same characteristics which render a civilization immune to intellectual decay and stagnation, if there be such characteristics, are accompanied by a repugnance to inflict harm on others, in particular to other "emerging" civilizations more primitive than its own. In that case, "They may be out there, but they're kind enough to keep quiet."

•

I confess to a humanitarian bias in the design that I suggest. Technological revolution is a powerful force for social change, and in choosing among several technical possibilities I have been biased strongly toward those which seem to offer the greatest possibilities for enlarging human options, and for breaking through repressions which might otherwise be unbreakable. Yet I offer no Utopia; man changes only on a time scale of millenia, and he has always within him the capacity for evil as well as for good. Material well-being and freedom of choice do not guarantee happiness, and for some people choice can be threatening, even frightening. Though I acknowledge that my study will be of the physical environment, and only indirectly with the psychological, I will still try to describe an environment which combines with its efficiencies and its practicality opportunities for increasing the options, the pleasures, and the freedoms of individual human beings.

I have argued that there is only one way in which we can develop truly high-growth-rate industry, able to continue the course of its development for a very long time without environmental damage: to combine unlimited solar power, the virtually unlimited resources of the Moon and the asteroid belt, and locations near Earth but not on a planetary surface.

Large tugboat driven by reaction engine seen in background has decelerated an asteroid sent in from the main belt and is about to process it for materials to be used in construction of new space habitats and Earth-orbital solar power stations.

Chesley Bonestell, 1976

L-5 News

Much the best source of news on Space Colony matters, the L-5 News is an enthusiast publication of admirable vigor, and it's no fanzine. The editors and contributors are in the thick of Space Colony design and speculation. They plan to live in Space.

— SB

L-5 News

$20.00/yr (monthly)

from:
L-5 Society
1620 N. Park Ave.
Tucson, AZ 85719

Also there is a West European Branch of the L-5 Society at 40 Lamb St., Kidsgrove, Stoke on Trent, ST7 4AL England.

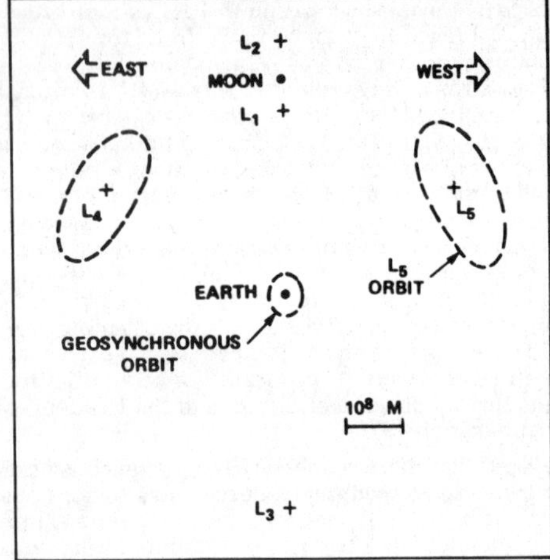

WHAT ARE LIBRATION POINTS?

T. A. Heppenheimer

Our organization name is the L-5 Society, and our newsletter has stated that L-5 is one of the libration points in the Earth-Moon system. But what are libration points? The answer is that libration points are locations where a spacecraft may be placed so as always to remain in the same position with respect to the Earth and the Moon.

Suppose the Earth and Moon were fixed in space and did not move. Then a single libration point would exist at the point where the gravity fields of Earth and Moon cancel out. A body placed there would feel equal and opposite attractions from Earth and Moon, and so would stay fixed in place. But if the body were moved slightly, it would feel a slightly greater attraction from either Earth or Moon, and so would fall down, moving rapidly away from the libration point. That point, therefore, is unstable.

In reality, the Earth and Moon are in motion about their center of mass. (It is the Moon, of course, which does most of the moving.) This means that, in addition to the gravity of Earth and Moon, we must take account of the centrifugal force acting on a body in orbit. Libration points are then the points where these *three* effects cancel out: the two gravity fields, and the centrifugal force.

The French mathematician, Lagrange, in 1772, showed that there are five such points. Three of them lie on a line connecting the Earth and Moon; these are L-1, L-2, and L-3. They are unstable; a body placed there and moved slightly will tend to move away, though it will not usually crash directly onto the Earth or Moon. The other two are L-4 and L-5. They lie at equal distance from Earth and Moon, in the Moon's orbit, thus forming equilateral triangles with Earth and Moon. These points are stable. It is a curious fact that they are stable because the Moon is only 0.01215 times the total mass of Earth and Moon together. If the Moon were greater than 0.03852 times the total mass, L-4 and L-5 would be unstable.

The situation, however, is even more complex than this. The Sun is in the picture, and it disturbs the orbits of spacecraft and colonies. It turns out (from an extremely messy calculation done only in 1968) that with the Sun in the picture, a colony should be placed not directly at L-4 or L-5, but rather in an orbit around one of these points. The orbit keeps the colony about 90,000 miles from its central libration point. The orbit is roughly bean- or kidney-shaped. It may seem curious to speak of an orbit about a point. Actually, the colony is in orbit about the Earth, but the simplest way to describe the orbit is from the point of view of an observer sitting always at the libration point.

In the colonization project, the colonies are to be located in the vicinity of L-4 or L-5. L-2, located behind the Moon, is the site of a catcher vehicle, which collects mass shot up from the Moon by the mass-driver. L-1, between the Earth and Moon, is the location of a satellite power station, to provide power for the moon base. No use has yet been found for L-3. However, at L-3 the Moon is permanently eclipsed by the Earth, so it could serve as an asylum for people suffering from lycanthropy (werewolf tendencies).

•

VAPOR DEPOSITION OF MASSIVE STRUCTURES

Work done this summer suggests the possibility of fabricating space structures directly from vaporized metals. If proven out by further design work and by vacuum chamber experiments, it promises to provide a textbook example of the use of space conditions in processing, and may cut the cost of a colonization program by some tens of billions of dollars.

Fabrication of seamless hulls or hull segments would be a simple and useful application. The solar energy flux, vacuum, and lack of gravity in space make it easy to vaporize metal and direct it as a conical beam; metal plate of the desired strength and thickness can then be built up on a balloon-like form made of plastic film.

This technology seems applicable to both aluminum and steel alloys, to structures many kilometers in diameter, and to structures more complex than smooth balls. Because it requires little equipment and negligible labor, it promises to reduce the cost of some space structures to little more than the cost of raw material. It seems a fruitful area for further research.

Eric Drexler

•

A recent study at the Hudson Institute entitled The Next 200 Years in Space (NASA Bicentennial Planning) by Herman Kahn and William M. Brown, predicts the building of space communities and manufacturing facilities.

Those interested in the complete report should write to:

Hudson Institute
Quaker Ridge Road, Croton-on-Hudson
New York 10520
1 Rue du Bac, Paris 75007, France
1-11-46 Akasaka Minato-ku, Tokyo, Japan

National Space Institute

This is a public lobby set up by Werner von Braun shortly before he died to help build a broad Space constituency and help keep Washington informed of what that constituency wants. Apparently it's pretty effective and is riding adroitly the burgeoning new public interest in Space. Good newsletter!

— SB

National Space Institute Newsletter

$15.00/yr (monthly)
(includes membership)

from:
National Space Institute
1911 N. Fort Myer Dr.
Suite No. 408
Arlington, VA 22209

Comments on O'Neill's Space Colonies

Is Balance Really Possible Where Even Gravity Is Manufactured?

Something about O'Neill's dream has cut deep. Nothing we've run in The CQ has brought so much response or opinions so fierce and unpredictable and at times ambivalent.

It seems to be a paradigmatic question to ask if we should move massively into Space. In addressing that we're addressing our most fundamental conflicting perceptions of ourself, of the planetary civilization we've got under way. From the perspective of Space Colonies everything looks different. Choices we've already made have to be made again, because changed context changes content. Artificial vs. Natural, Let vs. Control, Local vs. Centralized, Dream vs. Obey — all are re-jumbled. And Space Colonies aren't even really new. That's part of their force — they're so damned inherent in what we've been about for so long.

But the shift seems enormous, and terrifying or inspiring to scale. Hello, stars. Goodbye, Earth? Is this the longed-for metamorphosis, our brilliant wings at last, or the most poisonous of panaceas?

I've done a couple of things to help drive the question as deep as I could, for most original yield. Along with the presentation in the Fall '75 CQ of several angles on O'Neill's vision I stated all of my enthusiasm for the project and few of my misgivings — the usual brainstorm approach. I was partially aiming at environmentalists, including me, who have become too predictable of late, too smug, certain, convergent, uninquiring and unimaginative. We have come to love our famous problems (population, inequity, technology, etc.) and would feel meaningless if they went away. That's a lousy design posture.

This winter (1975-76) I wrote personally to a number of notable people inviting their comments (which follow beginning overleaf). I said I would pay them $30 and print whatever they sent, and would they hold it please to 1,000 words (some didn't). Response was amazing; at least half of the list wrote something — divided almost equally pro and con, somewhat more on the con. For this I'm grateful. Space Colony planning is going ahead full-bore and one-eyed. It badly needs intelligent criticism.

The readers at large were invited to remark, telling whether they think Space Colonies are a good or bad idea, whether they'll gladly emigrate, and why. 170 people responded, which seems a lot for our modest circulation (17,000 or so in 1975). Results have been tabulated in the graph. Excerpts from the readers' letters are scattered through the following pages (they were paid $10 each).

There's one thing within the response perhaps worth mentioning. The colleges these days are having another Silent Generation like the one I was in during the late '50's.

Nothing seems to make them jump — not politics, drugs, heroes, projects, or any special sense of themselves as a generation. The 122 people who wrote us enthusiastically approving of Space Colonies — and most of them wanting to go — are nearly all college students.

—SB

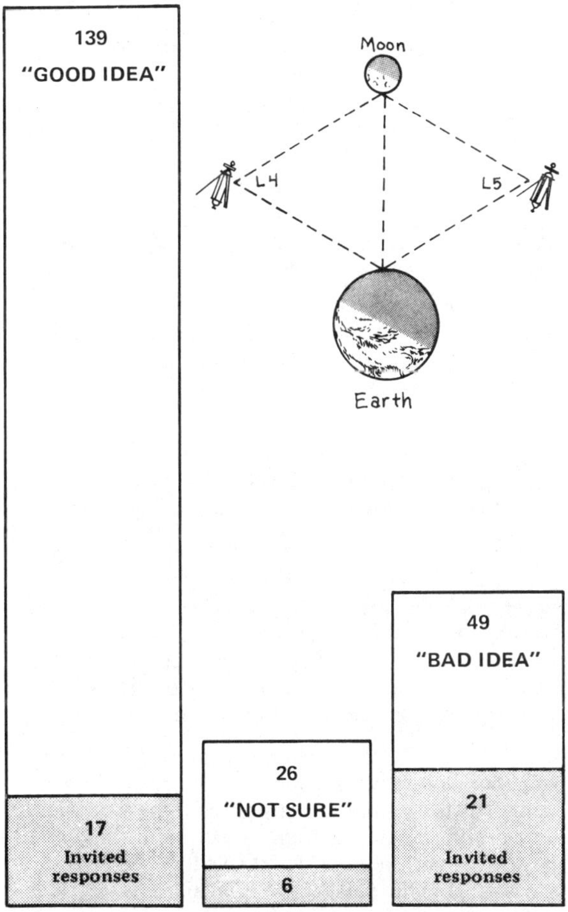

139 "GOOD IDEA"
17 Invited responses
26 "NOT SURE"
6
49 "BAD IDEA"
21 Invited responses
Moon — L4 — L5 — Earth

Response to Space Colony questionnaire by readers and invited commentators. Of the 139 who liked the idea, 104 said they were ready to go aboard. Women accounted for 13 yesses, 7 no's, and 1 maybe. The most universally favorable group was artists.

Inside a Model III Space Colony. This painting by Don Davis, a portion of which is on the cover, has aroused as much ire as admiration. Viewing from a colony end-cap mountain it depicts weather, a running river, mature ecosystem, "night" falling, and agricultural pods outside the skylight. Real Space Colonies may never be so nice, may never be at all, or may be much nicer.

KEN KESEY

Novelist, pioneer, author of One Flew Over the Cuckoo's Nest; Sometimes a Great Notion

A lot of people who want to get into space never got into the earth. It's James Bond. It's a turning away from the juiciness of stuff. That's something that's lost its appeal for me.

RUSSELL SCHWEICKART

NASA astronaut (Spacewalk, Apollo 9).

Gerry O'Neill is my hero. At a time when hair-shirting is the style and immediate utility the password to success, O'Neill dares to open the door again to man's destiny. A yes of fresh air in the no of a closed and stuffy room. This is the elixir on which mankind grows toward a deeper understanding of his nature and purpose — a mountain to mountain climbers.

Many of us, on returning home from space, brought back the perspective of a lonely and beautiful planet crying out for a more responsible attitude from its most prolific partner. Strangely, we didn't talk about the stars much. Perhaps with O'Neill's "seed pods" now emerging the perspective is enlarged again. Not that we should care less for the earth, for it will remain the principal home for most of us for a long time. But now, mother earth need no longer remain barren and generations of diverse offspring can continue to ask why.

LEWIS MUMFORD

Cultural historian, author of Technics and Civilization; The Story of Utopias; The Pentagon of Power; *etc.*

Dear Stuart Brand:

If you were familiar with my analysis in 'The Pentagon of Power' you would know that I regard Space Colonies as another pathological manifestation of the culture that that has spent all its resources on expanding the nuclear means for exterminating the human race. Such proposals are only technological disguises for infantile fantasias.

Lewis Mumford

LYNN MARGULIS

Microbiologist, author of Origin of Eukaryotic Cells, *co-deviser of the Gaia Hypothesis*

Sorry to be so late to respond to your invitation, but of course things on the surface of the earth, things here and now, always take precedence. That, alas, may also be the fate of the planning for Space Colonies that you ask me to comment on.

The effect of colonialization on the Mother Country self image is immense and incalculable. King George III was accused of "enslaving" his embittered Virginia and Carolina colonists 200 years ago. (It perhaps would have been more accurate to have called the gentlemen colonists slave masters.) However the analogy helped feed the flames and led to irreconciliation with the colonies. Can the effect of the American ex-colonies on the British self-image be assessed?

The Great Britain 4 p.m. tea service, white gloves and all are seen in perspective when Jeremy Button of Tierra del Fuego tried to retain it in the rugged offshore rocky islands of his forefathers. What seemed desperately important in 19th century drawing rooms looked ridiculous in Tierra del Fuego, the Gold Coast and the logging northwest of Canada. The values of the Mother Country must take on a new perspective from the distance of the colony. How small, idiosyncratic, isolating, anti-rational, parochial and repressive seem all tribal and national and imperial customs seen from a distance. I can't think of any exceptions (the American housewife lamenting Nicaraguan attitudes towards time as she waits for the hairdresser in Managua).

Of course Space Colonies are worthy of investigation and investment, in my opinion. (Why do some sun-requiring algae actually live inside carbonate rocks? Why do you find small blind arthopods scurrying at the backs of caves? Why do giant luminescent female fish (carrying their tiny males parasitically) inhabit the abyss? Why do red and green microorganisms cover the newly fallen arctic snows and multiply on its surface? Why do certain funny poorly known fungi (examples in the group Laboulgeniomycetes) live only on the left anterior appendage (read left front toe) of its insect host? The answer is the same as the one to the question why do people like O'Neill and his students imagine Space Colonies and advocate the move out. There are two parts to the answer: (1) the environment exists; and (2) the populations of organisms in question have the capacity to adapt to the environments. No fancy explanations are required. If there is space and if organisms can internally regulate utilizing the sources of energy at hand well enough to insure their replication, organisms will fill the space. This is the evolutionary pattern. It began over 300 million years ago, it still goes on. Steadily more and rougher parts of the earth's surface and near-surface have been colonized. There is no reason to believe the pattern will not continue to go on, at least in the near future.

What is obvious (and you already have said so) is that the John Todds of the World (e.g., holistic biological thinkers and doers) must connect with O'Neill and his crew to help stop the handwaving. Many details are not easily worked out simply because it is said that they are easy. Delivery of all needs, removal of all wastes, transport of the right things to the organisms in the right quantities with the right timing. Easy to say but perhaps incredibly complex to realize. I am not qualified to comment on the engineering difficulties except to perceive that they must exist. Furthermore, I think the working out of the details might be frightfully boring. But then I don't find Space War very interesting either, but (as I know from reading Brand) these issues are a matter of taste and indeed the working out of details might excite lots of differently temperamented people with inventive and exploratory natures. People who liked soldiering and rampages not because they are cruel but because they love excitement, for example, might go in a big way for Space Colonies. (Certainly such types have been highly successful in recent human history – evolutionarily speaking.)

Then there are the priorities. If we invest in Space Colonies from what other budget lines do we take the funds? I wouldn't mind a wholesale transfer from most of medical research to exploratory research on space colonies. Much of our illness is in the spirit anyway, and I like to see people permitted to be sick and die in peace. Although attainment of agreement on budget priorities will hardly be easy, perhaps it can be done.

What we need is a sober assessment of the technical feasibility by those qualified to make it, a nearly intrinsically impossible feat. But it is good to see **CQ** in the forefront of the search, as usual.

RICHARD RAYMOND

President, Portola Institute (which parented the Whole Earth Catalog*)*

If twelve people become passionately committed to a mission, the mission will succeed. That is an axiom. My friend Virginia Baker says two are enough, and I am inclined to believe whatever Virginia says on this subject, having observed the continuing growth of the Zen community with which she is affiliated. The New Planet mission appears already to have acquired a number of passionate sponsors, and I concluded several months ago that the question is no longer "if" but "how soon."

Another friend whose opinions I trust is Don Michaels, who recently wrote **On Learning to Plan and Planning to Learn**, in which he convincingly argues that the underlying justification of an organized society is to encourage and nurture the ongoing learning of its citizens. Thus an enduring society is not one that sees itself as primarily an aggragation of economic institutions – its economic stability will be a derivative outcome of its capacity to learn and learn and learn, endlessly.

There is no question but that the New Planet is a leading challenge to many of the world's most elite scientific and philosophical thinkers. Unless our culture produces some other more engaging intellectual challenge for this exploring contingent, their project beautifully fulfills the purpose of a human society. And that is not to say the world has no other deeply significant problems to solve; it is equally appropriate that a society encourage its citizens to become involved in each important issue that generates greater understanding. (Here at the Portola Institute, this is called "learning how the world works.")

I, personally, am challenged by some of the mysteries of this planet and am not arguing in defense of the morality, the economic benefits, or the political consequences of the New Planet. Yet I acknowledge it already exists as a project, it expands the understanding of society's members, and I wish godspeed to those who dig it.

The real backers

DEAN FLEMING
Artist, co-founder of Libre Commune

Space colonies are potentially the greatest creative focus in human history. If we may assume the highest achievement for humankind is to become totally, fruitfully human; responsible for being conscious of every detail of the living environment, heart and mind used for the mundane manipulation of matter while soul abides receiving continuously from the More then clearly the way now to reach such states is to take on the task of a totally integrated colony in Free Space. Every being so fortunate as to be able to participate will immediately be called upon to come up spiritually and psychically to handle the overload of unbridled forces (which may be ecstatic according to some astronauts) and still live the simple human life watching the vibes

WENDELL BERRY
Poet, novelist, farmer, teacher, author of
The Long-Legged House; The Memory of Old Jack;
A Continuous Harmony; *and* The Unsettling of America.

Mr. Gerard O'Neill's space colony project is offered in the Fall 1975 **CoEvolution Quarterly** as the solution to virtually all the problems rising from the limitations of our earthly environment. That it will solve all of these problems is a possibility that, even after reading the twenty-six pages devoted to it, one may legitimately doubt. What cannot be doubted is that the project is an ideal solution to the moral dilemma of all those in this society who cannot face the necessities of meaningful change. It is superbly attuned to the wishes of the corporation executives, bureaucrats, militarists, political operators, and scientific experts who are the chief beneficiaries of the forces that have produced our crisis.

For what is remarkable about Mr. O'Neill's project is not its novelty or its adventurousness, but its conventionality. If it should be implemented, it will be the rebirth of the idea of Progress with all its old lust for unrestrained expansion, its totalitarian concentrations of energy and wealth, its obliviousness to the concerns of character and community, its exclusive reliance on technical and economic criteria, its disinterest in consequence, its contempt for human value, its compulsive salesmanship.

The most striking feature of Mr. O'Neill's testimony is his lack of doubt, what I would take to be his unscientificness. He does not speak to us as the appropriately skeptical or dispassionate student of a possibility, but as its salesman. He sees no good alternative to his plan. He has no reservations. Or, rather, he has one reservation: he thinks that "it's very wrong to assume that something like this is going to promise happiness to all people. . ." – a point that will not greatly discomfort the plutocrats at whom ultimately this pitch is aimed.

As a salesman, Mr. O'Neill faithfully utters every shibbolith of the cult of progress. If we will just have the good sense to spend one hundred billion dollars on a space colony, we will thereby produce more money and more jobs, raise the standard of living, help the under-developed, increase freedom and opportunity, fulfill the deeper needs of the human spirit, etc., etc. If we will surrender our money, our moral independence and our judgment to someone who obviously knows better what is good for us than we do, then we may expect the entire result to be a net gain. Any one who has listened to the arguments of the Army Corps of Engineers, the strip miners, the Defense Department or any club of boosters, will find all this dishearteningly familiar.

The correspondence between the proposed colonization of "the high frontier" of outer space and the opening of the American frontier is irresistible to Mr. O'Neill. I find it at least as suggestive as he does, and a lot more problematical. The American prospect after, say, 1806 inspired the same sense of spatial and mental boundlessness, the same sense of limitlessness of physical resources and of human possibility, the same breathless viewing of conjectural vistas. But

it is precisely here that Mr. O'Neill's sense of history fails. For the sake, perhaps, of convenience he sees himself and his American contemporaries as the inheritors of the frontier mentality, but not of the tragedy of that mentality. He does not speak as a Twentieth Century American, faced with the waste and ruin of his inheritance from the frontier. He speaks instead in the manner of a European of the Seventeenth and Eighteenth Centuries, privileged to see American space and wealth as conveniently distant solutions to local problems.

That is to say that, upon examination, Mr. O'Neill's doctrine of "energy without guilt" is only a renewal, in "space-age" terms, of an old chauvinism: in order to make up for deficiencies of materials on earth we will "exploit" (i.e., damage or destroy) the moon and the asteroids. This is in absolute obedience to the moral law of the frontier: humans are destructive in proportion to their supposition of abundance; if they are faced with an infinite abundance, then they will become infinitely destructive. Mr. O'Neill sets it down as a false premise "That any realistic solutions to our problems of food, population, energy and materials must be based on a kind of zero-sum game, in which no resources can be obtained by one nation or group without being taken from another." That is the lesson that the closing of the earthly frontiers puts before us; it calls for an authentic series of changes in the human character and community that, if made, will afford us the spiritual resources to live both within our material means and with each other.

Mr. O'Neill proposes to learn no such lesson and to make no such changes. He proposes to outflank the lesson entirely. What we obviously need – as any old buffalo hunter or strip miner would also tell us – is a "new frontier:" some place, that is, where the mentality of exploitation may proceed without restraint, "correcting" the ruin of one place by the ruin of another.

But it seems to me that, in essence, Mr. O'Neill's false premise is true everywhere in the universe. All we have to do is re-phrase it in terms of what we know of our history and of the insights of the ecologists, so that it reads this way: Whatever human beings use they must take from the rest of creation, either temporarily or permanently. So phrased, this allegedly false premise is seen to be a practical truth with profound cultural implications. Instead of the moral escape valve of yet another "new frontier" to be manned by an elite of experts, it proposes certain limits, restraints, and disciplines that apply to all of us. It avoids the corporate and governmental big-dealing that will be bound to accompany the expenditure of a hundred billion dollars. And it avoids the open-ended chauvinism, perfected by the strip miners, that is always willing to advocate the destruction (for money, jobs, and the general amelioration of the human condition) of some other place. It has always been possible to export bad character and disrespect, but that can be understood as a solution only by misunderstanding the problem.

This brings me to the central weakness of Mr. O'Neill's case: its shallow and gullible morality. Space colonization is seen as a solution to problems that are inherently moral, in that they are implicit in our present definitions of character and community. And yet here is a solution to moral problems of the rock and the pond. Every conceivable act would have to be as the Hopis' dream an act of consciousness and inevitable worship. If being creative without ceasing builds humanness in humans imagine how sprung loose God will feel when relieved of the burden of creating the "skies" and the "waters"!

In May 1970 I attended a NASA symposium on habitability in Venice, CA. During which time I lobbied for an experimental community deliberately living as if in space. (Forget about the Arctic! Maybe on some Caribbean Island!) They thought I was nuts. But now they would have 5 years of solid info at their fingertips. Do you think they could go for it yet? Sign me on! I swear all this time I've been painting for that kind of space!

that contemplates no moral change and subjects itself to no moral standard. Indeed, the solution is based upon the moral despair of Mr. O'Neill's assertion that "people do not change." The only standards of judgment that have been applied to this project are technical and economic. Much is made of the fact that the planners' studies "continue to survive technical review." But there is no human abomination that has not, or could not have, survived technical review. Strip mining, fire-bombing, electronic snooping, various forms of genocide and political oppression — all have been technically feasible, and usually economically feasible as well.

Stripped of the glamour that we associate with adventures in "space" and of the romantic escapism left over from our frontier experience, Mr. O'Neill's project is clearly not a solution, in any meaningful sense, to any problem. It is only a desperate attempt to revitalize the thug morality of the technological specialist, by which we blandly assume that we _must_ do anything whatever that we _can_ do.

Mr. O'Neill's testimony is littered with the evidence of his moral bewilderment. His concern for the environment leads him directly to a plan to strip mine the moon. He says, "I have a deep suspicion of governments," but he does not hesitate to promote a scheme that would vastly increase the power and influence of government. He apparently sees no chance of political corruption in an expenditure of a hundred billion public dollars. He says that he "would far prefer to see a cooperative multinational program formed" to carry out his project, but only five paragraphs later he speaks of the project as a way to return to the "traditional role [of the U.S.] as a generous donor of wealth to those in need." Nowhere does he see the absurdity of trying to solve on a grand scale by expensive technology a problem that can probably be solved on a small scale, and cheaply, by moral means (see E.F. Schumacher's article, same issue). Nowhere does he see the absurdity of trying to solve with existing technology a problem that, as Schumacher suggests, "has been produced by the existing technology."

Mr. O'Neill predicts readily that his scheme will promote diversity and freedom. But he neglects to consider that the machine is already a renegade concept that sees people as spare parts, and uses them as such. Exactly how, one wonders, is this to be corrected by building an even bigger machine and causing people to live _inside_ it, in absolute dependence on it? What, exactly, would be the effect of a completely controlled environment on human character and community? What, exactly, would be the influence of space colonization on earthly political and social forms? Mr. O'Neill does not know, and he has no way to know. He is not, then, merely asking for a public subsidy in the amount of a hundred billion dollars. He is proposing that he and his colleagues should be permitted to _experiment_ with fundamental values. This is the violence of the _specialist_. This _kind_ of thing is familiar enough. What is new here is the scale.

Perhaps most important of all is Mr. O'Neill's failure to see that the so-called energy crisis is a moral crisis. He assumes that it is simply a matter of scarcity, which can be remedied by the time-honored method of getting more from somewhere else. But it has been obvious for some time that the energy crisis has at least as much to do with the _uses_ of energy as with its availability. The world will tolerate the use of even less energy than it can supply. The question is not of how much energy we can get, but of how much we can use without destroying, at a minimum, our ability to enjoy the use of it. The question of restraint is much more pertinent to the problem than the question of supply. And Mr. O'Neill has apparently never thought to ask what good might be accomplished by the proliferation in space of a mentality that cannot forbear to do anything at all that is possible.

Mr. O'Neill has failed to think of these things because temperamentally he is a scientific super-star. His ambition can comprehend only the grandiose. He is a professional mind-boggler. With the apparent simple-mindedness of the true-believer, he sees himself as the evangelist of the next "giant leap for mankind." He makes the overwhelming presumption of the evangelist — that he knows better than we do what is good for us. And he is asking for an influence over the material means of our lives which will require our spiritual capitulation. Like an evangelist, he wants both our faith and our money.

Finally, I would like to raise the question of what may be meant by this advocacy of space colonization by **The Co-Evolution Quarterly**.

Evolution, as I understand it, is a slow process. It develops gradually, by seasons, like a plant. It paces itself according to the capacities of organisms and the changes of the environment. It involves a profound mutuality of response between an organism and its environment. It is deliberate, meticulously attentive to details. It does not proceed by coup or decree; it does not often risk the wholeness or coherence of its systems. It would not allow one man or any few men to take "a giant step for mankind."

Coevolution I understood to mean a concept of human change modeled on evolution: changing together slowly, coherently, feeling for the right ways with some mutuality of consciousness and regard. Fundamental to it, I thought, would be a suspicion of change by technological or governmental coup. I thought that it grew out of the new awareness of material and moral limits that Mr. O'Neill's project is designed to repudiate.

I admit that I am bewildered. Perhaps I will have to admit that I have been wrong. It is certain, however, that the Fall 1975 issue displays a potentially ruinous split between what I at least have thought to be coevolution and what I think the energy lobby would unhesitatingly recognize as Progress — between the mind of Gerard O'Neill and the mind of E.F. Schumacher. On the one hand we have an admirer of Mr. O'Neill's project saying that if it should be implemented, "maybe humankind could walk gently in the Universe." And on the other hand we have an article by a twenty-year-old scientist at work on Mr. O'Neill's project, in which it is proposed that we should send out to the asteroid belt "a work crew equipped with about one thousand 100 megaton hydrogen bombs. . . ." The editor's implicit approval of both statements makes of the first a vacuous sentimentality. The other, in any context, would be monstrous. □

WILSON CLARK

Author of Energy for Survival, *advisor to California's Governor Brown.*

Perhaps the greatest significance of O'Neill's proposal is that it is being taken quite seriously, particularly by the technologically-oriented members of our society. The idea has been rooted in science fiction since the industrial revolution's adolescence, and it has a certain charm, in light of the fact that the very scientific "advance" which we have produced has rendered our remaining earth environment a battleground, pocked with the evidence of technology's increasing sophistication in war, environmental destruction, etc.

I am reminded of the wisdom of Oliver Wendell Holmes, who commented in 1872 that, "Science is a first-rate piece of furniture for a man's upper chamber, if he has common sense on the ground floor."

Notwithstanding the palpable evidence that our science lacks common sense, it is conceivable that we could ignore the implications of this, and build the space colonies – to prove that we could do it. Why? O'Neill's primary rationales are (1) Future energy needs to be met by space-generated solar energy for terrestrial and space use; and (2) Exploitation of lunar materials. From an energy and materials standpoint, the production of metals from space ores would have to exceed the demands on terrestrial energy and materials to construct and maintain the space colony/industry. Given the extraordinary demands on earth materials and energy supplies to initiate the project, it is doubtful that it could meet even this basic criterion.

From an energy standpoint, there is no compelling need to demonstrate solar energy in space. Solar radiation reaching the earth's surface is ample in most areas of human habitation to provide essential energy needs, at far less cost than a space system.

One practical rationale, which also meets the requirement of common sense, is to proceed with construction of the colony, thereby occupying the time of thousands of technologists and planners, who otherwise would be spending it inventing new genetic manipulation techniques, plutonium warheads, nerve gases, etc. If the colony were actually built, they could be sent there *en masse*, relieving earth of an annoying burden.

In proposing this idea, O'Neill has done us all a favor, by forcing us to ponder our philosophies and approaches to life on (or off) this planet. What if we were now living in a space colony, debating the possibility of colonizing earth? Would we be asking the same questions? I think not. From the viewpoint of a creature in some metal and plastic isolated environment, the key question would probably be seeking the companionship and cooperation of the diverse population found on earth, rather than avoiding it for the "safety," or "high technology" of a space cocoon.

Stewart Brand's incisive remark in the Winter '75 CQ, that self-sufficiency has done more harm than good, reflects a far more important view of the future than space fantasy. By taking stock of our precarious existence on earth, we can begin to rethink our relationship to technology and value, and hopefully find some approaches to help us redress many of the wrongs already accomplished by our species. What we need is nothing less than a quantum leap in our approach to science, relieving ourselves of the onerous image of *Homo absurdus*. O'Neill speaks in terms of a "first beachhead in space," evoking the image of the greener grass on yonder hill. Unfortunately, we have little time in which to prevent the elimination of the vegetation altogether.

Space for peace

. . . O'Neill is as convincing as the Atomic Energy Commission was advocating the benefits of Nuclear Power. . . .

Barry Hughes
Eureka Springs, Arkansas

E. F. SCHUMACHER

Economist, author of Small is Beautiful *and* Guide for the Perplexed.

Yes, Stewart, I'm all for it. I am prepared to nominate, free of charge, at least five hundred people for immediate emigration. For every one of these emigrants, once they are well and truly gone, I am prepared to donate $1,000.00 US dollars for the furtherance of the work that <u>really</u> needs to be done, namely, the development of technologies by which ordinary, decent, hardworking, modest and all-too-often-abused people can improve their lot. With the above-mentioned emigrants out of the way, it will be a great deal easier to obtain support for this work.

> P.S. "As for those who would take the whole world to tinker with as they see fit, I observe that they never succeed."
> – Lao Tzu

STEPHANIE MILLS

Editor of Not Man Apart *(Friends of The Earth)*

Apropos of space colonies, the sainted Walt Kelly expressed a basic qualm better than ever I could. To wit:

"I feel that we have done little enough with our present world to warrant our going off and putting soft-drink signs all over Mars and cluttering up the moon with oil rigs. These places are not ready yet for such advanced achievements. We should really go a little more slowly. We don't expect kangaroos to whistle 'Dixie,' do we? Not the first try, anyway. Let the universe come here when it's ready. Don't rush it."

Me for tending to home sweet home, that's frontier enough.

STEVE DURKEE

Artist (back cover), co-founder of Lama Foundation,
book designer of Be Here Now; Seed; Toward the One.

The positive feelings that I have for such a project rest on the assumption that we will get more out of it, in terms of net energy gain, than we put into it. In order to make such an assessment, I feel that we need to do further research & exploration of leads turned up by Skylab & other missions. **Items like:** promising fishing grounds indicated by observation of ocean currents, surface & geologic indications of subterranean water sources in West & Saharan Africa, surface hot spots indicating geothermal energy sources, location of ore & oil deposits suggested by aerial photography. If it could be shown that some of these leads pan out it would become much easier to convince the people & those who control the purse-strings that funding of such a project could result in manifest gains. Also I would think that there would need to be future Skylab related missions which would specifically undertake research & development work in Zero-G conditions in the fields of crystalography & metallurgy since it seems that much could come from this in terms of computer based technology perhaps paving the way for a new generation of computers freed from planetary restrictions & hazards.

Going down the line of positive feelings; 1) Planetary Consciousness: Furtherance of whole earth thinking provided by daily input to the planet surface of planetary phenomenae affecting local & global interests. At present we have such feedback (weather, re-con, etc.) but what I envision is more at the level of World Game, International Design Decade, Club of Rome kind of material (weather patterns, tidal flows, crop patterns, inventory & mapping, etc.). People would become used to the whole earth, seeing it they would not see all those funny lines denoting political divisions, arbitrary in many cases, but would rather learn to think in terms of regions rather than nations, biotic provinces rather than NATO or SEATO, geo-oceanographic reality rather than trade embargoes with island republics, watershed reality rather than states rights. When the blood & guts of Viet-Nam began pouring out of the TV News onto the living room floor the peace movement in America began to extrapolate; similarly I feel that the whole ecological awareness of people & their understanding of inter-relatedness would be vastly accelerated by information (visual & sensory) that could be provided by the stations.

2) Energy Transmission Stations: Tho this seems the most promising, cheap, clean power I have yet to understand from reading the articles what the proposed means of transmission would be. So far as gathering the solar energy this is not so different from what we have already been doing with the various probes, missions & skylab experiments. I can also see that crystals necessary for conversion of solar energy into electrical energy could conceivably be produced with a greater rate of success than on the planet (the high cost of such crystals I understand comes from the fact that there is a fail rate of 90%+ in the manufacturing process). But the big question I am left with is HOW it gets from there to here? Do you convert solar energy to radio waves, beam to earth, radio waves convert to electrical energy? Seems like a lot of steps & how much is lost in each step? Further I can't quite see in economic terms why, when we are told that it is not feasible here on earth, to use large areas of Aztlan (Sonoran Mexico, New Mexico, Arizona, So. Cal., Nevada, etc.), Saharan Africa, The Hedjaz, The Afghani-Iranian Desert, The OutBack of Australia (to mention a few possible places) to produce solar energy, why would it be more feasible to try & do the same in space? But it seems as tho there are possibilities in the area of Transmission so this remains a positive area for me.

3) Cosmic Ship or Ark: I dont feel it's necessary to elaborate on this. **Cities in Flight** by J. Blish as well as countless sci-fi novels plot this scenario. For one reason or another, a station decides to go off the line & the rest of it I leave up to your imagination. Pirate Stations & Rogue Stations, Interplanetary & Intergalactic(?) Travels. Generations growing up in space with only rare visits earth side; that should breed a whole new type of being. This is probably my favorite plus reason.

My negative feeling about such a project revolves around the kind of "Pie in the Sky" vibration which I feel from the project. Given the problems we are facing from population overload & its attendent horrors: famine, land rape, water pollution, dwindling resources, etc., etc., I wonder as to the wisdom of putting before people the idea or even the possibility of "getting out of here," when here is where the work is, here is where the problem is. At that level it seems like escapism, and as such the last place where we should put our energies.

As for recommendations; I feel that the project should become an international one from the start. That all phases of the project should be based on the ideals of co-operation & co-elaboration both on an international level & inter-disciplinary level. That people chosen to work on such a project would blend together rigorous scientific methods with ecologic awareness & whole systems mentality & hopefully that the mystical & transcendent implication of the project would be deeply considered.

Within this project there is the germ of a possibility in creating something which would serve to focus the aspirations & consciousness of many people the world over. What Sputnik did for the last generation this could do for the next. When I told my nephew what I was typing he said, "I'm ready, when's it happening?"

In short even if economically it was a borderline case there are other payoffs which might seem ephemeral at first but which could provide more important, if subtler, payoffs. This will be in the hands of the inceptors & the kind of values they stress & put forth. If it were handled wisely it could gain somewhat of the aura that the ancient Olympic games had & as such might serve as a model or operating symbol for world & planetary evolution. One could, for instance, move the UN into space. That would change the game.

In closing I would like to offer a cautionary tale;

One sunny spring morning leaning up against the east wall of an old adobe, soaking in the warmth, the fields around beginning to green, the apple trees in fresh bud, the freshly turned earth of the vegetable patch steaming, I was listening to this old man, can't remember if he was a Sioux or Kiowa, talking to a group of us "younger boys." He was talking about the red people & the green people & what he called "the hurry up & let's go" people. He was saying something like; "You know them hurry up & go people. They always want to go someplace else. Never like it where they are. I betchyou them people they just going to hurry up & go out there (pointing heavenwards with his pursed lips). Yeh, that's all right & God bless them, they don't like this one here anyway (patting the earth with his hand). Yeh, they just gonna up & go. But one thing you should tell them, those boys they gotta be careful, real careful with that moon."

That's all Folks, BE CAREFUL WITH THE MOON.

Extra energy

 Extra energy, only adds to the positive feedback of the very system which created all of the problems which O'Neill wants to solve. Aside from the damage that all this energy will do when put to work (what will the "limitless" amount of electricity do besides create more of the same?), the thermal effects can only create new, unforseen problems. Claude Summers (Sci. Amer. Sept. 1971, "The conversion of energy") has demonstrated that, assuming all pollution problems associated with energy production were totally solved, "waste" heat would inevitably provide overriding negative feedback and bring limits to growth. All energy beamed to earth must end up as heat. Only invariant systems of energy production (systems that add no energy to the earth besides naturally received solar energy) can avoid adding excess, and potentially damaging, heat to the biosphere. . . .

Eric Alden Smith
Ithaca, New York

STEVE BAER

Inventor, writer, president of Zomeworks, author of
Dome Cookbook; Zome Primer; Sunspots.

I can't keep my own car running, can't fix a broken
radio or light bulb — who am I to recommend anything as complicated as a space colony?

The project is spoken of as if it were direct as our
stepping over and grabbing a large rope, giving it a
huge pull and flinging people into space. But I know
that instead it consists of orderforms, typewriters,
carpets, offices, and bookeepers; a frontier for PhD's,
technicians and other obedient personnel.

Once on board, in my mind's eye I don't see the landscape of Carmel by the Sea as Gerard O'Neill suggests.
(He must keep his eyes on his shoes and not breathe
too deeply in order to suggest duplicating the pacific
coast aboard the space colony.) Instead, I see acres
of airconditioned Greyhound bus interior, glinting,
slightly greasy railings, old rivet heads needing paint —
I don't hear the surf at Carmel and smell the ocean —
I hear piped music and smell chewing gum. I anticipate a continuous vague low-key "airplane fear."

There is something especially boring about car nuts
talking endlessly about different model cars. I am
suspicious that the space stations are the next step for
these people, for the whole world would have a manu-
facturer, a model number, etc. Who would ever be
able to shut them up once one got started on the
"AJAX 8" or O'Neill's "Island One"? Absorbing
yourself in all that is to me just barely second best to
putting in your time looking at your own asshole with
a mirror and a magnifying glass.

I admit, though, that there is something fascinating
about the Space Colonies. They might be the perfect
cure for the "car nuts" — probably after a term in
such a place you'd never want to see another piece
of machinery again — just go off and roll in the dust
for a year.

Of course the positive side is that Space Colony
activity is less dangerous than multiple war head
guided missile activity — but I'm sure others will
describe such advantages. Space Colonies =
Methadone for technology junkies?

The SF vote

. . . What a project.

I'm hooked.

My 17-year-old brother who's also quite a fan of your
magazine (and very much into the Whole Earth mentality)
was super-cynical, predicting that it wouldn't work technically, that the ecology would fall apart in "a big bottle,"
and that the government would be a totalitarian autocracy,
with absolutely no personal freedom allowed. My question:
why did we react so differently? Was it, perhaps, because at
his age I was reading 5 - 10 science fiction books a week,
while, to the best of my knowledge, he's never read one? (I
myself haven't read more than 2 - 3 in the past four years,
something that will no doubt change now that dreaming
isn't so painful anymore.). . . .

Jose Garcia
Naperville, Illinois

*How would the skylights be kept clean? One envisions
thousands of scrubwomen, their role in society still
preserved intact.*
— Larry Todd

DENNIS MEADOWS

Social technical-systems analyst, co-author of The Limits
to Growth; Dynamics of Growth in a Finite World. *This
one was a conversation.*

"I have a mixed mind about L-5.

"My impression is that there are cheaper ways, ways less
demanding of capital, to satisfy any goal put forward by
the L-5 effort — to do that on Earth rather than to do it
in Space. You cannot justify the L-5 effort in any sort
of physical terms.

"It plays a function, which may be negative or positive,
of giving us another frontier, when we've used up all the
ones we have on Earth. I'm not sure if we should want
to have another frontier. It seems to me to block constructive response to problems here on Earth. If you look
around at societies to find those that have come into some
sort of a peaceful harmonious accommodation with themselves, many of them turn out to be on islands, where the
myth of a new frontier vanished long ago."

SB: How did they get to the islands in the first place?

Meadows: Well they migrated. I'm not sure that we want
a new frontier. If we do, this is a nice way to do it.

"But now let's turn to things like providing food or energy
here on Earth. Of all the ways I can think of doing that,
L-5 is very costly, risky, and long-term. The thing that it
doesn't involve is institutional and value change. Our current values and institutions will give us an L-5, conceivably.

"What are needed to solve these problems on Earth is
different values and institutions — a better attitude towards
equity, a loss of the growth ethic, and so forth. I would
rather work at the root of the problem here."

Larry Todd

Living in a big rotating cylinder in the absence of true gravity might play havoc with the instincts of migrating birds.
—Larry Todd

HAZEL HENDERSON
Co-director, Princeton Center for Alternative Futures; advisor, Office of Technology Assessment

As you asked, here are some of my thoughts on Prof. Gerard O'Neill's space colony proposal. I have also discussed them with Alice Tepper Marlin and they reflect her views also.

Firstly, I have little doubt that a model space colony could be initiated and I will accept, for the purpose of this discussion, O'Neill's trust in the availability of technology to accomplish this. How long it and its human inhabitants could function according to plan, and what surprises they would encounter are largely unknowable at this point.

I also have no doubt that the money could be found to fund the program, since NASA is still desperately casting around for new, politically-sexy projects and they have a good deal of clout when they lobby in concert with their prospective corporate contractors. Furthermore, the high-tech community is currently drumming up concern on Capitol Hill about a new and terrifying "Sputnik Gap" which will justify hyping U.S. science and technology so as to "improve our competitive position in the world" against all comers. They implicitly assume that the health of the U.S. scientific enterprise is coterminous with the health of the country, so a new WPA-type project to keep them fully employed will not be hard to sell.

It would be nice to believe that the money for the space colony could be carved out of the military budget, or even out of the Highway Trust Fund, the Corps of Engineers, or half a dozen other well-heeled Washington-based, make-work schemes. But we must assess realistically the political mood of the country that "big government must be trimmed," and "the Federal Budget must be cut" etc. You know as well as I do that it is always social spending that gets axed, while "military gaps" and "space gaps" and the fear thereof are always sure-fire stampeders of legislators and voters alike.

Given this sad reality, O'Neill's certainly innocuous space proposal must inevitably compete with other priorities in our national budget. Therefore, less benignly, it will very likely displace programs advocated by weaker societal interests or those less appealing to the voters' imaginations. I suppose it is hard to dramatize food stamps and all the other dreary, ameliorative and income transfer programs that are made necessary to cope with the human casualties and other consequences of our high-tech, industrialized society.

So there is a temptation I have noticed in O'Neill's statements to try to justify the space colony as a hope for the Third World, to ameliorate the population crisis, to solve the energy "problem" and to provide "immediate jobs of the high-technology kind which economic studies have shown generate wealth throughout the economy." O'Neill can't be naive enough to believe such studies, which are based on per capita averaging, and are discreetly mute about how that wealth generated will be distributed. Such studies are usually based on neoclassical economics, now rapidly being discredited within the discipline itself, since such studies rely on now clearly unrealistic assumptions of the free market, where buyers and sellers are supposed to meet each other with equal power and equal information; where these markets are always cleared at the equilibrium point where supply and demand are equal, at however astronomical a price, and a whole string of other myths which I have examined in my own writings. (Ecologists versus Economists, **Harvard Business Review**, July - Aug, 1973, New Models for a Steady State Economy, **Financial Analysts Journal** May 1973, Limits of Traditional Economics in Making Resource-Allocation Decisions, Transactions of the Fortieth North American Wildlife and Natural Resources Conference, 1975.)

In fact, these justifications, based on neoclassical economics and borrowing pragmatically from John Maynard Keynes, seemed to be the main pitch of the group of "true believers" in O'Neill, styling themselves as the L5 Society, who so fervently buttonholed participants at the recent Club of Rome meeting in Houston, Texas. So it seems that this is to be the latest, most baroque elaboration of Keynesian "trickle-down industrialism." The poor are always told that one day, when those at the upper end of the income scale really have <u>enough</u> then you can get yours. Meanwhile they are told they must accept inequality in economic distribution so that some of us can continue to accumulate wealth for capital investments so that these lucky few can create jobs for those so "unfortunately" dispossessed. This myth too, is unravelling, in spite of the massive barrage of corporate image advertising about how they must have profits since this is the only way to accumulate capital for investment for jobs. We see all too easily however, that corporations given generous tax credits for capital investment in the hope that they will create jobs, quite frequently use their capital investments to <u>dis</u>-employ people, (as the new automated check-out systems in supermarkets, and electronic funds transfer systems in banks will do) or to export the capital to set up factories in the Third World to exploit cheaper labor, or even to go on acquisition shopping sprees, such as Mobil Oil; while pleading for more profits to plow back into developing new energy, it promptly acquired Montgomery Ward and the Container Corporation instead. So far, only the **Wall Street Journal** has jumped ship on this set of myths. In an August 5th editorial entitled "Growth and Ethics," the **WSJ** finally admitted that the American people would have to <u>choose</u> between economic growth and redistribution! The <u>latest escalation</u> in corporate propaganda on these issues is due to begin soon, when the U.S. Department of Commerce is to launch with our tax dollars a $5 million Advertising Council campaign of "economic education" of the American people.

In his less euphoric moments, O'Neill does more soberly assess such issues of concentration of economic and political power, which will so materially affect his space colony's chances of ameliorating our planet's social ills, and indeed, may cause it to exacerbate them. He states, "If the new option is taken, it would be naive to assume that its benefits will

41

be initially shared equally among all of human kinds. The world does not work that way, and since people do not change, there is no reason that it will work that way in this case." So the moral seems to be, "If you can't beat 'em, join 'em."

I will leave assessments of the biological feasibility of the space colonies to those more qualified: Lynn Margulis, Paul Ehrlich and others. Paul Ehrlich suggests, "A biologist like George Woodwell, who really knows ecosystems inside and out, will have to sit down and explain to O'Neill why his ideas are complete nonsense." At least, I sense some sweeping dismissals of possible biological constraints in such phrases as "the space community residents could enjoy a per capita usage of energy many times larger even than what is now common in the U.S., but with none of the guilt. . . ." It seems to me that in the U.S. we are already consuming more energy than we can digest, and we have all the pathologies of high-energy industrialism to prove it, from the current epidemic of cancer to hypertension, mental dis-ease, drug addiction, heart ailments and obesity.

Similarly the blithe dismissals of long-term effects in O'Neill's statements: "People can breathe pure oxygen atmospheres perfectly well. The Apollo astronauts were breathing pure oxygen atmospheres for days at a time. I've done it for hours at a time." and "Plants couldn't care less." One wonders at this typical attitude of scientific over-optimism. I have just read a rather shattering paper on "Scientific Optimism and Societal Concern," by Gerald Holton, Professor of Physics at Harvard (**Hastings Report 5**, Dec. 1975). Holton comments on the psychology of scientists as well as their acculturation. The scientific and technological imperative are both powered by the same optimistic dynamism. The prevalent "other-orientation" of scientists leads them to attempt to transcend people-oriented problems and escape to a world filled with order, rationality and neutrality. Problems with a high degree of non-falsifiable and non-verifiable content are avoided in favor of problems submitting to quantification or instrumental manipulation.

A. NORCIA '75

In the last analysis, for me at least, the space colonies appear as simply a linear extension of the technological fix, instrumental rationality so beloved of this culture; another case of too much hardware and not enough software. Perhaps O'Neill would enjoy tuning in to some of his fellow physicists who are exploring the physics of consciousness and trying the indescribably more difficult task of writing the observer back into the equation (see for example, Jack Sarfatti and Fred Wolf, **Space-Time and Beyond**, Dutton, 1974). If our destiny is, as I believe, to strive for cosmic consciousness, I cannot see space trips in clunky, materialistic, tin-lizzie spaceships as the means for embracing the cosmos. Surely we will roam the universe with our minds and dream up far less materialist, more elegant means to re-structure and re-pattern ourselves. Surely we can leapfrog the current instrumental materialism and perhaps, even escape the prison of matter entirely. For heaven's sake, what difference does it make which arm of our spiral galaxy we are in? Will journeying across it in a spaceship really change anything? As O'Neill himself says, "people manage to make themselves unhappy in almost any circumstances." When will we stop trying to re-tool the planet and get on with the job of trying to re-tool ourselves? For me, not the High Frontier, but the Human Frontier. My question remains. "Why?"

Allure

I was a little surprised at your enthusiastic presentation of O'Neill's space colonies in The CQ. The concept seems to me well thought out, rational, very alluring, and quite mad. . . .

Martin Holladay
Sheffield, Vermont

Daughter

. . . Whatever I can do (contribute) — may help my beautiful daughter to slip away from this failing civilization here on Earth.

Capt. Jonas Caron
Watertown, Massachusetts

CARL SAGAN

Space scientist, exobiologist, author of Intelligent Life in the Universe, The Cosmic Connection, *and* The Dragons of Eden.

I think "Space Colonies" conveys an unpleasant sense of colonialism which is not, I think, the spirit behind the idea. I prefer "Space Cities."

One of the many virtues of the Space City proposal is that it may provide the first convincing argument for extensive manned spaceflight. The earth is almost fully explored and culturally homogenized. There are few places to which the discontent cutting edge of mankind can emigrate. There is no equivalent of the America of the 19th and early 20th centuries. But space cities provide a kind of America in the skies, an opportunity for affinity groups to develop alternative cultural, social, political, economic and technological life-styles. Almost all the societies on the earth today have not the foggiest notion of how best to deal with our complex and unknown future. Space cities may provide the social mutations that will permit the next evolutionary advance in human society. But this goal requires an early commitment to the encouragement of cultural diversity. Such a commitment might be a very fitting Bicentennial re-dedication to what is unique about the United States.

Rather

. . .I would rather have gone with Hank Hudson, Sam Champlain, or Lewis & Clark, but . . . now is yet. . . .

Timothy C. Cornwall
New York, New York

PAUL and ANNE EHRLICH

*Population biologists, environmentalists, author (Paul) of
The Population Bomb; co-authors of Ecoscience — Population, Resources, Environment. Paul co-conceived with
botanist Peter Raven the concept of co-evolution, our
founding idea.*

The prospect of colonizing space presented by Gerard
O'Neill and his associates has had wide appeal especially
to young people who see it opening a new horizon for
humanity. The possible advantages of the venture are many
and not to be taken lightly. In theory many of humanity's
most environmentally destructive activities could be removed
from the ecosphere entirely. The population density of the
Earth could be reduced, and a high quality of life could be
provided to all *Homo sapiens*. It might even make war
obsolete.

What can one say on the negative side about this seeming
panacea? At the moment the physical technology exists
largely on paper, and cost estimates depend in part on
numbers from the National Aeronautics and Space
Administration (NASA) — not necessarily a dependable
source. There appear to us, however, to be fewer barriers
inherent in the further development of the O'Neill technology
than in others in which society has committed itself to
large, open-ended and highly speculative investments —
fusion power technology being a prime current example,
the atomic bomb one from the relatively recent past.

On the biological side things are not so rosy. The question
of atmospheric composition may prove more vexing than
O'Neill imagines, and the problems of maintaining complex
artificial ecosystems within the capsule are far from solved.
The micro-organisms necessary for the nitrogen-cycle and
the diverse organisms involved in decay food chains
would have to be established, as would a variety of other
micro-organisms necessary to the flourishing of some
plants. "Unwanted" micro-organisms would inevitably
be included with — or would evolve from — "desirable"
ones purposely introduced. Furthermore in many cases
the appropriate "desirable" organisms for introduction
are not even known to us. Whatever type of system
were introduced there would almost certainly be serious
problems with its stability — even if every effort were
made to include many co-evolved elements. We simply
have no idea how to create a large stable artificial ecosystem.
For a long time it's likely that the aesthetic senses of
space colonists would have to be satisfied by artificial
plants, perhaps supplemented with "specimen" trees
and flower beds.

The problems in the agricultural modules would be less
complicated but very far from trivial. Since, according to
O'Neill, agricultural surface is relatively cheaply constructed,
it seems likely that early stations should have perhaps four
times as much as required to sustain the Colony, and that
it should be rather highly compartmented and diverse to
minimize the chances of a disaster propagating. A great
deal of research will have to go into developing appropriate
stable agricultural systems for space. The challenge is
fascinating — especially because of the variety of climatic
regimes possible, the potential for excluding many pests,
and the availability of abundant energy.

We can say, then, that although there appear to be no
absolute physical barriers to the implementation of the
O'Neill program, potentially serious biological barriers
remain to be investigated. What about psychological,
social, and political barriers? The question of whether
Homo sapiens can adapt to the proposed space station
environment seems virtually answered. Six thousand men
live for long periods on a Navy super-carrier orders of
magnitude smaller than a proposed space habitat, without
women and without the numerous other amenities
envisioned by O'Neill. Many city dwellers pass their lives
in a similarly circumscribed area and in much less interesting
surroundings (travel among stations and, occasionally,
back to Earth is envisioned). There is little reason to doubt
that most people would adapt to the strange situation of
access to different levels of gravity.

Whether or not society will support the venture is another
matter. Much may depend on whether O'Neill's calculations
on the profitability of the solar power generating enterprise
stand up under closer scrutiny and limited experiment. The
strongest objection that will be raised against space
colonization is that it cannot help humanity with the
problems of the next crucial decades, that it will divert
attention, funds, and expertise from needed projects on
Earth, and that it is basically just one more "technological
circus" like nuclear power or the SST. That Space Colonies
will have no immediate impact is recognized by O'Neill,
but he would argue that we should look to medium range
as well as short range solutions. Diversion of funds and
expertise also do not seem extremely serious objections.
There is, for instance, no sign that capital diverted from,
say, a boondoggle like the B-1 bomber, will necessarily be
put to "good" use. Equally it does not follow that money
for space colonies must be diverted from desirable programs.
The expertise needed is superabundant — many trained
aerospace engineers, for example, are not able to find
appropriate employment now.

The problem of diverting <u>attention</u> from immediate problems
like population control is much more serious and can only
be avoided by assiduous care on the part of O'Neill and
other promoters of the project. Some of O'Neill's associates
have done his cause great harm by not realizing this. At every
stage people must be reminded that for the potential of
space colonization to ever be explored, society must be
maintained for the next three decades.

Environmentalists, including us, had a strong negative
reaction to the O'Neill proposals when first presented with
them. They smack of a vision of human beings continually
striving to solve problems with more and bigger technology,
turning always away from learning to live in harmony with
nature and each other and forever dodging the question of
"What is a human being for?" But again O'Neill's vision
shares many elements with that of most environmentalists:
a high quality environment for all peoples, a relatively
depopulated Earth in which a vast diversity of other
organisms thrive in a non-polluted environment with much
wilderness, a wide range of options for individuals, and
perhaps time to consider those philosophical questions.
The price of this would, of course, be a decision that a
substantial portion of humanity would no longer dwell
on Earth.

Environmentalists often accuse politicians of taking too
short-term a view of the human predicament. By prematurely
rejecting the idea of Space Colonies they would be making
the same mistake.

Pioneer

. . . I'm fourteen years old right now, and I think in, say,
20 years, this planet is going to be pretty cramped. Not
only for humans, but for other living beings too. Resources
of every kind will probably be shot for the most part.

It's a good idea to have an alternative ready.

The people who are saying "no, no" now, will be screaming
"yes, yes" then — a little too late. . . .

I'd like to go — do my teensy bit for depopulating the Earth.
The idea of going scares me, but I think I could overcome
that. Yes . . . I'd definitely like to go.

Paula Read
Inverness, California

Down the road

A journey of a thousand miles does <u>not</u> end with the first step.

My mind is not Earth-bound,
My body doesn't want to be either.

Charles D. Walker
Bedford, Indiana

WILLIAM IRWIN THOMPSON

Author of At the Edge of History; Passages About Earth;
co-founder of Lindisfarne

We need to transform our civilization, not simply extend it.
If we extend ourselves as we are now, we will simply be
setting up metastases of the carcinoma of industrial civilization. If we are going to move out into space, we will have
to learn how to be inhabitants of the universe, and that will
require a transformation of consciousness. Such a transformation of consciousness was beautifully expressed by Rusty
Schweickart in the piece you published last summer. What
I am asking for is an exercise of imagination more profound
than the science-fiction fantasies of the comic books of a
generation ago. It is not an exercise of imagination to envision solar energy as the means of beaming down intense
concentrations of power to drive capital intensive economies
of scale. There is abundant solar energy on the earth for a
good life. It may not provide enough energy to fly Ritz
Crackers in jumbo jets to Venezuela, and if it doesn't we
need to rethink the whole kind of crazy economy we have
created. If that economy is now running out of energy,
perhaps there is good reason for it to do so.

I don't see anything wrong with setting up a colony in space
but I do see something wrong in thinking that one can create
wildness by placing it into a container. At the present time,
there is a battle going on in American culture between those
who are trying to surround management with Culture, and
those who are trying to surround and contain culture with
Management. If the space colonies are sold to the American
public as a way of escaping the juggernaut of apocalypse, of
escaping the internal contradictions of our industrial civilization, and of not having to face those contradictions but
simply to extend, extend, extend always to a new American
frontier, then I think we will overextend ourselves to a point
of a deserved collapse. I think the space colonies excite the
Faustian imagination of the managers and the technocrats
for it offers them a way of continuing their existence without going through the pain of a transformation of consciousness. Though I see nothing wrong, in and of itself, with the
idea of an experimental small colony in L5, I do see a lot of
things wrong with the hype that is being generated in order
to sell the American public on colonies, so that they will
encourage congress to pass the appropriations. You, yourself, are guilty of encouraging some of this hype by caption-
ing an article on the colonies as apocalypt goodbye to the
juggernaut of apocalypse. The apocalypse, in the fashion
of the Tibetan book of the dead, is but the malevolent aspect
of beneficent dieties. If we can in the face of famine, pollution, and war, remember our Buddha-nature, then we can go
through the terrifying initiation of the race to discover that
the apocalypse that we seek to escape is inside us, and until
we come to terms with it, it will follow us wherever we go
— to L5, to the moon, or to Mars. Since all of us have to
make some kind of choice as to what work will receive our
limited energies, I prefer to work to create a planetary, decentralized, meta-industrial village on the surface of the face
of the earth. Earth may not be the best place for a highly
technological civilization, but it is, as Robert Frost said,
"The best place for love. I don't know where it's likely to
go better."

(The rest of it)

I'd like to get away from earth awhile
And then come back to it and begin over.
May no fate willfully misunderstand me
And half grant what I wish and snatch me away
Not to return. Earth's the right place for love:
I don't know where it's likely to go better.

—Robert Frost

Sent by John Graham
Guernewood Park, California

GEORGE WALD

Biologist, pacifist; Nobel Prize 1967

CoEvolution has asked me to write my thoughts about
Space Colonies — O'Neill's or any others. Let me say at
once that I view them with horror.

I am a little late in getting this to you since I spent the
last eight days in Rome as a member of the Bertrand
Russell Tribunal II, hearing testimony on U.S. imperialism
in Latin America and the atrocities being committed by
the atrocious military dictatorships we have set up and
maintain there. That too was horrifying; and not altogether irrelevant to the present subject. If we took better
care of what goes on here, if we managed to live decent
human lives in decent environments, if we relieved the
exploitation and oppression of life on Earth, human and
animal, and of the Earth itself, living in space might seem
less enticing. And we could use the wealth and energy that
it would take to put a colony into space to do some of
these other things.

What bothers me most about Space Colonies — even as
concepts — is their betrayal of what I believe to be the
deepest and most meaningful human values. I do not
think one can live a full human life without living it
among animals and plants. From that viewpoint, urban
societies have already lost large parts of their humanity,
and their perversion of the countryside makes life there
hardly better, sometimes worse.

We are cultivating a race of fractional human beings, living
in fractional environments; machines for living for human
parts; providing mass produced and standardized unit spaces
for depersonalized human units.

The whole concept of space colonization carries this impulse
to the ultimate limit. What has already gone much too far
on Earth in technologizing all aspects of life — nutrition,
motion, medicine, birth and death and everything between
— will find its complete consummation in space.

A few years ago I had a strange experience at the University
of Vermont. The students had organized a symposium on
"Man-made Men." I had taken that title to be ironic, symbolic, hyperbolic. A few days before the symposium, I
learned with dismay that it was meant literally. Those students, brought up and surrounded by mass-produced objects,
were excited by the conviction that within another decade
human beings would be produced in that way. They were
looking forward eagerly to coming off an assembly line.
When I told them this is absurd, neither possible nor desirable, I had a fight on my hands. Their first response was
that I was old and behind the times.

So that my point is that the very idea of Space Colonies
carries to a logical — and horrifying — conclusion processes
of dehumanization and depersonalization that have already
gone much too far on the Earth. In a way, we've gotten
ready for Space Platforms by a systematic degradation of
human ways of life on the Earth.

A nice example of that is in the major developments of
architecture throughout this past generation. It's all gone
to designing monstrous machines for human existence in
the cheapest ways out of the shoddiest materials.

All around me at Harvard buildings are going up of poured concrete, surely the ugliest and least durable of all construction materials, gray and dead. Pier Luigi Nervi is their prophet. I am told with enthusiasm, "You see, we don't smooth the concrete. We leave the board marks. You can even see the grain of the wood!" That's just great! Wood made out of poured concrete! Like shingles of tar paper or extruded plastics.

Le Corbusier designed a building for art and design students here, the Carpenter Center, of poured concrete of course. Unfortunately, the building itself is the showpiece. It's a goldfish bowl – visitors can see from outside everything going on inside. There is no privacy whatever. Just the thing for an artist.

Walter Gropius, ending his career at Harvard, designed a major living space for students. Again, the shoddiest materials. Every room was a unit space for a unit student, small and forbidding, lined with cinder block. One couldn't drive a nail to hang a picture. In the open space outside is a Bauhaus objet d'art: a stainless steel Tree of Life. It looks like an umbrella stand. A young woman student once said to me, "On moonlit nights in the spring Radcliffe girls dance around it, dropping ball bearings!"

Mies van der Rohe: I gave a series of lectures in Chicago in his memory. Before the first lecture my hosts were so kind as to drive me about Chicago to see his buildings. They were huge, skyscraper apartment buildings, great "functional" concrete and metal and glass structures, each housing thousands of persons. And expensive: for the most part condominiums. That word has always bothered me; for me it carries an inescapable suggestion of contraception. After that trip I realized that the suggestion is apt. Those condominiums are no place for kids. They are the greatest contribution to birth control since the sports car.

Paolo Soleri: There he is, that gifted man, making bony constructions in the American desert. I haven't seen one that I would remotely want to live in. But he has larger plans: for single, integrated structures, each to house hundreds of thousands of persons, a city. All human needs fulfilled in one great block. And "functional": one of Soleri's models shows such a construction that is simultaneously the spillway of a huge dam. That might be fun for Soleri. But can you imagine trying to live, even to raise children, in such a place?

All that dehumanizing architecture is getting us ready for Space Colonies. So one last but not least consideration concerning them: Who is to go to them? The power elite of our over-developed society? The highly affluent? Who else? Perhaps, having made piles of money out of war, smart bombs, nuclear weapons, they can find in the Space Colonies the refuge from which to watch the rest of humanity killing and maiming and poisoning and mutating one another – deciding eventually when it is safe to come back down.

Ed Taylor, the physicist, seems to be a nice guy (see McPhee: **The Curve of Binding Energy**). His parents were missionaries and pacifists, so when he went into making atom bombs he had a problem. First he set out to make the biggest, most effective fission bomb ever. It was going to be so horrifying as to end all possibility of war. It didn't. Then Ed Taylor took a new tack; he designed the <u>smallest</u>, most effective fission bomb, the poor man's atom bomb. That would spread so widely as surely to end all war. It didn't.

Ed Taylor ended up wishing he'd had nothing to do with any of that. So what's he into now? You've guessed it! It's to design the biggest space vehicle ever dreamed up, one on which persons could survive for many generations. And after that?

How about a farm, Ed Taylor? How about a horse?

Our sun has about another six billion years to run on the Main Sequence. That's a long time. If we take to serving life rather than death, if we can come to realize that maximizing profits is not the primary aim of human existence, if we could begin to take care of life – human, animal, plant; if we cultivated rather than devastated the Earth – then it could be a great place to live on and to enjoy – for the next six billion years! It's worth a try!

DAVID BROWER

President, Friends of the Earth; initiator of "exhibit format" books such as This is the American Earth

Thank you for the chance to add my comment about the O'Neill Space Colonies to the comment of the brilliant people you have written to, even though I suspect you expected me to give a knee-jerk reaction, which I shall, because people who don't have good reflexes are in trouble, and my knee jerks when pounded.

Ever since I first discovered a fatal flaw in the logic of one of Edgar Rice Burroughs' Mars books (I once read them all), I have maintained a healthy skepticism ordinary people are supposed to maintain about attempts to achieve perpetual motion, or anything like it. That goes for the Mars lights that Burroughs invented that kept recycling their own light and thus needed no energy added. It goes for breeder reactors. And I am afraid it goes for the Space Colonies.

Owing to my preoccupation with problems on our friendly local planet, I have not had a chance to apply my sciolist's analytical capabilities to all the details of the colony scheme, but I am afraid I must remain in opposition unless Mr. O'Neill can guarantee that Murphy's Law will be inoperative on all his satellites, and that the Little Prince running each of them will never be overcome with what inevitably happens to leaders, sooner or later.

With Murphy's Law operative, I have grave misgivings about the colony gravity machine. Won't an awful lot of things start floating around willy nilly and getting hopelessly mixed up – people, crops, fertilizer, sidewalks, vehicles, and schoolteachers – when the rotating device that develops the gravity must be stopped for replacement of defective parts? I shall always remain suspicious of anyone who tries to mess with the Law of Gravity. Such a person might meddle next with the Second Law of Thermodynamics, and foul up the whole system.

As for the Little Prince, what happens if, upon your arrival at his colony, you ask him to take you to his leader, or his successor? Or he asks you. Colony politics needs to be thought through, because although engineers can build, politicians are at the switches.

Further aspects would worry me, I am sure, if the major argument against the scheme didn't make all other considerations academic. The major argument is this: In the last analysis, the scheme is born of despair, which, as C.P. Snow says, is a sin. Despair leads to Escapism. Let's not worry about what mischief we are wreaking with our one pass at the planet, Escapism says, because (a) we can get away for weekends, vacations, or sabbaticals, (b) we will be rewarded in heaven later on for putting up with the hell others have perpetrated for us on earth, (c) some remote island or continent will put enough space in between us and our tormentors, or (d) some other colony, on a (1) ready-built planet nearby, or (2) a custom-made contraption, pleasantly devoid of all the honing, perfecting forces of the very adversity that was solely responsible for making our present shape, senses, and spirit possible.

I would not wish to appear adamant. If Mr. O'Neill's colonies, after due energy accountancy and review of the environmental impact statement, prove more desirable than the present alternative, then let me be the first to place reservations, for the first colony, for all who would continue the atoms-for-peace/war experiment here. Let all of them, salesmen and customers, be aboard the maiden voyage, absolutely free of charge, with a bonus if they promise to stay away.

And let the rest of us stay here, on this poor old beautiful planet, plagued only by ourselves, and try in good heart to fix it. □

45

MICHAEL PHILLIPS

Co-founder Briarpatch Network, author of The Seven Laws of Money. *As POINT director, made the grant ($600) that paid for O'Neill's first Space Colony conference in 1974.*

My comments on the new planet are divided into two parts: (I) predictions that may be of interest to your readers, based on the three years of thought I have given this subject, and (II) current research directions of the International Committee for a New Planet.

Section I. Predictions

1. The milieu at the end of the end of this century: there will be two to five construction satellites in earth orbit with 10,000+ inhabitants on each, building a variety of space objects (such as power facilities and manufacturing stations), including a few planets for solar orbit. One or two of these solar orbiting planets will be functioning with populations of 40,000 or more. Most of these space objects will be oriented toward nation/cultures on Earth and will be affiliated with language groups such as English, Russian, and German, and also possibly Spanish, Japanese, and Chinese.

2. The scenario for the following quarter century, to 2026, will see the proliferation of planets in solar orbit based on a greater variety of peoples; some will be corporate creations — next-century equivalents of Xerox and Mitsubishi; some will be concepts groups such as Scientology, EST, and the Catholic Church; and some will be indescribable in current terminology, such as our own International Committee for a New Planet, which is made up of several cultural groups.

3. Interest in new planets will be primarily American. It fills our need for a new frontier. And as we did with the ecology fad, we will believe that the rest of the world thinks the way we do about new planets (they won't). Remember our "race" to the moon, which turned out to be us running alone.

4. The effect of the New Planet on people's minds will be of the same magnitude as were the adventures of Marco Polo and Columbus on the Western world. If there were a hierarchy of mind experiences ranging upward from opinions to attitudes to ideas to concepts, and higher to persuasions, religions, general philosophies, and cultural horizons, then there would be one still higher on the scale. I'll call it "human species perception." It is this level that was affected by Marco Polo and Columbus and it is this level that will be affected by the New Planet. I think a good case can be made for the position that these two men shaped the Reformation, science, and all subsequent Western political thought. I think it was their actions that fostered and developed the view we now refer to as The Perfectibility of Man doctrine. This doctrine of perfectibility was the basis of Bacon's and Newton's development of scientific technique. It also led to political doctrines ranging from Hobbes and Locke to Rousseau and Marx — and ultimately to the political experience we call constitutional democracy.

5. I think the nature of this new "human species perception" will be to understand and accept the findings of biology and archeology in the past century . . . so that if we are "perfectible," it must be in the context of our being an _animal_ which evolved over hundreds of millions of years and which has 3 million years of direct experience in its current physical form. In addition, this evolutionary view will be subordinated to a new scientific finding emerging in the next quarter century: that we are a species with a single common culture that is 60,000 (or more) years old and has shaped us far more than the recent 6,000 years of recorded history.

6. We will come to learn of ourselves as descendants of a small (maybe 20 million worldwide) population of genetically homogeneous people who had a continuous cultural history of more than 50,000 years. This culture was partially absorbed into our city/non-nomadic culture since it was translated from the oral tradition directly into writing in our first written documents (Old Testament, Baghavad-gita, I Ching, the Books of the Dead, Euclid, Plato, Upanishads and Arabic algebraic works, etc.). The part that was not absorbed may today still be around us but ignored or partially destroyed (American Indians, gypsies, Africans, Shintoists, so-called "island primitives," and probably some esoteric gurus). With future research the language and experience of this 50,000 years of our cultural heritage will unfold and may dramatically shape all of our personal directions. The power of this tradition along with the immediacy of its influence will so drastically shape our view of our "perfectibility" that we will no longer look to governments or contemporary social structures to "perfect" us as we do now. We will begin long, arduous, personal (almost monastic) endurances to achieve a harmony with our "old" cultural heritage.

7. Science and the scientific method will become even more preeminent. Within 25 years 70 percent of all literate people will understand statistical testing and statistical "significance." The power of 300 years of science, which was empirically validated (completely!) by the first nuclear bomb and by our bringing back rocks from the moon, will be so completely accepted, absorbed, and understood that its primacy will spread rapidly to other fields. The fields most affected will be medicine and the political sciences (within these two broad categories I include biology, anthropology, and economics). In these fields high-quality empirical testing will become the norm, not the exception.

Section II. Current Research Directions of the International Committee for a New Planet

1. The end of the next decade will see the beginning of one of the most momentous changes in human behavior since the development of agriculture. It will be the beginning of the era of the "great Network." Computers are the reason, but they won't be visible and we won't call them computers (we don't call the phone system a computer, or talk about the computer-controlled elevator, but that's what they are!). We will just become media freaks as the new voice-operated interactive network terminal (or whatever) becomes more usable. Information will become free. Information has two incredible non-economic properties (as Dr. Peter Sherrill has taught me) 1) it is more valuable as more people have it; thus, the value of a single piece of information is not diminished when it is duplicated and distributed (for example . . . this article). 2) It is the basic source of all economic (and political) power. This will drastically alter all human institutions, ranging from power structures to work patterns, and most important, people's behavior will change (I won't go into detail here). Peter suggests an increase in psychopathy, particularly its milder forms such as manipulative role playing and the detached bureaucratic personality. I think there will be more Briarpatchism, people living on lowest incomes, with the fewest possessions, sharing resources, and emphasizing service to others.

Our research consists of interviewing, studying and testing people who have lived partially in this "future milieu": telephone operators, phone freaks, and real-time computer operators. Anthropology of the future.

2. We will support and encourage research in areas described in Section I, No. 6 above: the culture of the period 50,000 years before writing. Much of this research will be anthropological and archeological, but some of it will be scientific analysis of specific remaining artifacts, ranging from pre-Columbian knotted ropes to Tibetan stones, Shinto paper structures, Tarot images, and American Indian petroglyphs. The time scale on this work will be long and our Committee's efforts will be mainly supportive, looking for the occasional Rosetta Stone that will encourage other researchers.

3. We're going to get started on the technical work related to building a new planet. In less than 15 years NASA will be looking for a prime contractor to build Gerard O'Neill's working station. The airplane companies that became aerospace companies are no more appropriate for this than auto or railroad companies would have been for building rockets. So we can start building such a prime contractor now. I think the businesses that have the potential skill are in the underwater diving field, especially the ones doing the most difficult high-tech underwater work on the off-shore

ANT FARM
Image technologists

SPACE COLONIES ARE TAIL FINS.

I've been doing some research into World War II technocracy and its legacy of pesticides, appliances, plastics, rockets, and other household items, especially the automobile, all in preparation for a book on cars by Ant Farm.

It seems almost everything we are living with today, from aerosol cans to McDonald's and the Interstate Highway System, has its roots in World War II. Harley Earl, who was the Vice President in charge of styling at General Motors until the mid-sixties once credited the airplane, particularly the slick wartime fighter planes, with being the greatest influence on car styling. The Cadillac tail fins, which Earl designed, were directly derivative from the Lockheed P-38 Lightning. (We found a good use for tail fins when we built the roadside Cadillac Ranch in 1974). G.M. used dream cars to introduce far out ideas like the tail fins, rocket ports, and wrap around windshields because they realized that the American consumer responded to IMAGES of technology in far greater numbers than they did to technological innovation. (Technological innovation was never G.M.'s best suit). But the images were a subtle way of selling an attitude that pervaded the war-time years, and has become synonomous with American Know-how. It's an attitude of machine idol, an infatuation with technology in general as a catch-all for problems both large and small. It can be seen in a little boy's love for cars, trucks, trains, planes, and anything shiny and fast. It was romanticized, mediaized, and sold American as the dream, a way of life,

during the fifties. Not until the mid-sixties did we first recognize feedback on some of the fallout from our frenzied technological growth. Our pollution, our poisons, our congestion and our cancer rates can be measured statistically, but our loss in terms of alienation, attitude, and style of life can not.

So here I am, finally aware of my addiction to machines, my love of cars and all that sparkles and where it came from. I'm trying to come to terms with it. I don't want to become a 'return to the woods' anti-industrialist and yet I must become free from a generation of techno-conditioning. In the midst of this dilemma I open up **CoEvolution** and find Space Colonies – the ultimate machine fantasy. It will solve, you tell us: "The Energy Crisis, the Food Crisis, the Arms Race, and the population problem – in that order." Like my LSD fantasies of the late sixties it will have: "a Hawaiian climate in one and New England in the other, with the usual traffic of surf boards and skis between them." It sounds like a dream come true! It sounds like General Motors Futurama at the New York World's Fair of 1964: "In the distance, our eyes make out a jungle metropolis – dramatic proof of how mobility can help master even this tropical environment and turn it into a productive contributor to the world's marketplace." A reference to the Giant roadbuilder that makes 'instant highways' in jungles and other 'alien' environments.

Hey, wait a minute. That gives me a great idea. Why not give the Space Colony contract to General Motors? G.M. could stop building cars (a solution to the energy crisis) and put a war-time-like effort into Space. The assembly-line workers could again feel they were doing meaningful work and G.M.'s tremendous industrial capacity could be put to work building next year's Space Colony for the good of all mankind (those who can afford it that is). G.M.'s huge network of dealers could become space travel agents and in their service departments citizens could learn how to live in space, in simulators.

Well, that settles it. If you give G.M. the nod we will support Space Colonies 100%. To give you some incentive we are enclosing a stylist's rendering of the 1991 Space Colony by G.M. And even if they don't work they will be totally entertaining.

Good luck in the future,

—Chip Lord

---

oil rigs. A large well-run company, multicultural, in this field would be a very appropriate prime contractor to build planets in space. So I'm going to start work on creating such a company.

We welcome help from **CQ** readers on any of these last three specifics. No philosophical treatises please.

—International Committee
for a New Planet
330 Ellis Street
San Francisco, CA 94102

JOHN TODD

*Biologist, co-founder of The New Alchemy Institute,
co-designer of The Ark on Prince Edward Island, Canada*

I have a perspective on the proposed space colonies that
might be of use in the debate. I am a biological designer
and for the past decade have been simulating a variety of
aquatic and terrestrial ecosystems in contained spaces or
"capsules." Some of them, like the tropical marine environment which my friends and I set up in a greenhouse at the
Woods Hole Oceanographic Institution, were difficult to
design. This "tropical" ecosystem was established to house
the blennies and gobies (fishes) Bill McLarney and I were
bringing back from Central America. Such a contained
environment had to be designed well enough to permit
the little marine fishes to breed, which is not an easy task.
Fortunately, we were able to raise several generations in
the north and learn something about them as I was interested in their sexual and social communication. We also
simulated conditions of a local pond and an inshore marine
area for studies of the effects of environmental perturbations on the social organization and behavior of a number
of fish species.

From all this we learned that the social behavior of fishes
and other animals can often be an extremely sensitive indicator of the health of a contained ecosystem and the
highly social creatures have an intrinsic ability to bioassay
their own environments. We also discovered that artificial
environments were often unpredictable even when established as diverse ecosystems. Occasionally under conditions
we did not fully comprehend, a particular alga species
began to predominate, producing an antibiotic which killed
the green and blue-green algae and a toxin which wiped
out the molluscs and fishes. Sometimes an ecosystem
became sick and we had to link it back up to the ecosystem from which it was derived originally. During my
period at the Oceanographic Institution I didn't know too
much about designing or using sub-ecosystems for self-
regulation and biopurification, and was highly dependent
upon technology. If a circulating pump blew or an air
system quit, the next morning brought putrid smelling pools
filled with dead animals floating belly-up on the surface.

At New Alchemy we have evolved biological design ideas to
create food producing ecosystems seeded with organisms
from around the world. They range in size from a tiny
Chinese-type polyculture fish-raising system contained in a
solar tube in my house to the Ark on Prince Edward Island
which is a wind and solar-powered bioshelter containing a
living area for people, a research facility, a family-sized
food garden and commercial greenhouse and aquaculture
components. The Ark, to be completed this summer, will
be as close to a contained living space and biological entity
as yet exists. It will be dependent upon its living elements.
The Ark on a smaller scale has many of the attributes of a
space colony, with one fundamental difference. Its gaseous,
climatic and biological health is created through its coup-
lings and linkages with the exterior environment. This fact
is important and should be kept in mind.

After a decade of living intimately with designed ecosystems I am coming to know that nature is the result of
several billion years of evolution, and that our understanding of whole systems is primitive. There are sensitive, unknown and unpredictable ecological regulating mechanisms
far beyond the most exotic mathematical formulations of
ecologists. When I read of schemes to create living spaces
from scratch upon which human lives will be dependent
for the air they breathe, for extrinsic protection from pathogens and for biopurification of wastes and food culture, I
begin to visualize a titanic-like folly born of an engineering
world view. At this point we don't know enough, being
totally reliant on knowledge as well as physical subsidies
from nature to survive on earth. In space there are no
doors to open or neighbouring ecosystems to help correct
our mistakes.

At New Alchemy we have established two backyard fish
farm-greenhouses for a comparative look at different biological design strategies. Three summers ago both aquaculture ecosystems continuously produced massive blooms
of green algae which, in turn, supported fishes as food, as

providers of essential gases like oxygen and as detoxifiers
of harmful ones like ammonia. Dense blooms, critical to
the success of both systems, were maintained with tiny
amounts of ground and roasted soybeans, a trick we picked
up from a 1922 issue of a biological journal.

Two seasons ago one of the systems would not produce
dense algae populations no matter what we did. It was a
puzzle. Fish growth was reduced and the system was vulnerable to extremely low oxygen concentrations in the
culture component. Some days the fish stuck their heads
out of the water in search of a breatheable medium. The
adjacent ecosystem housed in a geodesic dome produced
excellent algae blooms.

This past growing season both of these backyard fish farms
lost their algae blooms as well as their ability to produce
oxygen to sustain large, rapidly growing populations of
fishes. We had to use a number of management techniques
including splashing water and having it flow through beds
of oxygenating rooted aquatic plants to sustain these pond
ecosystems. The fishes required larger amounts of supplemental feeds to grow. In ecological-economic terms the
dilemma was pushing us in the wrong direction. Other
systems were producing algae so that we could not explain
away our problems as an "Act of God" although it's always
tempting to do so.

What was wrong? Three years ago all was well. Then the
next season one system failed and the following year so
did the other. We started to track down sequential changes
which might have been common to both. The explanation,
or at least partial explanation, illustrates how little we know
and bears on the space colony concept. Two years earlier
I had planted a small bed of macrophytic or higher aquatic
plants in the connected biopurification sub-ecosystem and
water lilies in the supplemental food chain sub-ecosystem.
The next year others did the same thing in the companion
fish culture complex. The year they both produced algae
in abundance was the year no higher aquatics were resident
in the systems. The disappearance of massive blooms coin-
cided with the introduction of the rooted plants and one
or perhaps several of the aquatic plant species produced an
antibiotic substance which severely curtailed reproduction
of the algae in the fish culture component situated downstream. The antibiotic travelled in the flowing water to
the main culture pond.

*"Congratulations, Judy! It's a 12 pound 5 ounce male
two headed calf!"*

The higher plants had been placed in the biopurification section to help remove toxins and to provide oxygen for adjacent carbon dioxide producing shell-bacterial filters which have high oxygen demands. They did their job well, but elsewhere within the systems the effects were negative. In the future either the higher plants will have to be eliminated, which is unlikely as they serve many critical functions and are used as feeds for the Chinese White Amur fish downstream, or we shall have to break down the antibiotics which inhibit algae reproduction through some as yet unresolved biological means.

I don't mean to overload you with detail, the point being that little is known about chemical competition between higher aquatic plants and specific species of algae. The numbers of combinations of interactions are endless. A number of biologists have studied antibiotic activities between a few algae species in laboratories, but it's all tip of the iceberg stuff.

When one adds the relationships between fish species and algae, the matter becomes more complex and the interactions between the two might have some influence on the biogenesis of oxygen for contained atmospheres. A recent study in Poland by Maria Janusko has demonstrated that the composition and densities of fish species affected both algae production and algae species composition. Bighead Chinese carps, for example, which have a preference for microscopic animals as foods, caused blue-green algae to predominate, while the Silver Chinese carps, which are phytoplankton feeders, resulted in a shift towards the predomination of diatom algae in the experimental lakes. The whole ecology of the lakes was changed by a shift in fish species composition. Here we have a single scarcely understood example of webs that influence not only plant populations but gas production as well. Prediction is almost impossible – but in a space colony where it would be prudent to have the bulk of the oxygen ultimately generated from algae in ponds and carbon dioxide from soil-bacterial complexes, I would consider it unsafe to attempt to simulate liveable environments from our present biological knowledge. Let me elaborate.

A few years ago Howard Odum and Ariel Lugo put together contained microcosms (terraria) made up of components of the forest floor in a tropical rain forest. They were seeded with mineral soil, litter, forest floor herbs, algae and small animals to simulate some of the properties of that ecosystem. Most of the capsules were left on the forest floor but others were brought to controlled environment chambers at the Universities of Georgia and North Carolina.

The biotic communities of the terraria varied in the ratio of litter and consumers to plant producers. As a result, the carbon dioxide levels and gaseous equilibria within these contained ecosystems were different.

This is what Odum and Lugo concluded:

> This may be an important demonstration of the control of the atmosphere of the planet by the biotic components existing in the system. The physical properties of the atmosphere are a result of biological evolution as much as vice-versa.

In the space colonies the only long-range solution is to create ecosystems which create atmospheres upon which the vital support components, including humans, will depend. I suspect it will take decades or even centuries of seeding and reseeding organisms and varying of ratios of litter, soil, lakes, etc. to achieve a liveable environment. Here on earth our oceans act as a buffer protecting us from rapid and harmful misuses of landscapes . . . but only to a point. In the future we will come to realize the importance of wilderness areas on land as biological regulators of the planet and learn to respect them.

The space colony designers have planned to handle the atmosphere question in a variety of ways. All are highly technical, costly, subject to failure, engineering solutions. Initially, they intend to bring aboard supplies of oxygen-nitrogen mixes, if I understand them correctly.

O'Neill (page 13, Fall 1975 **CoEvolution Quarterly**) stated:

> Nitrogen constitutes 79% of our atmosphere on earth, but we do not use it in breathing: to provide an earth normal amount of nitrogen would cost us two ways in space colony construction, because structure masses would have to be increased to contain increased pressure, and because nitrogen would have to be imported from earth.

I am not sure how the ecosystems illustrated on page 7 of the **Quarterly** are going to like a nitrogen poor atmosphere. Certainly it would make the creation of ecologically-derived food producing culture systems difficult, and atmospheric nitrogen fixing plants might have a hard time.

O'Neill is more optimistic than I am although he acknowledges that semi-closed-cycle ecology will need prior verification. I think he is off the mark when he states that "Isolation and heat-sterilization can halt any runaway biological subsystem."

A resident population of some 10,000 individuals are going to have a difficult time making it in an area circumscribed within a mile's distance. This whole question of carrying capacity in space has been looked at before, albeit in a cursory fashion. During the hey-day of interest in space exploration (summer 1962) a symposium on the ecological aspects of space biology was convened at Oregon State University. Several ecologists present, including Howard Odum and Jack Myers, argued that space biology was presently confined to rather narrow disciplines tending to be simplistic and reductionistic and that if humanity were to explore space for any length of time we would have to take complex ecosystems with us. They went on to say that ecosystems were elegantly miniaturized and perfected with biocircuits, control functions, repair mechanisms, biopurifiers, gaseous regulators and had integrating abilities far beyond the wildest dreams of electrical engineers. Under regimes of total or overall accounting, including hidden subsidies, ecosystems were also efficient.

At the symposium Robert Beyers described eight artificial aquatic ecosystems conveying the message that balanced, steady-state ecosystems indeed could be contained. But his miniature ecosystems were in no way intended to produce end products such as human foods, nor was he trying to prove that the wastes from 10,000 people could be utilized.

Odum went one step further. Being optimistic, he stated that space environments for humans were possible, but he didn't let it go at that. He calculated a free energy budget for a self-maintaining, light supported, closed ecosystem at climax (stable state). His discussion which appeared in the **American Biology Teacher** [Vol. 25 (423-443) 1963] is worth reading. He feels that multiple seedings will be required, but that biological support in space can be developed given enough time. His estimates of the size of support ecosystems necessary for stable conditions are based upon his energetics and efficiencies calculations at various levels in the biological realm. The space required to support humans artificially makes one damn respectful of planet earth. He calculated that it would take about 2.5 acres of ecosystems combining water and land to sustain safely one person in a space colony. If my crude estimate of the first colony's potential biotic area of one hundred acres free of structures, machinery, storage and what have you is correct, Island One could support 40 people, not the 10,000 proposed by the exponents. If Odum is right that means that the other 9,960 people will have to bring along their own gases, food, and waste disposal units, and even at that they might tragically overload the colony.

I think that when people talk of colonizing space they really don't have any genuine perception of what it will involve. All the present support for space comes from earth and until we learn much, much more about contained ecosystems it will continue to do so. It won't be the kind of knowledge that a crash program of space biology will generate, but the very thinking about ecosystems in space potentially has the ability to move part of biology in the direction the New Alchemists are exploring.

I should like to end by stating my bias. The idea of moving nature into the cosmos is staggering. It may be the ultimate human folly, or it may be life's experiment in using us to extend itself. I do not believe that we as a species have in any way earned the Right of Passage. □

BROTHER DAVID STEINDL-RAST

Benedictine monk, founder of "Houses of Prayer" movement

You've done it again! If the publicity you are giving to O'Neill's idea of space colonies catches on, it could have an impact comparable to those first photographs of Earth from Space. Isn't it fascinating how quickly we moved from recognizing our planet as "Spaceship Earth" to designing an Earth Spaceship?

Let me first tell you what the idea did to me personally. When I picked up this issue of CQ, I was at best moderately interested. When I put it down, I knew that I had read some of the most important information of my lifetime. Here was a challenge. It would not leave me.

In fact, it has been growing on me ever since. Not so much the idea of shipping off our surplus population (that'll take a while and have its problems), but the possibility of tapping a practically inexhaustible source of energy by putting electric power plants in space. That does something to me far beyond its practical implications. It renews my experience of the abundance of nature. Yes, I remember the time when we thought of ocean and atmosphere as inexhaustible. But as we realized the implications of the term Spaceship Earth we ran the risk of thinking small instead of living frugally, which is quite a different thing. O'Neill's invention restores my childhood confidence in Mother Nature. And only now do I realize how important that is for our spiritual wellbeing.

Also, I suddenly find myself looking with different eyes at machines. Anything more complex than a pencil sharpener still makes me feel a bit uneasy. Is it going to break down? If so, I know I won't be able to fix it. Machines just isn't one of my talents. But for the first time the other day I felt something like respect, affection almost, when I watched that computer the teller was consulting at our local bank. After all, down here we can still make do with simple tools, if we are willing to pay the price and live austerely. In Space that's different. And that again is spiritually significant. I had long suspected that simplicity could not be reached by going back, but only by going forward, beyond complexity.

Of course, I see problems, too. One set of problems springs from the fact that we are dealing here with a controlled environment. It must be highly controlled if it is to work. But this entails the double danger of its becoming boring and vulnerable. I don't know enough about the technical aspects to assess the material vulnerability of so complex a system. (I hope it's not too likely that some schoolboy could inadvertently shoot a hole in the vault of the sky with his slingshot, and the air goes out as from a balloon.) The vulnerability I have in mind is of a different sort. When control is necessarily extensive, power accumulates and thus the danger of usurping this power. If scores of people have hijacked airplanes, how about hijacking a little world out there in Space?

Boredom is another shadow following control. How are we going to make room for the unpredictable, the wild, the surprise, for the untoward even, to save us from boredom? How much architectural variety sprang from the resistance a not fully conquered environment offered in the past. Now, that we can adopt any style we want, will every imitation be tried out? It's bad enough to come upon a Dutch windmill next to the Damascus gate in Jerusalem. I hope this kind of thing and the resulting boredom will be spared us in outer space. How boring, to make a space colony look like an island in the Bahamas or a township in New England! Why not make it look as much as possible like a space colony, and discover what that will be?

Another area of concern may still be connected with control: the inevitably high socialization of a space colony. One has to have lived as a member of a village community (or of a monastery, in fact) to appreciate the impact of social interdependence. Of one thing I am sure: there will have to be a sociological safety valve built into a space colony, a breathing space, a place to be alone for a time. A return trip to Earth might not quite be feasible every time you develop the kind of global cabin fever one must foresee. But there may be room for temporary hermits, say, on that belt of agricultural cylinders. There won't be need for someone to do a little hoeing of those space potatoes, will there? At any rate, monks ought to be able to teach us something in preparation for space colonizing. They have for quite a while experimented with intentional, often self-contained communities, and with the creative tension between cenobitic (communal) and eremitical (solitary) life.

But the sociological aspect of the need for solitude is still not the most important one in this context. There is a spiritual dimension to it. A space colony, even the largest possible one, will entail limitations of a kind we never come to know on earth. Now, there's one aspect of limitation which many people enjoy. It's what we find attractive about a compact car, a snug backpack, a ship where every inch of space is used to advantage. But the other side of limitation is a threat to us: every limitation points to our ultimate limitation, to death. We'll need solitude and silence to come to terms with death. All the more so in a setting where the limitations are too pronounced to hush death up, to gloss over it as we tend to do here. The hermit, as I see it, brings into society the lifegiving energy that flows from a radical confrontation with death. Ask Gerry O'Neill if he needs a starter group of space hermits. I think I'd know some volunteers.

Meanwhile, what can one do to help? What a rallying point space colonizing could be for our whole human family! Not because it can be done "without any sacrifice on our own part" as we are promised. The frontier is always an area of sacrifice, and the High Frontier demands higher sacrifices, because the stakes are higher. Sacrifices won't scare people off when the cause is inspiring. On the contrary. They'll rise to the challenge.

Suppose you collected — not yet money — only pledges of, say, $1,000 per person, conditionally. No one has to pay a penny unless and until the pledges add up to the total estimated cost. Only then does everybody pay. You could even make your contributors shareholders in this profitable enterprise. But we better let Mike Phillips work that out. Meanwhile let's make as much noise about space colonies as we can. If Columbus had gotten a little more publicity he wouldn't have been lobbying (a one man lobby) from 1485 to 1492. We can't afford seven years in 1976!

Info isn't heavy

... I want to help. Each man sees things in terms of his own craft; as an information system designer it occurs to me that one element of space colony construction has a very small gravity well transit penalty — this element is information. The available resources in space are energy and materials; the 'cook book' for transforming these into useful artifacts can be sent up from earth at a very modest cost using mature data processing and communications technologies. These technologies are on a rapidly declining cost curve.

Money and Politics

Building space colonies will be a very expensive venture; the break-even point for the investment is at least tens of years in the future. How to get money out of existing governments? A most interesting possibility is linked to disarmament. The U.S.A. and the U.S.S.R. have given considerable lip service to disarmament and to cooperation in space technology ventures. Here is a chance to get their money where their mouths are. The combined defense budgets of both countries exceed 170 billion dollars per year. O'Neill estimates that the space colonies project could be funded conservatively at 12.7 billion dollars per year; about 7% of the defense budgets. Funding a joint space colony project represents an interesting form of disarmament. Verification becomes much simpler — no need to count ICBMs, bombers, etc. Expenditure levels on the space colony project would be directly visible in terms of progress toward concrete goals. ...

Dave Caulkins
Los Altos, California

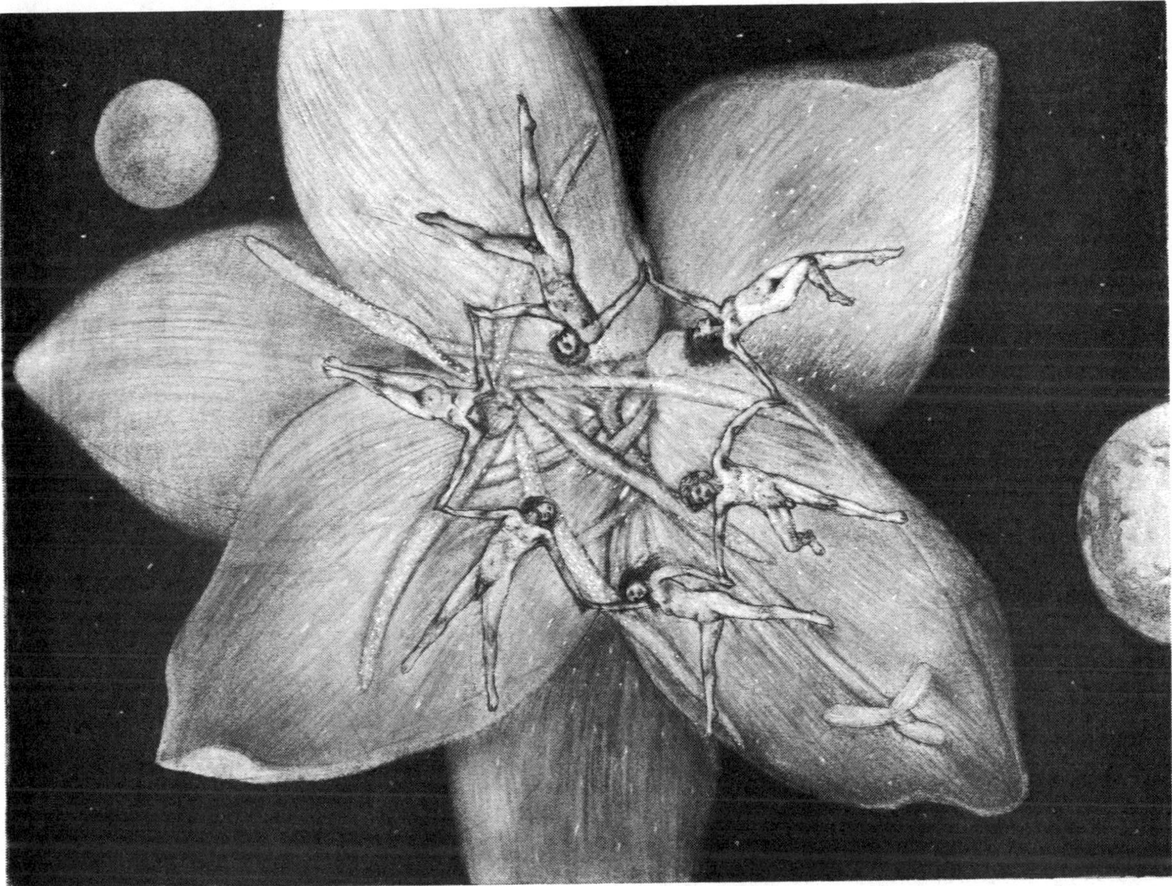

Arthur Okamura

Planet parenthood

. . . I keep getting the vision of the planet as a plant which has flowered and wants to go to seed. Perhaps all the insanities of our time come basically from a frustrated planetary reproductive urge. . . .

Ramon Sender
Occidental, California

RICHARD BRAUTIGAN

Poet, novelist, author of Troutfishing in America

OWLS *Hoot! Hoot! Hoot! Hoot!*

I think for the time being, the remaining years of this century, we should limit our exploration of outer space and concentrate our creative energy and resources on taking care of our mother planet Earth and what lives here.

Owls hoot in the early Montana evening when the air is very still and floats the scent of pine trees.

I like this planet.

It's my home and I think it needs our attention and our love.

Let the stars wait a little while longer.

They are good at it.

We'll join them soon enough.

We'll be there.

Juvenile space

. . . Dixie and I have been discussing the space cities in the light of Margalef's ecological theory. We've considered that a healthy ecosystem, human or not, needs a dynamic patchwork of juvenile and mature sub-systems; the juvenile systems to promote evolutionary experiments, and the mature systems to remember the best results. Of course the systems will always be moving towards "climax" through succession, while occasional rejuvination will occur due to cataclysm (i.e., change too swift for life to swiftly adapt). Now, the state of Europe in pre-Columbian times was of a climaxing human ecology with dwindling frontiers of juvenility. Had the new world not been discovered, Europe would have certainly calcified to the point where it could not change rapidly enough, and a disaster would have probably nearly destroyed civilization. (The Great Plagues may very well have been such a disaster at an earlier time of "over-maturity." Bateson's notion that unused flexibility disappears is also relevant here.) But the New World was discovered, and opened up a new frontier both to exploitation and evolutionary experimentation. Democracy was one of the (apparently) successful experiments exported back to the old world; super-industrialization was another.

The analogy to our present dilemma should be obvious. World civilization has spread to the point where frontiers are scarce and flexibility is being gobbled up. In the Last W.E.C., Stewart, your proposal for "outlaw zones" was a recognition of the need for juvenile systems for necessary experiments to take place. But space cities provide a more expansive and open-ended frontier of juvenility to restore healthy balance to our culture. Our alternative, as we see it, is Apocalypse. To O'Neill and company we say, "Bravo!" and "Let's get on with it!" . . .

B. Alan Scrivener
Laredo, Texas

51

PETER WARSHALL

Biologist, watershed consultant, **CQ** *Natural History editor,*
author of Septic Tank Practices

My curiosity focuses on soil, Earth. Rather than say "10,000 humans are impossible to feed in a Space Colony," I would insist that O'Neill has drastically underestimated the problems of space agriculture.

I cannot judge the physics, but O'Neill's understanding of plant growth leaves me totally uninterested in this project. No details are needed. Just consider:

Comments like "the main reason for anything but a pure oxygen atmosphere is just fire protection." An enzyme (nitrogenase) is responsible for taking atmospheric nitrogen and changing it into high energy ammonia (NH_3). The ammonia is necessary for protein production. Here's the catch: *oxygen must be kept away from nitrogenase.* Too much oxygen, then no ammonia and no protein. It's hard enough on Earth to keep nitrogenase active (see below) without upping the amount in the atmosphere to 80%. O'Neill doesn't even mention nitrogen-fixing bacteria – the creatures that make all this possible.

O'Neill doesn't mention sources for the trace elements. For instance, molybdenum is needed to make nitrogenase.

O'Neill doesn't even mention the crucial elements like phosphorus. The universal energy "currency" of all living organisms is ATP (adenosine triphosphate).

O'Neill doesn't understand the simplest chemical processes involved in photosynthesis. For instance, photosynthetic compounds tend to mistake oxygen for carbon dioxide. When oxygen is incorporated in place of carbon dioxide, plants cannot produce carbohydrates. Remember, the early atmosphere of Earth had much more carbon dioxide. Plants (like soybeans) adapted to these heavy CO_2 atmospheres and still produce more protein in artificially high CO_2 environments.

The two major elements needed for photosynthesis (carbon for carbohydrates) and nitrogen-fixation (nitrogen for proteins) are the scarcest mean elements according to O'Neill. How many times do we have to be told that life means organic life and organic life is just one great theatre of carbon molecules.

Not to sound too sour grapes but these **CQ** articles are prime examples of contemporary American schizophrenia: a technological romanticism totally removed from agricultural practicality.

The only benefit of Space Colony research might be the tiny slough off of relevant information about semi-closed-system agriculture. More sensibly the billions should be spent directly on Earth agriculture. (P.S. I am totally unimpressed by the scare tactic that we must exploit space or starve. Just another way to dilute energy into the BIG & TECHNO rather than the small and planetary.)

My own romance would make O'Neill and all potential space cadet(tes) serve five to ten years in a major Earth watershed (like the Hudson or Brahmaputra or Congo or Danube). After acquiring intimacy with the Earth, then, maybe, we should allow exportation of planet consciousness in outer space. To send techno-romantics as ambassadors of Earth Mind is abrasive to my Taoist soul.

I have included a simplified organic mandala of a small part of the nitrogen-fixation process. O'Neill might want to meditate on its implications for outer space food production.

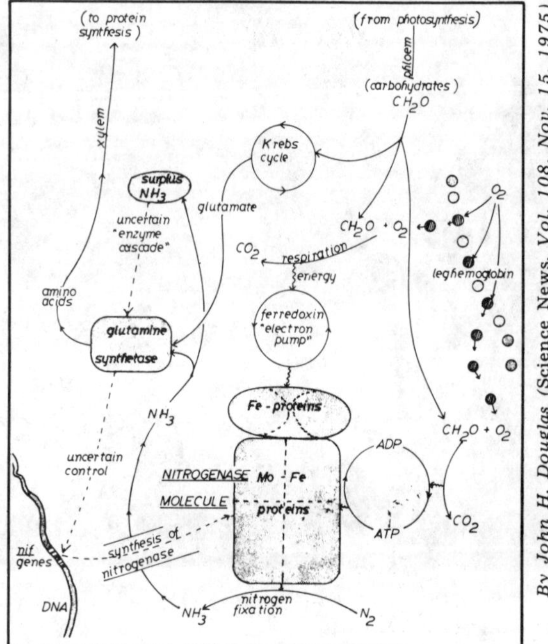

By John H. Douglas (Science News, Vol. 108, Nov. 15, 1975)

Note that the whole process is a buffer system. Starting with water and carbon dioxide, fuel is made by photosynthesis. The fuel is used to alter nitrogen to ammonia. The genes are genes of nitrogen-fixing bacteria that live symbiotically with the plant. Any alteration in atmospheric conditions will change the total process. Four crucial elements (Carbon, Nitrogen, Hydrogen and Phosphorus) are scarce to nonexistent on the Moon.

Space virus or child?

. . . Space colonization isn't something to rush into, and then be locked into, as a means to profit. Developed with care, as an end in itself, the space colony can be a most fertile inspiration, forcing/leading us to learn the meanings in the structure of communities, the balance of biological sufficiency, and the qualities of a sane technology, lest we export the amorphous, imbalanced insanities we live with now.

Consider that the building of worlds is the work of gods. Evolution is metamorphosis; when you leave Earth in a world we have created we will be literally and metaphorically leaving the cradle. The question is whether we break out of a ruin like a virus exploding from a shattered cell, or leave Earth as a child leaves a playpen, no longer needed but ready for the next sibling. The virus is a degenerate fragment that just replicates on; the child grows towards limits we haven't yet found.

Doug and Missy Mink
Brookline, Massachusetts

Arab Space Program

While I am as potentially patriotic as the next person to his country of origin, I don't think we here in the U.S. have the surplus to put into such a project right away. And while we don't, OPEC does.

The oil rich nations have billions more than they themselves can spend, and are looking for some capital intensive scheme to start paying off about the time their oil runs out. They have the most to lose and the most to gain from a competitive, and clean, energy source. When you throw in abundant, exponential growth of both resource and demand, and long-range schemes, it becomes all the more inviting. They of course know first hand the economic-political clout in controlling an energy source. Big investment, Big profits. . .

If your idea about their space station program and our shuttle program is correct, what could follow is an alliance by U.S. and U.S.S.R., racing the first OPEC nation or nations who get started on this. A hassle over nationalism may come when unemployed aerospace engineers head in droves to the jobs waiting for them in Araby. I wonder if they would live in domed cities, many square miles in area, which had been built not only as protection from the desert clime, but also as models of the Colonies to be constructed in space.

M.A.W.D. Hoffman
Sturbridge, Massachusetts

ANNE NORCIA

Cartoonist

This cartoon is the summation of my commentary on the subject... I tend to feel abandoned faced with scientific jargon (the necessity of it aside) and become punchy when some obvious questions − to me − get passed over. A few of the remarks are from friends but most are mine. I realize some of the questions have already been discussed, but I threw them in anyway purely for compositional balance. By the way, the scientists are caricatures of no one in particular − just composite stereotypes.

MARC LE BRUN

CQ *Personal Computers editor*

Is it a good idea? Where we rub up against the limits of our environment it begins to pinch, to form a blister. By your "Law of Paradoxical Effects" the combination of two more-or-less closed systems (Earth and Colonies) results in a more-open system, both materially and in more ephemeral ways. A blister is a self-healing process. The question is: Do the Space Colonies lance the blister (and thereby interrupt the healing) or do they provide a bigger (and presumably better fitting) shoe? Is the immediacy with which we are beginning to feel environmental pressures an important factor in our growth towards a more mature world civilization? Will our expansion into solar space delay our coming to grips with important social and ethical questions for perhaps millenia? I believe the Colonies can provide some respite from some pressures, but humanity will always be grappling/growing with itself and the rest of the universe. Between frontier and homeland the issues will vary, but never the vitality of the struggle.

Metallurgical paradise

. . . you seemed to me to play down the possibilities for heavy metallurgy out there. Now, this is the basis par excellence for industrial civilization. It uses more energy, water, more of earth's mineral substance, her oxygen, & puts out more heat, than any other branch of human work. (1) All that heat, expanded into the solar wind, would be nothing. (2) If ores could be obtained cheaply out there, they would start out at the top of earth's 4,000 mile well; going to market would cost only potential energy plus friction. (3) Thin, reflectively-coated mylar films, billowed out into big solar sails, could be used to fuse, smelt, refine such ores. (4) Zero-gravity would render massive castings easy to move about and work. (5) Vacuum-purity in metallurgical processing yields the strongest alloys; these would sell at a premium on earth, and if they were produced in sufficient quantity, they would amount to corners in the international market; just as English steel "cornered" the European steel market by the 18th Century − nothing comparable available. (6) Liquids in a vacuum (e.g., molten alloy) tend to form spheres. Therefore, just as shot was formed by free-fall in "shot towers" back in olden times, so could ball bearings be produced in zero-gravity. How about − all the ball bearings needed in a single year? Mucho $$! (7) Also, electromagnetic forces could be used to work the metals in zero-g.

Americans would love to get out there & pull off an economic coup like that!

J. F. Muggs
San Francisco, California

Hot Gaia

. . . If O'Neill is talking about a 5000 MW energy producer the energy just doesn't get from space to earth through the 5th dimension, it has to pass through the atmosphere (Gaia-breath, remember?) and while I'm sure you can select a region of the microwave spectrum that's relatively transparent, you can't get one that's perfect (you realize this when you speak of Arizona only getting a tenth of the space-flux) so some of that energy is going to be absorbed in transit. Even if it's only one per cent, that is stil 50 MW per station. Now by spreading out the beam of microwaves you can reduce the heat per volume of atmosphere, but you would likely lose efficiency, not to mention requiring greater area (=$$$) on earth. Liberating this quantity of energy into the atmosphere at small spots would do bad things to the weather, I'm almost certain. . . .

Another objection I see is funding. Do you really think the govt is going to let a funky rabble on their new space colony? It is a politically open ended sort of thing now, but open ends have a way of being closed rather quickly if you're talking abt POWER (polit., energy, $$$ or whatever). . . .

John Beutler
Philadelphia, Pennsylvania

ERIK ECKHOLM

Planetary Analyst at World Watch Institute, author of **Losing Ground.**

Justify it as a worthy extension of the imagination's playing field, as an interesting experiment for the mere price of a puny portion of a gross and misshapen national product, or even as bread and circuses. But not as a solution to the demographic, environmental, nutritional, and other sundry attributes of the earthly predicament. Anyone who knows the simple arithmetic of exponential population growth knows the irrelevance of unearthly migration to the reversal of our current downward spiral. □

GARRETT HARDIN

Biologist, author of Nature and Man's Fate; Exploring
New Ethics for Survival; Stalking the Wild Taboo.

Sure and it's an intoxicating vision Gerard O'Neill has given
us, the dream of creating a shiny new world all of our own
out toward the moon. How nice it would be to escape earth's
population problems! But we had better be wary of intoxi-
cation, even by 100 Proof Technology. The trip may not be
worth the hangover.

The image fails. This hangover would come before the trip —
which probably would never take place. Let me explain why.

I'm not going to spend any time on the technical details,
though I think Brother O'Neill has overlooked a few things.
But he's done a pretty good job. An exciting job. Let's not
carp at trivia.

On the economic side I think his vision fails. We must always
measure proposals like his against Hitch's Rule, which says
that a new enterprise always costs from two to twenty times
as much as the most careful official estimate. The more exotic the technology, the greater the cost over-runs. O'Neill's
space colonies are so exotic that the cost will surely go beyond Hitch's Rule.

So what? We're rich, aren't we? Yes, but not infinitely rich.
For awhile, the cost of mammoth public works can be met
by normal (though painful) adjustments in the economic
system. But at a certain level, corrective feedbacks fail and
the system goes into the destructive positive feedback mode.
Uncontrollable inflation takes over; prices and taxes spiral
upward out of reach. Attempts to evade the flopover point
of the economic system introduce new evils.

History has something to tell us. In the 16th century the
Papacy became intoxicated by the dream of building a
monumental new Saint Peter's in Rome. Hitch's Rule soon
ran the cost out of sight, and the Church had to finance the
project by the sale of "indulgences" — advance forgiveness
of sins yet to be committed. Parish priests pushed indulgences with all the subtlety of second-hand car salesmen on
Saturday night television. Resentment of the hard sell led
to the Reformation, and the Church never recovered its
temporal power.

O'Neill says his space program will cost hundreds of billions
of dollars. Applying Hitch's Rule we can be sure it will
cost thousands of billions. Would such a venture push the
economic system past the flopover point? Would O'Neill's
space stations be civilization's Saint Peter's?

But there is a more serious criticism to be made. Let us, for
the sake of argument, grant all of Professor O'Neill's technological and economic assumptions. The space station has
now been completed. It is ready for occupancy. Question:
Who is going to be permitted to move in?

Because of our powerful (though recently developed) tradition
of integrating minority groups it is obvious that the comple-
ment of the spaceship would, if it were U.S. controlled, have
to include blacks, whites, Puerto Ricans, Chicanos, Indians
from Wounded Knee, Wallaceites, American Legionnaires,
Weathermen and members of the Symbionese Liberation
Army. If the emigrants were drawn from the whole world
they would have to include Moslems, Hindus, Irish from
both Belfast and Dublin, Greeks, Turks, Israeli, Arabs,
Lebanese and Palestinians.

Some of the groups just mentioned are races, some are
religions, some are political groups. It doesn't matter. Gener-
ically we can call them all <u>tribes</u>, where a tribe is defined as a
group whose members pursue one code of ethics in their in-
group relationships, and another code for their out-group.

A libration point spaceship is a precision instrument, far more
delicate in its construction and far more vulnerable to sabotage than is our massive earth. How could such a fragile craft
withstand the buffeting of warring tribes?

Paradoxically, the creators of such a spaceship would be
psychologically least suited to be its permanent inhabitants.
The Professor O'Neills of the world might make brief visits
and inspection tours, but they could not tolerate the sort of
life that permanent residents would have to pursue there.
People of great originality and independence of spirit would
be intolerable in the spaceship community, particularly if
they belonged to different tribes.

For a libration point colony to survive it would have to have
only one tribe on it. (This is a necessary but not sufficient
condition, for even an initially uniform tribe may differentiate in time.) This means that the political system of the
spaceship must include progress-stopping features from the
first day people go on board. This means totalitarianism.

What group would be most suitable for this most recent
Brave New World? Probably a religious group. There must be
unity of thought and the acceptance of discipline. But the
colonists couldn't be a bunch of Unitarians or Quakers, for
these people regard the individual conscience as the best guide
to action. Space colony existence would require something
more like the Hutterites or the Mormons for its inhabitants.
Scientists and college professors would, as residents, be
disastrous.

The peopling of a spaceship creates an ironic problem for a
society like ours. We worship "integration" and consent to
forced diversification via "affirmative action." But integration could not be risked on this delicate vessel, for fear
of sabotage and terrorism. Only "purification" would do.

How could we possibly sell a purification program to our
people? If residence on a libration point colony was regarded
as a plus, then every tribe would demand the right to live
there. If it was regarded as a minus, no tribe would consent
to be made the sacrificial goat. It seems unlikely that precisely one tribe would view residence as a plus, and all others
see it as a minus. Yet that is what it would take to make a
selective residence system work.

Let's go back to fundamentals. What was the motivation for
this space colony proposal anyway? It was just this: to solve
earth's population problems. But there is another way to do
this: institute political controls of population here, setting
and enforcing limits to the size of families. Technologically,
this would be easy; politically, we haven't the foggiest notion
how to do it. (We all are appalled by the thought of "a
policeman under every bed.")

The principal attraction of the space colony proposal is that
it apparently permits us to escape the necessity of political
control. But, as we have just seen, this is only an apparent
escape. In fact, because of the super-vulnerability of the
spaceship to sabotage by tribal action, the most rigid political
control would have to be instituted from the outset in the
selection of the inhabitants and in their governance thereafter.

R. BUCKMINSTER FULLER

Design scientist, author of Synergetics; Nine Chains to the Moon; Ideas and Integrities

Conceptualizing realistically about humans as passengers on board 8,000-mile diameter Spaceship Earth traveling around the Sun at 60,000 miles an hour while flying formation with the Moon, which formation involves the 365 revolutions per each Sun circuit, and recalling that humans have always been born naked, helpless and ignorant though superbly equipped cerebrally, and endowed with hunger, thirst, curiosity and procreative instincts, it has been logical for humans to employ their minds' progressive discoveries of the cosmic principles governing all physical interattractions, interactions, reactions and intertransformings, and to use those principles in progressively organizing, to humanity's increasing advantage, the complex of cosmic principles interacting locally to produce their initial environment which most probably was that of a verdant south seas coral atoll — built by the coral on a volcano risen from ocean bottom ergo unoccupied by any animals, having only fish and birds as well as fruits, nuts and coconut milk. First the humans developed fish catching and carving tools, then rafts, dug-out canoes and paddles and then sailing outrigger canoes.

So the whole project fails by reason of a pair of paradoxes. (1) The people who can conceive of this clever solution cannot be part of it. (2) The reasons for seeking the solution — refusal to accept political control — require that the solution be rejected.

———

What has just been carried out is an exercise in futurology. Every discipline has its distinctive techniques. We have just uncovered what is — or should be — a basic technique of futurology. Let me spell out the details.

In Euclidean geometry there is a technique called the *Reductio ad Absurdum* proof. A question is settled once and for all if it can be shown that the necessary assumptions lead to a logical absurdity (as that A both is, and is not, equal to B at the same time). A *Reductio ad Absurdum* proof is of overriding power; it puts an end to further investigation. (The only exception: one can look for errors in the proof itself.)

In futurology we have just seen the workings of a *Reductio ad Paradoxum* — let's call it RAP for short. If the very means of "solving" a problem thwarts the reason for using those means, then the "solution" is no solution. RAP overrides all other approaches — fancy technology, computer readouts and what-not. O'Neill's colonies run right up against a political *rapout*. There is no need to look further into problems of technical feasibility once we understand the political rapout.

Will this explication of the rapout put an end to the dream of libration point colonies? Most unlikely. Near the end of the 20th century we still have the Flat Earthers with us. From now on we will no doubt have the Librationists too. O'Neill may have given birth to a new religion.

People don't like to have their dream-balloons punctured. The rapout here explained was first presented (not quite so explicitly) in a paper I published in the **Journal of Heredity** fifteen years before O'Neill's proposal. In my 1959 paper I criticized an earlier escapist proposal that was rather similar to O'Neill's. The way my paper was noticed was significant. My cost estimate, a minimum of three million dollars per emigrant from the earth, was frequently quoted. But the *Reductio ad Paradoxum* analysis was (so far as I know) never mentioned. Yet any cost estimate is only tentative, whereas a rapout is final and decisive.

Why should the least decisive result be cited while the most decisive one is ignored? I suspect it is because of our rather decent underlying love of "fair play." A decisive argument stops the game; so we pretend we never heard it, thus permitting the argument — now pointless — to go on. Our behavior does credit to our hearts, but not to our minds. If embarking on a hopelessly escapist program leads to the downfall of a civilization, a mere sense of fair play will be a poor excuse for having closed our eyes to the practical implications of a rapout.

Reaching the greater islands and the mainland they developed animal skin, grass and leafwoven clothing and skin tents. They gradually entered safely into geographical areas where they would previously have perished. Slowly they learned to tame, then breed, cows, bullocks, water buffalo, horses and elephants. Next they developed oxen, then horse-drawn vehicles, then horseless vehicles, then ships of the sky. Then employing rocketry and packaging up the essential life-supporting environmental constituents of the biosphere they made sorties away from their mothership Earth and finally ferried over to their Sun orbiting-companion, the Moon.

Employing principles of optics, chemistry and electromagnetics, humans have now gained celestial information at the range of 11.5 billion light years in all directions around our Spaceship Earth. They have photographed equi-deeply into the microcosm. Macrocosmically they have located and photographed a billion galaxies of hundreds of billions of stars each. They have photophed atoms. Humans are now operating successfully in such vast and minute realms of scenario universe that 99.99% of their realistic activity is "invisible" to humans' limited range of direct sensing. Clearly, human beings are designed and equipped to operate in both ever larger and more incisive manner in respect to local universe and will for some time base their operations on their Mothership Earth.

I hear many people use my expression, Spaceship Earth, which I invented at the University of Michigan in 1951, yet I find almost all typesetters, editors and authors spelling Earth as "earth" with a small "e." I realize then that they are not thinking realistically about our world as being that of a planet whose name is equally well qualified to be capitalized as are those of Mars and Venus. People are going to keep on writing, earning their livings and enjoying kudos by glibly discussing space activities. I am confident, however, that they will also keep right on seeing the Sun "going down" in the evening and using the words "up" and "down" instead of using the words "in," "out" and "around," as used by the few who are working in cosmic realism. The word cosmic is frequently used to indicate non-realism. The vast majority are conditioned to think of the sky only as disorderly scenery. The only realism is cosmic. Cosmic includes all — macro-micro, — you and I.

I have now traveled around the world 39 times, never as a tourist but only in the course of my work. For the last quarter of a century, I have spent 9/10ths of my time away from my official home. People often say to me, "I do not see how you can stand so much travel." I answer, "You obviously don't know what you are doing. You and all of us are making 60,000 miles an hour around the Sun, which makes my kind of Earthian travel of utterly negligible magnitude." People ask me, "Where do you live?" I answer, "I do not mean to be rude or facetious, but I live on a little planet called Earth. I never leave home. My back yard has become greater and greater until it has proven to be a big sphere, and I can travel in any great circle direction and eventually find myself where I started, ergo: I never leave 'home'." If anyone asks, "How was the trip?" or "Where do you live?" they are not living in cosmic realism — they are "grooved" like an L.P. disc.

To all who are living in cosmic realism, the immediate inauguration of additional Earth-Moon, around-the-Sun flying formations of our team could not be more humanly normal. It is just as normal as a child coming out of its mother's womb, gradually learning to stand, then running around on its own legs.

Seed

. . . Analogies: Asimov's **Foundations**; the medieval Irish monks & others who preserved Latin learning while Europe fought over the Roman remains; the Atlantean legends; any plant packing its essence into tiny pods & scattering them on the verge of its autumnal death. Tell the colonists to carry an ark-load of life more than their support systems will require, & as many whole libraries as they can microfilm. I think it would make the rest of us feel somehow less threatened. . . .

Pierce Butler
Placitas, New Mexico

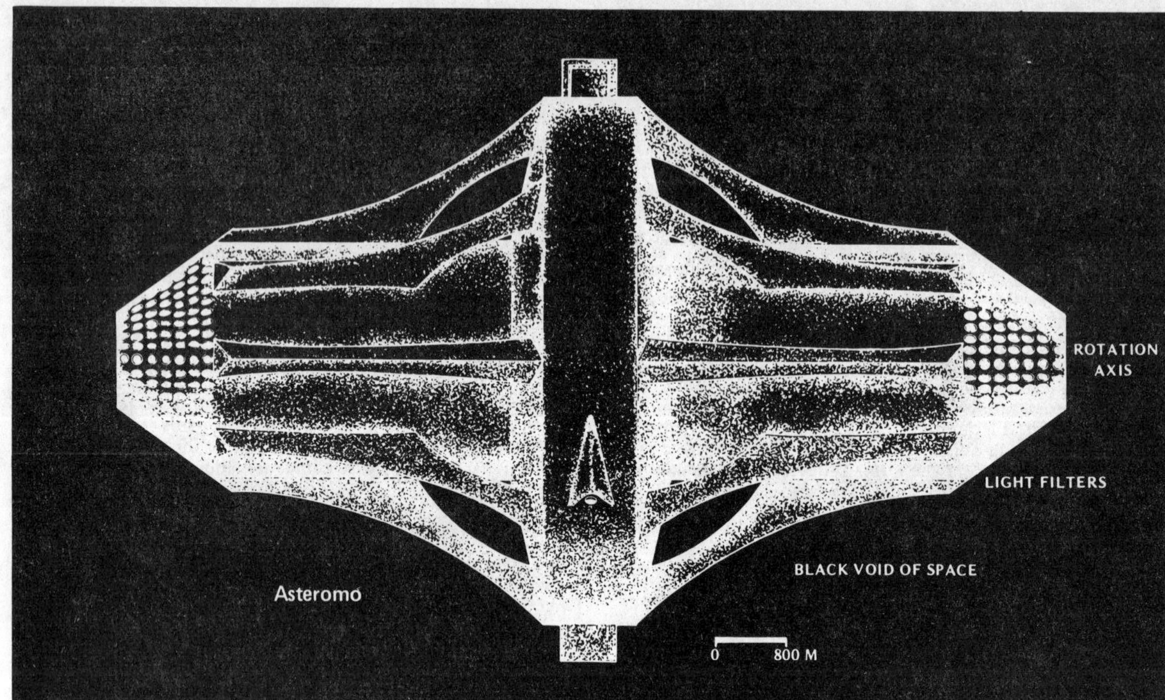

PAOLO SOLERI

Urban visionary, author of The City in the Image of Man; The
Bridge Between Matter and Spirit is Matter Becoming Spirit

On what can be defined the threshold to space "infinity," the
human specie is going to make momentous decisions. Some
of them will be unconscious and irreversible, and some will be
conscious and "crucial." For one, if the venture is to be developed, we had better seek a consensus. In this size and kind
of undertaking, it has to be a trans-national consensus and we
must try to have a knowledgeable one. This will be impossible
if the pioneers and promoters are themselves less than clear
about the scope and impact of the enterprise.

What follows is a tossing in the air of a few points in the form
of concerns. Consensus might have to be sought about them,
and the sooner the better.

A. The technological concern. It is such that I am not well-
equipped to tackle its technical underpinning and I, therefore,
assume for the sake of what follows that we are capable of
doing what the scientist and the technologist say can be done.
At the same time, since its end-product will be a habitat, the
technological concern is very close to my interest. But ultimately the technological concern is subservient to the eschatological concern and the habitat will have to be imprinted
by it and imbued of it.

B. The politico-economic concern. It escapes me in many
ways. Besides being not sufficiently knowledgeable on the
matter, it is per se a tangled knot to which we respond or
react in more and more "empirical" ways because more and
more the inertial stresses pervading it seem to be beyond our
limited wills and wisdom.

C. The eschatological concern. To my limited understanding,
there will be a renewed religious unrest caused by the space
probe and it will be an eschatological* concern which will
embrace the following: the social, the environmental, the
cultural, the ethical, the esthetic concerns. They are all directly operating upon the human condition together with the
questions of health and genetic "preservation."

This concern will be mostly unspoken of, but will be also
intentionally brushed under the rug of hard facts and techno-

---

*ESCHATOLOGY? A study or science dealing with the
ultimate destiny or purpose of mankind and the world.
(Webster's Third New International Dictionary).

political imperatives. And yet, dear "fellow travelers," the
stakes are frightfully high and we must, we ought, face what
we are about to plan and to implement.

Under the pressure of scientific and technological "progress"
stimulated by the space venture, the eschatological concern
will give rise to new or pseudo new theological models.

Thus, as I can perceive it, the probe of life into space is ultimately not a technological or a political or economic problem but a theological one.

The eschatological implications of "space colonization" are
most fundamental and critical and could be considered under
the 3 following, but not necessarily equally important, titles:
1) The eschatological concern. The question itself of ultimate aim, the purposefulness of life, that is to say, the
eschatological paradigm as such. 2) The genetic concern.
The splitting of the human specie into "sub" species (see
the pre-historical precedent) as a direct consequence of a
space "invaded by humanity." 3) The urban concern. The
space probe is the urban probe on "new grounds," therefore,
the urban question looming every more large on the destiny
of the specie.

Depending on how we will sense the 3 questions, we are in
for hope or despair.

In despair:

1A) We see ourselves as the (well-worn) apprentice sorcerers
incapable of halting our plunge into a technological "hubris"
which will bring upon us more and more forcefully the wrath
of our indignant Father, the Lord, and/or the merciless ex-
pulsion from nature's bosom.

2A) The human specie, abandoned by such Lord, or by
Providence, or by instinctual wisdom, which under the stress
of new (evil) environs will tear itself apart into inimical sub-
species foreign to each other. Those will find their own neme-
sis in specialization, genetic and otherwise.

3A) The human specie will make an ever more compromising
step into the urban syndrome, seen as the sum of all evil's
syndrome, since space colonization will be directly informed
by those conditions which are per se the definition of the
urban context.

In hope:

1B) We are making (remaking?) a promethean commitment
to the spirit, by unleashing it concretely from the gravi-

tational vise of the earth and, by so doing, opening the cosmos to "urbanization," that is to logos (see further).

2B) The human family recognizes its own genetic (and other) limitations and willfully seeks new ("morphologically") cognitive forms for the end of "outfitting" itself for the immense journey into the spirit via the flesh (mass-energy), and in the process going through a lengthy series of "transcendences" of its psychosomatic self.

3B) By stepping off the earthly landscape, man is turning by necessity (to be made into virtue) toward a frugality of environs and "hardware" which are specific to the urban condition to, and with one must add, ever more crucial transphysical longings. In sum, man will opt for the self-containment of his habitats, the inward orientation of them, the cooperative and inderdependent nature of the social and cultural texture, the high density of performance, the imperative of integrity and self-reliance and finally, the complexity and miniaturization of the milieu. The space city will, therefore, be unequivocally a test of how ready we are for the vertigo of a new momentous step toward the spirit.

I will deal with the hopeful triade, since there is where I stand and because by dealing with it I will also asystematically deal with the despair triade.

THE ESCHATOLOGICAL CONCERN

I look at it through a critique of the lifeboat metaphor and the carrying capacity thesis. I would offer the notion that their weakness does not come so much from a relativism peculiarly anchored to the consumerism ethos (gross national product) but that the scientific theory on which they stand is quite possibly unscientific, or better, a science which is basically a verification of facts (past) and which shies away from expectations that (to it) appear unscientific since they are only of the realm of the possible (feasible?).

If this were so, if the "science" of the carrying capacity (and the lifeboat ethic) is unscientifically applied, then its use is needlessly vicious. It is the viciousness of inflicting pain and death by the incongruous application of a paradigm.

Let's look at the carrying capacities as they seem to have developed.

The first living thing on this planet had, as a necessary support system, the whole existing cosmos, since only the existence of a specific if unknown cosmic balance made possible a specific if unknown solar system balance, that made possible on earth the appearance of a specific organism (one of them had to be the first). The cosmos in toto was the "territorial imperative" of a bacterial-like organism. No cosmos as such, no bacteria as such.

But at that moment already, or a bit sooner, things were getting "autonomous," that is to say, not purely deterministically generated. (Hopelessly prisoners of a totalitarian cosmic dictum). The solar system was "coming" to life and on earth, for instance, physical balances were interfered upon by physiological counterbalances. Oxygen was freed into the forming atmosphere by the initiative of living organisms, etc.

A first massive imploding of "territoriality" from cosmos to solar system, and specifically sun-earth-moon, was taking place. So it came to be that the "territoriality" for each organism was to be that amount of earth bulk which was the proportional fraction of the total "belonging" to it, to which was to be added a similarly defined fraction of the sun energy falling upon the earth and, in addition, relatively "infinitesimal" influences as moon-tides, cosmic radiation, etc.

If now one takes a look at the contemporary scene and one simplifies the model, one sees that for every human being is "needed" the presence and the "use" of a cone of matter defined by the radius of the earth and with a base measured by a circle upon the biosphere, atmosphere included, of let's say 100 acres. This gigantic mass to which must be added its share of sun energy is but an infinitesimal fraction of the "original mass" necessary for the advent of the first bacteria-like organism . . . and we begin to feel crowded!! But we begin to see also the "Providential" tendency of life toward frugality. Providential, since if it were to be otherwise, we would soon be face to face with the fatal dialectic of a self-parting universe (expanding) and a conscientization process that must abide to the fierce rules of shrinking carrying capacity and, face to face with lifeboat ethics, triage, et al. That is to say, that there would not be the rationale of evolutionary expanding capacity (extension of reach, contraction of needs), but an ontological wall made of undisputable and ultimate stop signs; a dying sun, the size and resources of the earth, and consequently life as a short fireworks of arrogance and opulence.

What if an even more incredible explosion of frugality is in store?

It is in the most frugal (and crowded) mode of all, in the brain of man, that powerful cognitions are working out more miracles of contraction and frugality (the Urban Effect is in full swing . . . see later) and, lo and behold, few hundreds of thousands of years after the invention of divinity, that prophecy of utter economy, the mind conceives a fully-lived, extra-terrestrial existence. The implosion in the brain of countless operations goes, at least potentially, for the numerical explosion of "ecologies" eventually unlimited in number, capable of sustaining and developing life. . . ad infinitum.

If for a moment, we assume that we will make the step and do it without diminishing man, then what do we have? An utterly new relationship with the world of "matter." We will literally mine the universe, the solar system at first, rearranging and processing matter into hollow urbis of all kinds of sizes and types and populations.

The order of the Christian God, "go and multiply," would see an unimaginable degree of realization and the carrying capacity of the cosmos would grow exponentially. With it, the lifeboat theory for all realistic purposes would be blown to bits since at best it would serve an acciduous and self-righteous society unwilling to get down to tasks and construct new boats one after the other, on and on "forever."

But why such explosion of life? For what purpose? The answer is theological. Before going into it, let's repeat what would happen to the "territorial imperative."

With the space venture the bulk of matter necessary for each person would dramatically shrink from the individual earthly cone, together with the corresponding ecological veneer and the sun energy, into a bulk on the scale of a home attached to an urban landscape to which would be added an open space of some acres or a fraction of an acre (and some energy from the sun, stars, to which the city might depend).

In other words, it would be as if the earth or any other celestial body were to be peeled off into successive skins,* each one of which would contain an "interiorized landscape" (cityscape) of minimal physical bulk. A characterization of the human environ as exponentially frugal. Thousands upon millions of hollow worlds, inner-oriented worlds because of locally produced gravity, would invade the universe (from as many points of it as there are conscious centers of it similar to earth.) Eventually, each galaxy would have the carrying capacity for four thousand millions of consciences (the earth today) billions of times over, a true explosion of consciousness throughout the physical universe.

What is so desirable in such model? On the personal level one could ask oneself if one would choose not to be born. If the answer is no, then to negate the birth of others, provided there is a carrying capacity, is rude to say the least. On the ontological level, it would seem clear that it is only with the intensification of reality by the presence and action of life, that "we" might eventually bring compassion and grace to the whole cosmos, the integrated universe.

The Theological Answer:

What is implied in this kind of process, insofar as life and consciousness can find as normative. It is implied that. . .

1) Consciousness is an unbelievably interiorizing stress.
2) Consciousness is an unbelievably complexifying stress.
3) Consciousness is an unbelievably miniaturizing stress.
4) Consciousness is an unbelievably frugalizing stress.
5) Consciousness is an unbelievably animating stress.
6) Consciousness is an unbelievably transcending stress.

*The earth as the last to be so treated, since the biosphere is "original," unique, precious and beautiful.

7) Consciousness is an unbelievably urbanizing stress.
8) Consciousness is an unbelievably divinizing stress.

It is only half true that such are only the potential powers of consciousness, since a cursory survey of life's evolution is already, if coarsely, a demonstration of such power. Those are all "versions" of the fact that the eschatological imperative is pressed out from reality as an inescapable command: Do unto matter (the mass-energy universe) what you do unto yourself. Make it into "conscious matter" into logos. That is, demand and force out of an ephemerally conscious universe that which ultimately will be a child God of infinite conscience, infinite integrity, infinite love.

Once more, let's take a step back and then a jump ahead.

Once the biological has reached dimensional limits, the size of organic molecules, and furthermore, the biological has not been as yet supplanted by a "better" media, and since only that which can congruously plan and operate,* intellection, can also put more understanding, design and will in ever smaller amounts of mass-energy, space-time (miniaturization), it is in the "space city," the urbis et orbis in one, which is to be sought the next step toward logos. This is also and at the same time a statement of "feasibility" and of desirability, since to work against the process is ultimately to work against logos itself (and consequently against the spirit). This feasibility is imperatively demanding implementation. The open question remains. . . when?

It would then be for the sake of logos that life must free itself from the "earthly prison." And if it is possible for life to free itself from the earth, isn't it then a "mortal sin" not to do so? And isn't the fact that life could not even conceive of leaving the earth before the appearance of consciousness, and that thousands of years ago the first step toward this leaving expressed itself in metaphorical, that is religious form, a proof of sorts that it is the task of life and specifically of consciousness to do just so?

From a "territoriality" of the whole cosmos necessary for the appearance on our earth of the first living cell to the ecological cone necessary today for each creature populating the earth, to the minute "territoriality" of a space city, we can measure the powerful trend toward frugality and concurrently the not-less powerful opening of the whole universe itself to the spirit. Once upon a time a whole cosmos for one infinitely puny life, now eons later, now that intellection is grasping at the alchemy of matter and makes it deliver its latent energy and potential conscience, comes the possibility of mining the cosmos, of making a moon into "large numbers of earths," etc.

The ape, leashed by gravity to a tether pulling down toward the center of the earth, has its "territorial" imperative defined by the ecological capacity of the earth. With the appearance of the human mind the tether is (potentially) cut. The ecological capacity is (potentially) transferred from the limited earth to the endless universe.

But for this potentially quasi-infinite growth of liveliness, two things seem to be indispensable: an eventually quasi-infinite growth of will and a quasi-infinite growth of reverence. Then divinity is expectant there in the future.

<u>Two Considerations.</u>

1) How much or how little we make of liveliness and consciousness-spirit is in a way not relevant since no matter what opinion we have of them, to have an opinion, any opinion is per se an identification of ourselves with them. One option we do not have is to give them up without giving up life itself. Then the fostering of them turns out to be a "law of nature" which tautological "block" is that the ultimate meaning of the ensuing process will not disclose itself before the exhaustion of itself. But this late disclosure is not due to secrecy but to "incompleteness." To reveal itself (to itself) before the "end," the Omega condition, would be an anticipation and as such they are an important exercise in futurism. But by necessity, those exercises fall short of target since the target is not there as yet, since it will be there only at the "end."

---

*That it can also operate incongruously is a well-known fact, therefore, our constant condition of emergency. But what if we were to loose our intellectual capacity?

2) A word about "compatibility." Each generation has its own environmental compatibility, that is to say, a person is hard put if he is asked to reject those conditions with which he has grown up. But each generation is piggy-backed on the preceeding one. The individual cycle demands a return to the infancy grounds (re-entry into the mother, we call it the father), but for each infancy such grounds are the maturity grounds of the parents (the preceeding generation). This is providential since it secures both change and continuity. In fact, it would be difficult, if not impossible, to come up with a better "scheme." This dialectic of change versus continuity makes conflict inevitable. It follows that it would be not only unrealistic but down right unjust and cruel to want to thrust the "space move" upon the present generation as if it were a pleasure trip or a palatable prospect. But that could be so for our grandchildren. The fact remains that notwithstanding the incompatibility parameter, the parameters of territory-carrying capacity and environmental development are in constant flux (and growing).

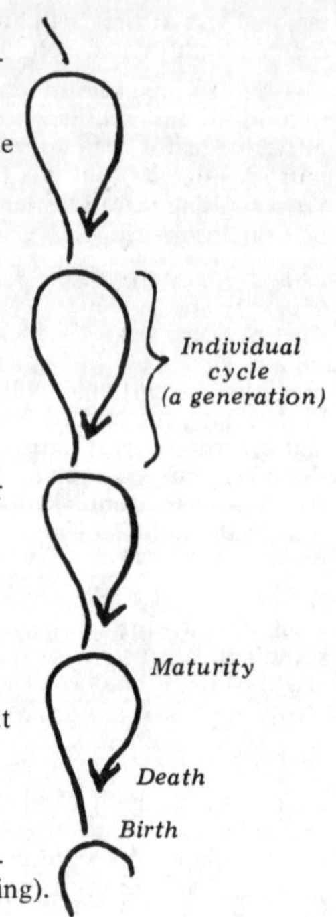

2) THE SPLITTING OF THE HUMAN SPECIES INTO SUB-SPECIES

One of the most momentous outcomes of the space colonization successfully carried on will be the appearance of human mutants that will more fittingly perform in not just a new environment but in a variety of new environments. (We are children of our environment).

A most critical area of decision (or non-decision) is, therefore, the genetic pool area, and with it new conceptions of human values, justice, equity, hierarchy, fairness, etc. Are we able to withstand the thought that quite possibly, I would say inevitably, the human kind might become fractionalized (fractured?) into for instance, living fossils (the earthlings?), psycho-techno man (cyber), superman (an intellect-relayed multitude constituting a single creature?) similar but on a different ledge of evolution to the insect colony? (See further).

Since the future is not a process of deployment like the unrolling of a (sacred) scroll but instead a process of creation, nothing that we might conjecture or plan for, will ever turn out to be the future. Therefore, as it is quite possible that we never leave the earth, it is also quite possible that a polarization on this planet might eventually force upon us some "strange" variances from the human kind we are now accustomed (or resigned) to. That is to say, a genetic schism might be not only a consequence of, or peculiar to, a space colonization, but could be in fact the cause for a space colonization.

Isn't indeed what is going on now with the mystique of technology a kind of pilot work for a mutilation in the making, which in turn urges man into its space probe? Technological man, the *enfant terrible* of the 21st century, would be the mutant that as yet has to find a sound justification for his appearance. Now or soon, as a form, a container filling itself with purpose, he would cause life to extrude itself into a new set of parameters. A new universe authored by consciousness and its prowess in understanding, guiding, transforming, metamorphosizing. Are we in for a new radiant creature or are we in for life's abjection? The question is open and will be open for eons. In fact, for our earthly prospective, from our present, that possible radiant creature would be a "monster" anyhow since we could not withstand its sight (Bible). Whatever the case, we cannot escape the future and thus we might as well become conscious of some of its possibilities.

A) Since it will not make sense eventually to copy earth's environment "over there," environments more congruous to the space situation will eventually define new morphological characters of fitness and response. Eventually, the physiological make-up might end its own usefulness, but well before such definitive and "radical" dismissal of the organic (see "Mass Transit - Mass Delusion"), we will probably mutate ourselves into cybers of sorts (we do already in the push-button syndrome).

B) Will there be in the meantime a transformation of the mother-lover. . . son-lover drive* in favor of a "scientific" intrusion in the immensely chancey but immensely rich genetic pool afforded by the bisexual mechanism? But might not this powerful and unpredictable (as yet) genetic pool be the best if not the only insurance policy against any cosmic emergency? Should we then dare to take such a step?

C) What will eventually make spaceman, besides the ability to think, like earthlings? And what if the thinking apparatuses become extravagantly different?

D) Will there be a "change of hearts" about slavery since sooner or later the notion of superior and inferior might be fostered among genetically different sub-species and peer groups? A "benign" version of such discrimination would make way for the earth as the living museum. The earthlings and their beautiful earth visited now and then by chartered space fraternities, a benign hypothesis in the restricted sense of doing away with violent confrontation, bloodshed and genocide. . .

We must beware and not make humility into insularity and eventually into bigotry. If it is true that man is unique, it is also true that he is uniquely isolated and confined (possibly isolated because confined). But what if he were to explode at first into the solar system, then into the Milky Way? Would not there be then good reasons for the specie to venture into new, unthinkable explorations of the psychosomatic potentials we are carriers of, in view of encounters with other centers of consciousness and grace?

If then one has to admit pure terror *vis a vis* the opening of a galactic trap door under one's humble but arrogant, venturesome but cowardly, compassionate but bigotted self, one must also not be blind to the possibly infinite radiance of the future on which such trap door is opened.

3) THE SPACE PROBE IS THE URBAN PROBE

Are we then really working at the creation of a Son God, the "masterpeice" of evolution, or are we deviated in the process and are we becoming the authors of a monster? To get an answer I think we have to consider that normative mode I call the Urban Effect.

It states that the transformation of the cosmos in the direction of the spirit is the Urban Effect. That is to say, the Urban Effect is an eschatological imperative. It states also, as a corollary, that divinity is the Urban Effect and third, that since the Urban Effect is as yet at an embryonic stage of development, neither is the city any closer today to what it will be (ought to be) nor is God any closer to being the monistic and absolute center of love we (religion) anticipate it to be. But there is an immanent Urban Effect which incarnates a limited divinity, the immanent God** (The God of the present).

This choice of terms, Urban Effect, divinity, God, is not casual or sensationalistic. It puts an eschatological thrust into the human condition by going at the tick of the human performance (where it becomes social, cultural, civilizing), the civic man. The civic man is not an experiment from which life might or will derive some advantages. It is instead the manifestation of the same kind of thrust that throughout evolution has and will keep the living stuff working upon raw matter at the cutting edge of conscientization.

The Urban Effect is the casual stress which forces the production of enclaves of consciousness, starting with the most ephemeral of them, the original micro-organism which

appeared on this earth thousands of millions of years ago. Those enclaves of consciousness are to incorporate eventually all of the universe in ways and expressions as multiple as are the world involved in the process and the time in which they perform. (There will occur then a further synthesis into the oneness of Omega). The frugal (miniaturizing) character of the process insures that the "end" will not find itself short of means by seeing that the equation end-means will exponentially grow to the final point where the nominator is near infinity and the denominator is near zero.

The enclaves of consciousness would eventually join together, thanks to the implosive force of the complexity-miniaturization paradigm which will be enabled by cognition to warp the cosmic laws into the "divine law."

It could be said that it is not so much the case of increasing the amount of spirit-consciousness that might exist scattered throughout the cosmos (the Taos) as the endowment of each particle, but it might be the case instead of the desegregation and conjunction of each particle of consciousness to one another by way of the Urban Effect and of the creative process which ensues from it.

This original atomized spirit, the atoms of consciousness, which can be conjectured to be composing the universe, this Alpha $1/10^{80}$ God is prisoner of its own iron cage of determinism together with the other $1/10^{80}$-1, and it is one of the 10^{80}* (polytheistic) deities of the pristine universe.

What is the logical thread that causes the Urban Effect to be the eschatological thrust? It is the common rule that any discipline (the eschatological thrust in the present case) has to be such as to foster that which has originated the conditions that have allowed for its birth. If the Urban Effect was the *causa prima* for the advent of consciousness-spirit, then the enhancement and the impregnation of the cosmos by conscience-spirit is dependent on how forcefully the Urban Effect goes on in multiplying and reinforcing itself throughout the cosmos.

There is a sort of premonitory situation which must be read even though it justly (but unjustly) sends shivers down one's own spine. It is written in the invertebrate, on land and sea societies of insects, in slime mold slugs, in the various (and infamous) ants or termites colonies, wasp and bee nests. They form aggregates of acting matter and why not thinking matter, perhaps comparable more to the mammal brain and its own specialized hierarchy than to anything else.

Up or down, side up, side down, gravity itself seems for the time being forgotten so much is the living matter inner-oriented, totally absorbed into itself, pure living matter, flesh, thinking flesh. They are the proto-conscious cities, anticipatory, if brutishly so, of urbis where the dependence on matter becomes less and less measured on bulk and more and more measured by arrangement and discrimination, complexity and miniaturization-intensity and transcendence.

If there is an eschatological guide to turn to for assistance in our decisions and actions, and if this guide is predicated on the emergence, via creative genesis, of the divine, then the breaking away from the earth's bondage is indispensable. It is not indispensable in the sense of being also sufficient, in fact this break away could end up being a run away from responsibility and "grace," but in the sense that there is no access to "full divinity" without the "concrete" intrusion of consciousness into all corners of the universe. Call it, if you like, redemption of matter whose sin is not to be as much as it ought to be: spirit. Or call it, as I prefer, the creation of a new and divine universe where each and all things are a radiant and transparent synopsis of each and all for each and all of all times and all places.** This assertion is not a breaking

---

*I refer to what I consider to be the Christian hypothesis about the fatherless birth of man presented in the myth of the Virgin Birth.

**This whole eschatological argument is absurd or blasphemous unless we are able to make a distinction and keep it constantly in mind, between the potential and the actual.

*The number of particles in the universe according to some calculations.

**A state where time and space have collapsed and, therefore, a condition of total resurrection. And resurrection is total or it is not.

of one's own bridges from (ecological-minded) "salvation" but it is a reinstatement of the concept that the bridge between matter and spirit is matter becoming spirit and the process cannot be halted at any accidental moment, the present.

That the notion of the breaking away from the earth is contemplated in most religions shows the anticipatory power of theological thinking. But usually this anticipation is not seen as entailing a physical migration of this life from the planet to other places. But if this is the interpretation, then the anticipation is wanting and the prophecy does not fit the eschatological grid. It would not suffice that the cosmos might sit in "adoration" around earthly magnets of intellection, as many of them might there be, scattered throughout it (thousands of millions)?! Intellection must move into the cosmos and consume it into consciousness, make it suffer the intellection of itself and so cause it to transcend itself and create the "not yet," the divine.

The reaching for the divine (the work of the spirit) demands this "inexhaustible" bridge of matter, all matter consuming itself into its own entelechy. But this is essentially the imperative of intensifying the performance of matter (mass-energy), that is to say, it is to cause the Urban Effect to become the universal concern, the rule and not the exception and, ultimately, cause not so much the City of God but instead and indeed the God-City, the Omega Urbis et Orbis . . . It might indeed be indispensable that our anthropomorphic god be reinvented into the God in the likeness of the city with the massive reservation that both the Urban Effect and the ultimate (entelechy) expression of it in divine terms are inconceivable, unimaginable for our limited conscience and grace.

To make a pale metaphor: to think of Omega *urbis et orbis* as a redeemed Detroit is comparable to the mystification one would be in for if one were to be presented with a tiny blob of tissue, an embryo, and were told that one beheld a beautiful, mature person. But the mystification would be even greater if the embryo turned out to be not of a human but of a fly. Detroit stays to the God-City as the embryo of a fly stays to the full-grown person. Between the two is the demonism of the evolutionary metamorphosis with its own cul-de-sac and its own triumphs.

Therefore, what we will weave in space is going to be a series of cities that will be adding new force, new degrees of eventfulness, new situations to the Urban Effect. One, quick to come to mind, is the riddance of the gravity burden (and as a consequence, genetic alterations). Another is the physical interiorization of the environs (and consequent psycho-social-cultural alterations), and another is the "crowding syndrome" coming to fruition with new force.

Crowding in this use of the term is not the bunching together of people, things and time, but the highly selective, discriminating coming together of disparate and per se less intense elements, in similar but not identical ways as it is done within an organism (plant, animal) or associations of organisms. "Crowded," living, durational events incredibly full of interdependent processes responding to the needs and the "hopes" of the organism. Since crowding is a divine attribute, co-presence, co-creation, understanding, knowledge, reverence, fullness, radiance,. . . then the Urban Effect (the crowding effect) is inexorably present whenever and wherever there is a thrust toward such condition of grace.

What then?

1) Space "migration" is in the human agenda.
2) Is space migration now a responsible act, a diversion, or pure escapism? A careful scheduling of our means, intellectual, ethical and otherwise, might show that the best defense against the squalor of the lifeboat theory and the carrying capacity "miscalculations" is the ultimate frugality of a life thrust, which can and if it really can, must also pervade the cosmos.
3) True frugality is the antithesis of mediocrity. (See "Relative Poverty" Summer '75 **CQ**). Therefore, a vivification of the cosmos which cannot but be frugal, is bound to move man on a higher ledge of the evolutionary pyramid.
4) To this end there is no escaping the need for a more reverential, urbane, civi-lized sense of the human experiment and, ultimately, the need for the eschatological vision of a universe in the process of self-divinization, the Urban Effect.

JULIA BRAND
CQ editor's mother, 72.

Space Colonies? Feasible, certainly. Inevitable, soon.

As you know, I thought years ago (20, 25?) we would have a space station before going to the moon. Gravity from centrifugal force, solar heat, algae for oxygen, recycling, etc., etc. O'Neill has extended & refined all those old ideas, added so many new ones, envisioned a beautiful and productive colony.

Remember the exhilaration of the moonshots & the chagrin when that marvelous assembly of trained and talented men was largely disbanded & the whole space endeavor diminished? Now at last a space shuttle is under way. I hope all will move faster and that at least one of O'Neill's colonies will get off the drawing board & into space. What marvelous possibilities!

Meanwhile my fascination holds for the far out probes to & beyond the planets. What wonderful surprises, what lovely mind stretching. Keep me posted. Remember what I asked you and the grandchildren. When I'm long gone, try saying to the wind, "Grandma, guess what!" & maybe . . . .

Better foolishness

Space Colonies are a good idea because they're bold but not Utopian. If my taxes are going to be spent on foolishness, this is the kind of foolishness I want them spent on. None of this trying to develop a frisbee that'll fling antipersonnel grenades or training porpoises to perform kamikaze missions.

I want to go. I need a home for my imagination. If I didn't go, how could I bear to sit down at the Space Cowboys (Cowpersons? Asteroidpunchers?) Bar and Grill in my old age and listen to the other geezers reminisce about slipping up to the end of Island One with Mary Lou to stare at the cosmos and doing it weightless for chrissakes. Sure did make your Betelgeuse, eh podner? Then they'd ask me what I did in the olden days before there were Golden Arches on the moon, and I'd have to hang my head and mumble, "I coulda gone, but my roots was too deep."

I want to help, particularly on food production/waste disposal or anything else my talents might be used for. I have my doubts about the immediate usefulness of lunar materials for growing food. Perhaps the best bet would be to start out depending on hydroponics and/or dehydrated food along with some algal reactors. (See Shelef, et al. "Algal Reactor for Life Support Systems," **Journal of the Sanitary Engineering Division**, ASCE, Vol. 96, No SAI, Proc. Paper 7105, February, 1970, pp. 91 - 110.) This algal reactor looks pretty good for use at L5 because of the availability of sunlight at L5. (Or the cheap energy available for powering electric Lucalux lamps as in the model used for this study.) Graham Caine's algae cycle might be useful if it works. See **Stop the 5 Gal. Flush!** [EPILOG, p. 487.]

If water is used for washing, or some kinds of manufacturing; perhaps a ground water recharge system or trickle filters would be useful for purifying it. See Bouwer's article in the previously cited journal for one recharge system design.

Maybe composting privies could be installed to provide for enrichment of the lunar "soil." There is also the question of bringing up the microorganisms (bacteria, viruses, nematodes, yeasts, etc.) necessary for establishing an Earth-like cycle. It might be wise to screen out some of the nastier pathogens, but I'm for keeping some garden variety diseases around to keep us on our toes.

Incidentally, a soils professor (Prof. Hole) who is a friend of mine once told our class, "There is no soil on the moon." His definition of soil is, "the portion of the environment providing water and nutrients for plants." Under this definition soils depend on their vegetation (or lack thereof) for their characteristics, and the oceans and atmosphere are soils. The fungi which provide water for the algae in lichens would be soils, hmm, maybe the algae are soils too, providing nutrients for the fungi. Until plants are grown on it the lunar sand/silt/clay isn't a "soil," as O'Neill called it.

Jonathan Beers
Madison, Wisconsin

DAVID SHETZLINE

Novelist, author of DeFord *and* Heckletooth 3.

No technical suggestions as yet, but may I (for the many who will choose to be left behind) offer some constructive — I trust — objections before feathering onto natural optimisms? (Or in these times, what remains of one's nat. ops.) First a confession:

My maternal grandfather was a second generation German-American whose eldest son signed up to be gassed in the First World War and whose youngest volunteered to be wounded in the Second. Grandfather saw virtually all modern devices Researched/Developed; and he believed, bought and personally used near every one. My father worked four decades as research engineer in the Bell Labs. Encouraged by such patriarchal models, I studied sciences at my first university. So this amateur is confident O'Neill's proposal will go. NASA seeks larger federal draw, detente is our metaphor, the Economy needs a bellwether other than arms, the Pentagon has been superbudgeted, etc. The time is over-ripe. One would have to be a rabbit, imagining Choice lies in the widening blackness between the oncoming headlights, to do anything but follow along behind the road proposed by the Space Colony energies. And in a sense this is exactly what I lament. An immediate future has been introduced which will fascinate and occupy at least one generation of Engineers, lock several generations of Laborers into steady toil, generate ongoing Capital Profits, maintain American Leadership, and open still another Damn Frontier.

Some fiction writers (O'Neill's scientist/novelist predecessor Tsiolkowsky) dream outward utopias while others (the prisoner/novelist Solzhenitsyn) reshuffle soap-operatic routines, seeking contemporary reforms. Certainly all are fascinated by any Grand Plan designed to keep us from further sliding down the socio-economic razorblade of life. Otherwise we are no closer to establishing the kingdom of god here on earth than two hundred years ago when some of our upperclass white male forefathers wrote in their diaries that they had. (Actually it was true for men of their persuasion.) We all need practical visions; the generalist in everyone is warmed by knowing at last Fulleresque System Structures will be given — literally — space to develop. I'm not sure what this promises the inner man, but no doubt divorce/bankruptcy/suicide will be amiably affected by the construction of human-made worlds proceeding from vestigial international cooperation. However, to begin anew, it

may be necessary to confront our past and literally live it down within ourselves. Now and here. This is terribly frightening for everyone, as it suggests the grimmest possible course, politically, personally.

My own experience counsels that Engineer/Scientists are eminently practical folks who get things done. So it must seem extremely unkind (if not perverse) to insinuate a further Mechanical/Technological/Brave New will suck off a lot of energies, beckoning another generation outward toward quality equality and peace of mind. Nevertheless O'Neill's vision is quite elitist. Although he declaims the elitism of the Apollo Project and offers "the possibility of direct participation by large numbers of ordinary people," no matter who his workers are, the concept is elitist, no doubt has to be. If that is its only handicap, then in the profoundest sense, amen.

The Good Engineer, I suspect, from those who gave us the Great Pyramids, through da Vinci to Fermi, tended to shy from political considerations (although Fermi was very unsettled when the First SuperBomb's fireball seemed to overreach). O'Neill could be whistling when he suggests further development of such colonies would relieve the earth of exploitation by the industrial revolution and consequently open new frontiers to challenge the best and highest aspirations of the human race. His closing testimony reports: "Many correspondents refer to space colonization by analogy to the discovery of the new world or to the settlement a century ago of the American frontier." I feel much the yokel in underscoring that the discovery of the new world led to the virtual extermination of every one of its indigenous races, several of its species and the massively unequal distribution of most of its natural resources. South America is largely under fascist or military dictatorship. North America gave us the model for the first modern war — our Civil Meatgrinder — and has — since the closing of its American frontier — continued to expand that frontier anywhere CIA/Marines etc., ad nauseam.

Obviously Gerard O'Neill and all the good people he has drawn to his practical vision know that finding a new route around these old habits is the game. And somewhere in that game the past must be confirmed, its lesson understood. Then the present can be taken, seized. Before we go on to the future. Or at least at the same time. Big money always seems to ignore the lesson of the past and skip the present by insisting the future is our present.

Forgive me for offering what might seem to be merely some rustic lament yawped from some province of the mind. Of course space is antiseptic. There are no indigenous cowboys, Indians, Vietnamese, out there to be exploited. But when Project Independence goes so expensively/expansively into that virgin sweet light it will be its own indigenous race. And it will leave a hell of a lot of us behind, alone with existing mischiefs, quite powerless, unfunded, and perhaps in a poor position to help. Naturally, that is our challenge. And O'Neill has already done us great service: whether one will choose to be on the Colony or not, we are already out there or down here in our minds.

You invited a thousand words, I'm afraid I've run over without offering a single technological contribution. But try this: May Project Independence be sure to include at least one dozen each: excellent historians, crack economists, keen sociologists, to assist all aboard in matters of fine print whenever contracts are made with those folks the Project has had to align and bankroll itself. And may it never forget to write home.

Partial atmospheric pressure

. . . Cooking at 3 p.s.i. is going to be a slow affair. Some PV = nRT work to be exact, but water's going to boil at maybe 140°F.?? . . .

Jonathan Feiman
Riverside, Illinois

Use pressure cookers.

—SB

JOHN HOLT

Author of How Children Fail; How Children Learn;
What Do I Do Monday?

It would take a book to discuss fully the many flaws and errors in Professor O'Neill's proposal for space colonies. Let me here mention a few.

First of all, the basic design of his cylinder is unworkable, for this reason. If the cylinder is rotating at a speed sufficient to produce centrifugal forces (*not* gravity, as they carelessly call it) equal to 1 g on the inner surface of the cylinder, then there will be equal or greater centrifugal forces on everything outside the cylinder which is rotating with it, i.e. the movable vanes which will supposedly reflect sunlight into the inside. Folded in flat, these vanes will experience along their entire length a centrifugal force equal to 1 g. Fully extended they will experience forces averaging about 4-5 g. But these vanes are supported only at the hinge end. To see how far this is from being possible, we have only to ask what is the longest structure that we can make on the earth's surface, parallel to the surface, and supported or cantilevered at one end. Hardly more than fifty yards, if that. O'Neill asks us to believe that with present technology we can make equivalent structures many miles long. Clearly, we will not do this in ten years, or a hundred; the odds are great that we will never do it.

If, then, we were to build one of O'Neill's cylinders and start it spinning, it would not be long before the vanes would begin to bend, fan out, and soon break off. We must ask ourselves, how is it possible that the by now tens or hundreds of thousands of "scientists" who have read of this proposal have not pointed out this elementary flaw? Either they have not seen it, because they did not want to, or because they are, quite literally, <u>out of touch</u> with reality, have lost the feel of real things. Or they have seen and not spoken, perhaps from fear, perhaps so as not to impede a project that would be good for "Science." Either way, they have shown that we dare not take their word about the benefits, feasibility, or costs of such a project. Despite their credentials they have shown themselves to be incompetent.

This defect in the design cannot be cured by supporting the vanes with cables. No cables are anywhere near strong enough to support many miles of their own length against such forces, to say nothing of the weight of the vanes. And if super cables, ten times the strength of any we have, should be invented, they could still not prevent these vanes, a mile or so wide, from sagging in the middle under the stress of centrifugal forces. Of course, the problem could be solved by having the reflecting vanes fixed in space, independent of the cylinder and not rotating with it. But there is no way with such an arrangement to have the day-night cycle which is the heart of O'Neill's plan.

Let me be clear at this point about what we can and cannot do. We can build cylinders in space as big as we want, if we do not try to put earth "gravity" and atmospheric pressure in them. Or, we can have earth "gravity" and pressure, if we keep the cylinders very small. But not with any technology we have now or are likely to have for a very long time, if ever, can we build the kind of cylinders O'Neill describes. We can make – have already made – human habitations in space. But they are, and must and will be, like submarines: they will never be like the surface of the earth.

This fundamental error of O'Neill & Co. – hereafter ONACO, though we could as well say CRASCO (Crackpot Scientists & Co.) or even MASCO, is only the first of a great many. Among others:

It is impossible, using available technology, or any likely to be available in the next generation, to refine metals and do other heavy industrial operations in space, or on the moon's surface, and any technology we ever develop to do it will be enormously expensive. The reason is that in a vacuum there is no medium to carry away excess heat. A rough figure often given is that it takes about 80 tons of water (and no one has estimated how many tons of air) to make one ton of steel.

The task of this water is simply to take heat, first, away from the steel, and then, <u>away from the steel mill</u>, and to dump it into the heat sink of the earth environment. What weight of radiating surface, and how disposed, would be required to dissipate heat at this rate into a vacuum? (It is another careless mistake to speak of space as being "cold"; it is neither

cold nor hot; objects in it may be cold or hot, depending on how much energy they receive and radiate.) The amount would be immense; at the temperatures at which humans can live and work, and at which presently existing machines are designed to operate, radiation is a very inefficient way to dissipate heat.

At any rate, the technologies to do such metal refining and shaping in a closed system surrounded by a vacuum <u>do not exist</u>, not in a pilot model, not on a laboratory scale, not on a drawing board, and probably not even in ONACO's imagination. In like manner, the technologies do not exist that would, in O'Neill's airy (no pun) phrase, "unlock" oxygen from the rocks on the moon. On earth's surface we break oxygen loose from metals by heating the metals to very high temperatures and then giving the oxygen an abundant supply of carbon to react with instead. On the moon, where would we get the carbon? How would we then free the oxygen from the carbon? And, in what sort of containers and heat exchangers, and with what sort of pumps, would we contain, cool, and finally compress these huge amounts of enormously hot gas?

Nor do we have, even on drawing boards, the technology to build on the moon's surface the materials launcher which is another vital part of ONACO's plan. We do not have linear motors capable of applying to a tracked vehicle the proposed accelerating force of 25+ g's and a decelerating force of 75+ g's, nor do we have speed measuring and controlling devices of the required sensitivity that would, and for long periods of time, withstand such forces, nor do we even have the vehicles themselves. Nor have we learned how to make a track level enough so that a vehicle could run smoothly on it at the required speed of about 4,000 miles per hour; the record on earth, and that over a very short distance, in a rocket powered sled, is only about 600 mph. Such a track would have to be heavily ballasted, not perched on flimsy supports as in ONACO's drawing. Nor have we any idea how these problems, difficult enough on earth, might be made more difficult by the moon's lighter gravity, different surface conditions, and, what is most important, enormous fluctuations in temperatures.

In these and many other respects ONACO have underestimated, by factors surely as great as ten and probably very much larger, the difficulty and expense of devising, building, testing, perfecting, and maintaining the devices needed to do the things they want to do. We do not yet know how to build a lunar habitat for even a half-dozen people; ONACO's plan will require a habitat that will house, and for long periods of time, hundreds and perhaps thousands.

In like manner, ONACO have grossly underestimated the weight of material that would be needed to build the space cylinder. From their words, and the artists' drawings, they seem to have no idea of the degree to which changing the scale of a problem changes the nature of the problem; they are like people who would try to build a 747 out of the same materials as a model airplane. Scale makes a great difference; to go from the 707 to the 747 we had to invent and make not only new metal alloys, but new machines to work those metals, among them forges many times larger than any that had existed.

Consider, for example, the effect of atmospheric pressure on the end of the cylinder. It is very much like the problem of building, on the earth's surface, a water tank, to be suspended from its upper rim, and to hold a 32 foot depth of water. If such a tank was to be ten yards wide, we could build it with fairly conventional means, heavy steel plates welded together. But if we try to imagine such a tank 50 yards wide, or 100 yards, and get some sense of the forces on the bottom, and their bending moment, we can see that much heavier construction, with massive stiffening beams, would be required, but ONACO are talking about such a tank more than a mile wide! In like manner the side walls of ONACO's cylinder would be subject to immense forces, both atmospheric and "gravitational;" it would take enormous amounts of reinforcing beams, both longitudinal and annular, to prevent the cylinder from bulging out into something more like a sphere, or even a disc.

Scale is also important in the matter of the microwave projection of energy that is another central part of ONACO's plan. No doubt we can transmit power through microwave energy on a laboratory scale, but that is not at all the same thing as doing it on an industrial scale. The technology to do that does not yet exist. And if it did, or when it does, how big will be the target area on earth to which this energy is projected? More important, at such distances, through what kind of feedback mechanisms will the projector be kept on target? And more important yet, suppose these mechanisms break down – things do break down – and this energy beam, probably close to what we might call a Death Ray, starts to wander around on the earth's surface. What then?

As to the cylinders themselves, even if we could solve the vane problem and so get a cycle of day and night, which we can't, we would still not get an earth-like environment. What about rain? To get rain at night, the relative humidity would have to be very close to 100%, certainly far above any level of comfort. What about balance? The weight in the cylinder would have to be kept in balance, not only around the axis of rotation but also down the long axis. Otherwise the cylinder would begin to rotate eccentrically, or to wobble. There being no correcting forces the cylinders would be in unstable equilibrium, and these motions, once started, would tend to increase. For that matter, an object at L5 is itself in unstable equilibrium; there are no forces tending to keep it there, and since any movement will bring it into the gravitational fields of earth or moon, once it starts to move it will keep moving. And what about wind within the cylinders? Since, if the vanes could be made to work (which they can't), the end of the cylinder nearest the sun would always get much less sun than the far end, and would hence be colder, what sort of air currents might be set up? And how long would it take to establish a stable biosphere, and how would it be done? O'Neill's brief remarks about doing away with unwanted pests show that ONACO is thoroughly ignorant in this area. And even if some miracle technology of the distant future, as yet undreamed of, could solve all these problems, it would not produce an environment like the surface of the Earth. Living there would not be like living on Earth, only nicer; it would be like living in the inside of a big rotating cylinder with mirrors outside reflecting in sunlight. Who would choose to spend the rest of his life there? Not me, for sure. Only the starving and desperate – and for them, no such palatial accomodations would be needed.

Space is not Heaven. It is not even Disneyland. It is an environment as hostile and deadly as the core of a nuclear reactor or the inside of a tank of nerve gas. In time, we will probably learn how to move around in it a bit more and do a few more things in it. But Earth's major problems will have to be solved on Earth.

Technical debate

John Holt's above letter to us has been answered hotly by one T.A. Heppenheimer of the Center for Space Science in Fountain Valley, California, writing to someone named Cheston at Georgetown University in D.C. John Holt sent Heppenheimer's letter to us, along with his own (Holt's) counter-remarks, which were addressed originally to Senator Edward Kennedy. A peculiar form of private publishing, all this, but a fascinating debate. I have trimmed preambles and shuffled the retorts and counter-retorts together.

Heppenheimer begins:

Before proceeding to address the technical points, I believe it is first worth-while to consider points of theology. There are a number of places in the paper where the author makes assertions which can only be described as theological, as articles of faith. These are the statements that space is insuperably hostile, that it can be of no significant value to man, that we must solve our problems on Earth, and the like. There is an alternate position, which is equally theological. This is compounded of assertions such as "the earth is the cradle of man, but man cannot live in the cradle forever," or the "man's colonization of space is as significant as the colonization of the land by aquatic animals in the Cambrian Epoch," or that "man's imagination and daring can overcome any limits."

I feel the pessimistic theology is naive, and is irresponsible. In a time of challenge to the foundations of our industrial civilization, it ill-behooves us to dismiss major technologies out of hand. But non-theological pessimism is valuable if it leads us to examine carefully proposed solutions and to prove critically for weak points. The optimistic theology is also naive, and may lead us to underestimate the obstacles which obstruct a difficult project. But a type of optimism is useful in that it may lead us not to be daunted by initial difficulties, but instead to seek to apply ingenuity and resourcefulness so as to overcome difficult problems.

I personally feel that the most fruitful attitude is not one of optimism or of pessimism, but of what might be called "critical ingenuity." That is, one must seek to be critical, yet to buttress one's criticisms with solid technical reasoning drawn from a multitude of fields. At the same time, one must have sufficient command of the pertinent technical fields as to be able to recognize what problems are truly difficult, what problems will readily yield to intelligent design. Then, faced with such a true difficulty, one must seek to cut to the center of the problem, to lay bare the core of the difficulty, and to apply the pertinent sciences so as to propose a solution.

With these comments, then, I will try to attend to the dozen or so chief objections which the author has made.

(1) The mirrors of the cylinders. These are exposed to at most about 2 g, not 4 - 5.[1] Structurally, they consist of lightweight supports for extremely lightweight reflectors, for example of aluminized mylar.[2] There is no reason to build them as cantilever beams. On the contrary, one can easily arrange a system of tension-line supports to guy the reflectors.[3] Nor is it necessary to fold them in and out each day.[4] It may well be preferable to build the mirror support structures as fixed assemblages, on which mirror panels are mounted in the fashion of venetian blinds.[5] They would then be tilted so as to give partial or total illumination, or to illuminate only a part of the colony. In short, the problem of mirror design, so far from being an insuperable obstacle, is the sort of problem I would cheerfully assign to a sophomore course in strength of materials.[6]

Holt replies:

1. Heppenheimer is mistaken here. The centrifugal forces on the reflecting mirrors will depend on the design of the cylinders. Given certain designs, these forces might be as low as 2g; given others, they could be as high as 5g's or even higher. Most of O'Neill's articles about space colonies have described, and the accompanying illustrations have shown, cylinders four or more times as long as wide, with mirrors long enough to reflect sunlight into the full length of the cylinder. As the accompanying sketch shows, for

such a cylinder the forces acting on the reflecting mirror when in the fully open (i.e. 45 degrees) position will range from 1g at the end nearest the cylinder to 9g's at the extreme end, with a 5g average. If the ratio of length to width of the cylinder is greater, these forces will be correspondingly greater.

2. Heppenheimer has missed the point here. What counts is not the lightness of the reflecting material, but its rigidity. Since all of these mirrored surfaces will experience "gravitational" forces greater than 1g, they must be rigid enough to remain flat under these stresses, or they will be useless as reflecting mirrors. When we consider the size of these mirrors — for some cylinder designs they might be ten or more miles long and two miles wide — it is clear that the supporting structures which will be needed to give the necessary flatness will not be simple or light.

3. In the first place, none of the drawings and sketches which have accompanied O'Neill's articles to date have shown or indicated any such cables. In the second place, to support the kind of structures mentioned in No. 2 above, a veritable forest of cables would be necessary. In the third place, if we imagine the 2 mile x 8 mile cylinder that O'Neill often talks about, the cables to support the ends of the mirrors would have to be about eight miles long. If, as would be the case in such structures, the average force on such a cable was 5g's, we would require a cable strong enough to support, on the surface of the Earth, 40 miles of its own length, plus five times the weight of the much heavier mirror and supporting structure. The strongest cables we now have will support about 35 miles of their own length — but that's all. The four or five times stronger cables we need do not exist. It could be said in reply that there is no need to design the space colonies in this way. But this is the way that O'Neill, in article after article for a year and a half now, has proposed that they be designed. And I must ask again, if the scientific community cannot see, or seeing, will not speak publicly about a mistake as great as this, how much can we trust them to <u>tell us</u> about other mistakes?

4. No, it is not necessary. But this is what O'Neill was for a long time proposing, to give an Earth-like illusion of the sun rising and setting.   5. Yes, but if the mirrors are attached to the ship and rotate with it, the problem of the centrifugal forces remains. If the mirrors are not attached to the ship, a new and equally difficult problem arises — how to maintain their position with respect to each other. As for the venetian blinds, that would probably work, but at the cost of much of the supposedly Earth's-surface appearance of the environment.   6. Perhaps — as long as someone rather more skeptical, and with a surer feel for the reality of things, was there to correct the papers.

(2) At the top of page 3 is a comment, "But they are, and must and will be, like submarines. . ." I presume this means the colony internal design. Who sez? Sez you! Even a submarine (viz, the Beatles' Yellow Submarine) need not be like a submarine. It is a matter of architectural design and of interior layout.[6a] The prospective population densities are expected to be[7] similar to those of San Francisco or of other cities, and the pleasure of living will be enhanced by an absence of autos, highways, and urban noise.

6a. By this I mean that these colonies, whether small or large, spartan or luxurious, will look like what they are — artificial environments, large containers floating in space (perhaps with some windows). They might even be as luxurious as the lobbies of Las Vegas hotels, or the insides of luxury ocean liners, which some people like. What they will not look like is the natural environment of the earth's surface.   7. This is sales talk, or an unrealistic hope. If and when such habitats are built, it is almost certain that most of them, like military barracks or troopships, will be crowded. More on this later.

(3) Chemical processing for extraction of metals, oxygen, and glass. Before addressing this critical problem, let us first consider the defining parameters within which a solution must be found.

The difficulties attending this prospect, space ore-processing, must certainly be daunting. To begin, the ores are not rich concentrations of oxides or other simple compounds, such as we find on Earth. Rather, we will deal with typical lunar materials such as anorthosite, plagioclase, ilmenite, forsterite, and the like. These typically involve complex chemical compounds, and concentrations rather lower than we are accustomed to dealing with.

Then, it is entirely true that we have no free availability of air or water, or of cheap carbon for reduction. Recycling of these materials will be essential.[8] The problem of heat disposal will also be critical.[9] But while we must be aware of these constraints, we also must realize we are operating

under conditions which in other respects are more favorable than on Earth.

8. "Re-cycling?" Where are these materials to come from in the first place?   9. Very true. And, as I said in my earlier letter, here we have no experience whatever to guide us. We do not know how to dissipate very large quantities of heat without a cooling medium. Yet this is a problem that must be solved before any construction of colonies can begin.

The chief of these is the economics of materials production. On Earth, for example, it may be required to produce metals at fifty cents a pound. This is about right for steel, and somewhat lower than the cost of aluminum, I believe. But at the colonies, far higher costs are tolerable. The reason is that the metals are not to be produced for sale as raw ingots, but rather are to be used in space for construction of power satellites, and of similar projects producing a very high economic return.[10]

For example, let us suppose that the prime chemical plant costs $60 billion over a twenty-year period, for its design, development, construction, establishment in space, and operation. In this time it produces one million tons of raw metals and two million tons of byproduct oxygen, most of which is used for rocket propellant. The net cost of these products then is $10 per pound, which by terrestrial standards is uneconomic. But the alternative, in space, is to ferry them up by rocket transport, at $100 a pound or higher.[11] (This is the transport cost to L5.) Moreover, the power satellites built at the colony may by then be generating a revenue of $30 billion a year, thus amortizing the debt — and then some.

10. & 11. Here I must refer you to my point No. 39. For reasons I point out, there will <u>not</u> be a very high economic return, if any at all. And the costs against which these costs should be compared are not the costs of sending this material into space from the earth's surface, but the cost of obtaining a comparable amount of energy from the sun (or wind, tides, etc.) <u>on the surface of the earth</u>.

In addition to being aware of these economics, we should

also be aware of the opportunities for integration of the chemical processing with the rest of the colony. The colony's thermal design may be so arranged that waste heat from the ore-processing would be used for internal space heating.[12] Then, the entire surface of the colony would function as a radiator.

12. It is almost certain that the waste heat from the metals refining process will be far greater than the internal space heating requirements of the colony.

Having said this, let us consider specifics. What sort of methods for ore-processing appear of interest? We have considered methods for the extraction and refining of aluminum, of titanium, and of glass; of these, the smelting of aluminum is indicative of the processes. We begin with lunar anorthosite. This is melted and quenched[13] to produce a glassy solid. This substance is treated with sulfuric acid[14] to extract the aluminum in the form of its sulfate. The sulfate is then treated with chlorine[15] and carbon dioxide. The former is electrolyzed, yielding aluminum. The carbon dioxide is put through the Bosch process, to recover the carbon and to produce as a by-product, oxygen.[16] Also, the sulfuric acid is recovered through acid reformation.[17]

13. "Quenched." How? And with what? And how obtained? 14. How obtained? Is sulfur plentiful on the surface of the moon? What will be required to refine it? 15. Again, how obtained? Same questions as above. 16. What are the raw materials requirements of this process?

17. Again, what other materials are needed to carry out this process? My point is that to establish in space a smaller scale counterpart of our materials refining and processing industries on earth, a great many raw materials must be produced — not just aluminum, oxygen, glass, and a few others, but a host of metals, chemicals, etc. These processes are interlocked: to make A, we require B and C; to make B and C, we require E, F, and G — but also a great deal of A. To get the thing going at all would require that we lift out of earth's gravity and into space very large quantities of materials — hardly less than many tens of thousands of tons.

We have considered the individual steps required for ore refining[18] through such processes. The temperatures required are typically of a few hundred degrees Kelvin where caustic chemicals are present, and up to 2000° K for the melting. The chemical technologies involve such long-established methods as treatment with sulfuric acid or with chlorine; indeed, the carbochlorination and electrolysis steps represent a process patented by Alcoa, for use with low-grade ores.[19] Certainly, we do not anticipate a need for any technologies as advanced as the hexafluoride methods which were developed so successfully for uranium isotopic separation, over thirty years ago.

18. Aluminum only. But an industrial base cannot be made from aluminum alone. The key metal is steel. What about steel refining? And what about chromium, molybdenum, tungsten, vanadium, and other metals needed to produce modern steel alloys? And what about tin, lead, copper? Do these exist generally on the moon's surface, and in what sort of alloys, and in what concentrations? And what kind of refining processes will we need to extract them? In taking the example of aluminum, Heppenheimer has picked the easiest case. What about the hard ones? 19. Yes; but these processes are not carried out in a totally enclosed space surrounded by a vacuum. We do not know how to do this.

We have estimated the overall systems requirements for a production capacity of 150 tons per day of aluminum.[20] We find the plant mass required is 7600 tons. The chemical inventory is 650 tons; the processing equipment is 3500 tons. Powerplant mass is 2800 tons. We estimate the required energy as 76 megawatts for process heating, 115 megawatts as electricity. Of the latter, 70 is for electrolysis and 40 for carbon reforming (Bosch process). Some 600 tons is required for the space radiators, with area of some 100,000 square meters.[21] The associated radiator temperature is 600° K or less, which is quite conservative for proposed space radiators.

20. It is not clear whether Heppenheimer thinks this is all the capacity that would be needed, or whether he has just picked this figure out of the air for purposes of illustration. If the former, the figure is absurdly small, as I will later show. And in any case, what this plant would be producing would be raw aluminum. Later, Heppenheimer talks of making much of this into cable. What would be the materials and power requirements for the factories to do that? Or to make the other forms of finished aluminum that would have to

be used? 21. This is very speculative. In any case, there would almost certainly have to be some kind of closed, circulating coolant (like the water-system in an auto engine) to carry heat from very high temperature areas to the radiators.

I feel that this work represents a useful preliminary effort to assess the requirements for ore processing. This difficult problem certainly represents one of the critical areas in which a modest amount of research is expected to pay off in greatly enhanced understanding of the issues involved.[22] However, this work certainly has given an indication of the types of issues with which we will be concerned. It is far, far too early to say we have achieved anything like full understanding of these issues. But we are quite prepared to propose and to defend answers to the types of questions asked in the letter.

22. This is very modest and tentative language, very far removed from the language of O'Neill in his articles. He does not talk about understanding issues involved. He says we know how to do this, right now. I will return to this point again.

(4) Lunar mass-driver. I am particularly pleased to have the opportunity to address this issue, since it is one on which I have spent a great deal of effort. So far from this mass-driver being beyond the state of the art, I would be quite delighted to point out how it may be built with existing lasers, tracking systems, alignment controls, cryogenics, and the like.

The linear synchronous motor, proposed for the acceleration drive, is an application of classical electric engineering. It certainly is much less complex than such commonplace devices as computers, particle accelerators, or similar electronic systems. It has been extensively studied for its possible application to high-speed ground transport.[23] It is irrelevant to state that such linear motors as we require have not yet been built, for the development of such motors appears to be a straightforward exercise in systems engineering.[24]

23. One would think from these words that linear electric motors were in common use, their problems well understood, their bugs ironed out, as is generally true of computers and Heppenheimer's other examples. Such is not the case. The linear electric motor is in a state of early research and development. To my knowledge, it exists, at least as a form of vehicle propulsion, only on a few miles of test track in a few countries. No practical, operating, installations exist, and none are projected for something like another ten years. We have to ask, if it seems likely to take ten years before we have a linear electric railroad on the surface of the earth, operating at perhaps 200 mph, with acceleration and decelerating forces of perhaps ½g, how long will it take us to develop and perfect a 4,000 mph railroad, with accelerating forces of 25g's and decelerating of 75g's, on the surface of the moon? I have no doubt that given enough time and money, it could be done someday. But it certainly can't be done, as O'Neill has repeatedly suggested, within the next five or ten years. 24. This is advertising agency talk, not scientific talk. One might say the same of the development from the Wright Brothers' airplane to the 747. Such talk ignores the relevant factors of time and cost.

The accelerating vehicles ("buckets") are envisioned as being built around small, powerful cryogenic magnets such as are well understood by physicists.[25] The velocity measurement systems proposed involved laser doppler from fixed locations, with no delicate hardware carried aboard the buckets;[26] in our estimates of the achievable accuracies, we have used the state-of-the-art performance of existing mode-locked lasers.[27] The problem of track smoothness is largely overcome by arranging for the buckets to be suspended above the track, by means of magnetic levitation; again we here propose to rely on technologies developed for ground transportation.[28]

25. As it happens, they are well understood by me. For any who may not know, cryogenic magnets are magnets that operate at extremely low temperatures, not far above absolute zero. They are kept at these temperatures by liquid nitrogen or helium (perhaps other gases), liquefied by complicated and expensive processes, confined under high pressure, and heavily insulated from the outside environment. Such magnets are expensive, cumbersome, and fragile. They exist now in the protected environment of laboratories, not along the edges of a railroad track on the moon, where ground surface temperatures may vary as much as three or four hundred degrees. The distance, in time and money, from today's cryogenic magnets to the

lunar railroad O'Neill proposes, is comparable to the distance between the radio and airplanes of, say, the 1920's, and the color TV and airplanes of today.

26. I stand corrected on this point. I take it that Heppenheimer is talking about a gadget comparable to that which the lurking highway patrolman measures the velocity of oncoming cars. The difference is that instead of talking about an accuracy of perhaps one percent, we will be talking about an accuracy of something like one-thousandth or ten-thousandth of one percent — again, in an environment of wildly fluctuating temperatures.

27. Where do they exist? What velocities are they measuring? Under what conditions? Does Heppenheimer claim that, using existing equipment we could regulate the speed of an earth's surface vehicle to the above-stated degree of accuracy? Where has it been done?   28. Here again we take promise for performance. I have followed with some interest the developments in this area. The facts are that the companies, mostly German, that are doing research and development on magnetically suspended trains, are running into serious problems, so much so that within the last year the city of Toronto, which had a contract with one of these companies to develop for them a magnetic-suspension system of transportation, has canceled the contract. It is a serious error of fact to speak of these technologies as "developed."

The principal requirement for accurate track alignment arises from the fact that slight misalignments will give rise to vibrations, transmitted to the payload on the bucket, thus preventing release and launch with desired accuracy. The track thus must be aligned to high accuracy immediately prior to release. This is to be done by supporting the track upon screwjack actuators. An alignment reference is provided by lasers; detection of track misalignments is provided by means of track-mounted Fresnel zone plates.[29] The specifications of the alignment system have been taken as those of the existing alignment system of the Stanford Linear Accelerator Center.[30] We have used methods of classical control theory to estimate the associated launch errors and miss distances; we find that following a flight of 40,000 kilometers, the launched payloads should arrive within a circle of 100 meters diameter.[31]

---

29. Once again, we have the difference between what can be done under closely controlled, optimum laboratory-type conditions, and what would have to be done in a much more difficult, variable, and uncontrolled environment. The Stanford Linear Accelerator is, to my knowledge, underground, carefully shielded from shocks, temperature changes, and other disturbances in the environment. Moreover, this accelerator is, in effect, a railroad only for atomic particles; it could accurately be said to carry no load at all. The proposed materials launcher would be a railroad carrying loaded cars, weighing at least several tons, moving at speeds up to 4,000 mph. To maintain an equivalent degree of alignment and rigidity on such a railroad is not a task already accomplished, but a wholly new task.

30. Included above.   31. This 40,000 kilometer flight will be a spiral path around the moon. Neither O'Neill nor Heppenheimer say how high above the moon's surface will be these loads of lunar ore when they are "caught." To keep them within a circle of 100 meters in diameter will certainly require a very precise control of velocity, greater by a factor of 1,000 (I have read) than the control we have so far achieved over our space rockets.

Thus, the mass-driver appears to be well-understood,[32] in terms of its major features and requirements.

---

32. What this means is, "We think we know how we might go about trying to build one." O'Neill says we know how to build one right now. We might note here that the maximum speeds of railroad trains in active commercial service, even in those countries (Japan, France) that take railroads seriously and spend money on them, has increased only about 40 - 50 miles per hour in the last fifty years. Some of the problem is with wind resistance, which would not occur on the moon, but much has to do with the difficulty of making a sufficiently level and smooth track — and the requirements of the 4,000 mph moon railroad would be far more stringent in this respect.

---

(6) Lunar habitat. NASA has conducted systems studies for the definition of habitats housing several hundred men, and there is a large literature of studies for smaller lunar bases.[33] Indeed, had the U.S. space program continued at the pace of the mid-to-late 1960's (the pace of John F.

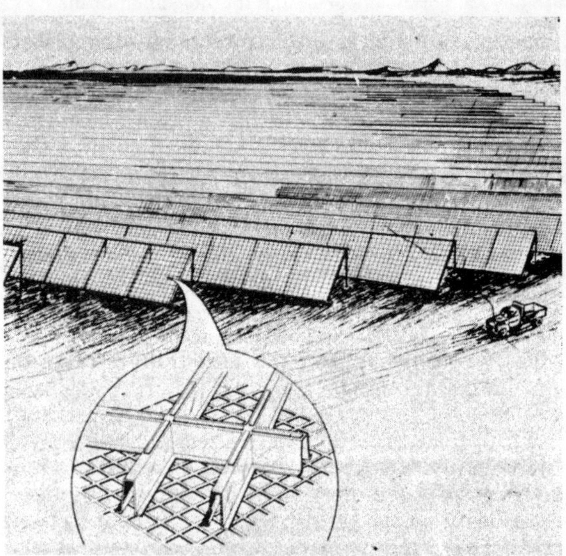

Microwave Receiving Antenna.

Kennedy), then by now we might well be on the way to building such habitats.

---

33. A study is not a habitat. No doubt we had studies for the F-111 or the C-5, and no doubt the studies showed that they would be wonderful airplanes. Events proved otherwise. Building self-sustaining habitats on the moon is to a very large degree a much newer and more uncertain enterprise than designing a new airplane.

(7) Design of large cylinders; structural considerations. The assertion is made that it would be very difficult to devise suitable structures for pressure vessels several kilometers in diameter. This statement misses the point. It is not true[34] that ONACO set the goal of a mile-wide cylinder and then tried to define appropriate structural supports. Instead, ONACO began with specific, rather conservative assumptions[35] as to the technologies to be used. These were principally the bridge-building structural designs similar to those employed in the construction of suspension bridges. ONACO then undertook to solve the problem: with such designs, how large a cylinder could one build? To their great surprise, they found that cylinders a mile or more in diameter, and tens of miles long, would be possible.[36] Thus, ONACO do not envision "massive stiffening beams"; rather, the emphasis is upon tension cables of conventional design. This involves the production in space of large quantities of wire and cable, rather than the fabrication of immense and massive structural members.[37]

---

34. I did not say it was.   35. That depends on who defines "conservative."   36. The proper comment here is GIGO, a computer maxim standing for "Garbage In, Garbage Out." Whatever O'Neill may have fed into his computers, what has come out is nonsense. The previously mentioned figure of 150 tons of aluminum ore per day shows that O'Neill has enormously underestimated the amount of materials that would be needed to build the kind of cylinders he talks about. It can be shown quite simply that if it takes a certain weight of materials to build a container of a certain size, to hold a certain pressure of gas, if we hold the shape and the pressure constant, the weight of materials needed will vary with the volume — double the volume, double the weight of materials. Or, in other words, the weight of materials increases as the cube of the linear dimensions. If we ask, how much material would we need to build, to use one of O'Neill's favorite shapes, a cylinder 100 feet long and 25 feet in diameter, to hold (with appropriate safety factor) a pressure of one atmosphere, a ton seems an optimistic answer, and two or three tons much more likely. But let us say a ton. Consider now a cylinder 2 miles in diameter and 8 miles long, not by any means the largest that O'Neill has proposed. Its linear dimensions are 400 times greater. This means, not taking into account the weight of its associated machinery, or soil, air, and water, or stabilizing ballast, but considering only the shell itself, we would need 400^3, or __64 million tons__ of material, to build it. For that little 150 ton/day plant, that would be something over 1,000 years worth of output.

37. So what. What is critical is the total weight of material.

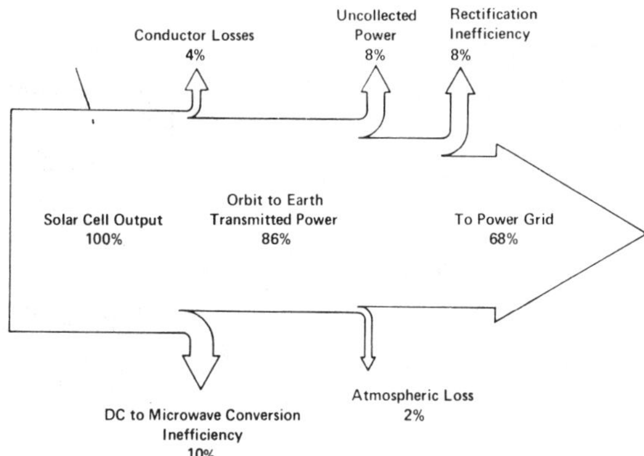

Summary of Microwave System Losses.

I realize now, too, that Heppenheimer, when estimating the weight of his 150 ton/day aluminum plant, did not take into account the weight of the dwellings required for the people working there, or the weight of the system required to raise the food that these people would eat, or the weight of the systems required to recycle their wastes. And this leads me to wonder, for every worker engaged in primary industrial production, whether of aluminum, steel, sulfuric acid, chlorine, hydrogen, oxygen, etc., etc., how many more would be needed to maintain and repair all these factories, dwellings, vehicles, "farms," etc., or to supply all the needs of the people? How many people would we need in space to operate a factory that required 100 workers?

---

(8) Power beams from the power satellites and safety assurance. The power beam is to be formed by means of a phased-array antenna involving microwave generators (Amplitrons) and ferrite-core phaseshifters. These exist and can be mass-produced by the electronics industry.[38] The beam will illuminate a target area at the Earth of some 50 square kilometers. Considerations of physical optics prevent the focussing of the beam into a smaller area.

---

38. No doubt they exist and can be mass-produced. But they have never been put together and used in this way and for this purpose, and least of all in space. We will have to learn, by trial and error, how to do that.

The associated power densities will approach a kilowatt per square meter at the center of the beam. This is somewhat less than the intensity of sunlight;[39] birds and animals will in no way be fried or cooked. However, they are likely to find the beam uncomfortably warm, thus avoiding it.

---

39. Here we reach the most extraordinary paragraph of Dr. Heppenheimer's letter. If his words mean what they appear to mean, and it is hard to see how they could mean anything else, we are seriously being asked to consider spending something like 100 billion dollars so that, from a point in distant space, we may beam to 50 sq. kilometers of the earth's surface somewhat less energy than the unaided sun (at least on sunny days) regularly delivers to the same area. This surpasses belief. And it totally destroys the myth of the eventual cost-effectiveness of such a project. If we were getting from space what we could not get anywhere else, there might be some grounds for paying these enormous prices for it. But the relevant comparison must now be against the cost of developing solar and sun-related (wind, etc.) energy on the surface of the earth. At this point the absurdity and wastefulness of this proposed project becomes clear.

Excuse me, John. As you must know, the microwave energy continues to arrive in all weathers and all night. It is far more easily converted to electric current than sunlight — the inefficient (and waste heat producing) part of the conversion having been done in space. The steady supply eliminates the need for storage, which remains the major problem with solar conversion to electricity on the Earth's surface. —SB

The beam will be formed using a pilot signal from the ground to provide a phase reference. This phase reference will serve to control the phase-shifters so as to form a tightly-focussed beam from the outputs of the individual Amplitrons. The pilot signal, in turn, will be run off power obtained at a fixed site (such as the nominal target area center) from the microwave beam itself. If the beam wanders off-target, this pilot signal will lose its power and effectively shut off. Then, lacking a phase control, the phaseshifters will fail to provide a coherent beam. Instead, the individual Amplitrons will radiate into the entire forward hemisphere of the antenna. The beam will spread out, and at any point on Earth the signal intensity will drop by some ten orders of magnitude, to levels such as are used in communications.[40]

---

40. I stand corrected here; I suppose such a control system could be made to work reliably. But No. 39 makes the question irrelevant; if we are not beaming energy to earth in enormously high concentrations, there is no economic reason or justification for doing it at all.

Thus, it is not the case that we will rely on techniques which can fail and produce disaster. We will not build a power beam transmitter which could swing about wildly, the beam being possibly quite hard to turn off or to control. Instead, we propose that it will be somewhat difficult to form the beam, and that a continuous control will be required merely to keep it properly focussed.

I am amused by the comment, "No doubt we can transmit power through microwave energy on a laboratory scale, but that is not at all the same thing as doing it on an industrial scale. The technology to do that does not exist." I suppose one would have made a similar comment about transmitting human voice or music through microwave energy, in the year 1910.[41] Actually, we are somewhat better off than that. The development of radio was immensely advanced through the invention of the vacuum tube, but we already have all the inventions[42] we need to do the job.

---

41. It would have been true. I return again to the point made in my earlier letter and many times in this one, that O'Neill and his supporters enormously underestimate the costs, in both time and money, of turning laboratory-scale inventions into industrial-scale technologies — costs certain to be far greater in the unfamiliar, radically different, and highly dangerous environment of space. 42. As above.

---

(9) Problems of rain and of dynamic balance in large cylinders. The largest cylinders will have atmospheres of radially varying density as well as clouds, so that rain may well be obtained much as it is on Earth.[43] If that is inconvenient, we will use sprinklers and give everyone a constant blue sky.[44] It may seem that a total lack of rain would be unnatural and even frightening. But to residents of southern California or Arizona, it is everyday reality. It all depends what you are accustomed to.

---

43. The conditions which generally produce rain on the earth's surface — variations in terrain, very large air masses of different temperatures, presence of microscopic dust particles (apparently needed for raindrop formation) etc. will not exist in the proposed space cylinders. By keeping the air at something close to 100% relative humidity, it might be possible to cause a little rain to fall when "night" cooled down the cylinder. But this would not be a pleasant environment to live in. 44. For years scientists have written that the blueness of our "sky" is caused by the diffraction of sunlight passing our upper atmosphere. No comparable conditions will exist in a space cylinder. People looking up will see one of two things: a) other parts of the cylinder on which people are living, or b) windows, through which they will see the sun (if it is being reflected in), perhaps some stars, and otherwise the black of space. In any case it will be clear to all that they are not in an environment similar to earth's surface, but inside a large cylinder (or sphere, or whatever).

As for dynamic balancing, that is readily accomplished by pumping quantities of water between holding tanks strategically placed.[45] Of course, it may be that the cylinders will be so massive that the mass-shifts will be small indeed. If we think of the rocking of an aircraft carrier as the men go about their duties, we may have the idea.[46]

---

45. As an old submariner, let me say that the balancing of a body in unstable equilibrium is not "readily" accomplished. But ballast tanks, as Heppenheimer suggests, might well do the job, though the mechanisms to do this would be large and complicated. 46. Yes, but if the sun were being reflected into the colony, that sun would be rocking in the "Sky," and it shadows would be moving back and forth on

the ground. Some might find this interesting or pleasant, others not; it would surely be unlike Earth.

---

(10) L5. L5 is a point of stable equilibrium, and not unstable equilibrium, in the restricted three-body problem. When one considers perturbations due to the Sun, it is found that there exists a stable orbit about L5. If the colony is moved slightly, it will not depart from this orbit, but will instead remain close to it.[47]

47. Perhaps; I will have to have this confirmed by some skeptical astronomers. For the moment, it seems that if L5 is a point where the gravitational fields of earth and moon cancel each other out, any movement toward either earth or moon would lead to further movement in that direction, there being no correcting or opposing force. The effect of these forces might be very slight, so that we could say of a 64 million ton cylinder that it would take many thousands or tens of thousands of years before it finally reached the earth. Still, it would be rather hard for those on earth when it did get there.

---

(11) The last page or so of the paper. These points are largely theological,[48] reflecting bias or intuitive dislike, rather than any semblance of reasoned assessment. It may make the writer feel good to turn up his nose and say, Ugh, I would never want to live there! But this is a poor basis for policy.[49] Certainly, one should not seek to deny others the possibility of what to them may be an important and exciting new type of life, merely because one would not himself choose that life. In this country, we provide for some people to live aboard aircraft carriers, others at Army bases, and still others along the Alaska pipeline, not from esthetic judgments as to whether we ourselves might like such a life, but from considerations of national needs.[50] It can scarcely be denied that large numbers of people will freely volunteer to live in space, even under austere conditions, when this becomes possible.[51] If it is in the national interest that they do so, then esthetic judgments lose much of their force.[52]

---

48. Again, "theological." My objections to this project are variously ethical, moral, philosophical, political, and economic. (I might add that, according to Gerard Piel, publisher of Scientific American, many scientists themselves oppose this project on moral grounds). To call such objections "theological" is imprecise, and has in it more than a whiff of Dr. Strangelove, or hard-nosed talk about "megadeaths" or "credible first strike capability" or "acceptable risks." And this may be the point to note that in all of O'Neill's and Heppenheimer's talk about space colonies there is no mention of risks. The risks would in fact be enormous. We have already lost three lives in space, and almost three more; the Russians have lost at least three. This is a death rate of something over 6%. But our ventures into space have been very modest, and surrounded by the most elaborate and expensive precautions. It seems altogether reasonable to assume that if we begin complicated mining and industrial operations on the moon and in space, our casualty rate will be even higher, perhaps much higher.

49. Not at all. "Do not do unto others as you would not have them do unto you" is a very good basis for policy, which would have spared us one major and recent national disaster. 50. Here the iron hand begins to slip a bit out of the velvet glove. So we "provide," do we, for people to live on aircraft carriers, army bases, and in other unpleasant and dangerous places. How do we do that? How in fact do we fill the ranks of the Armed Forces? To a large degree, poverty, unemployment, and boredom do the job; when that is not enough, we conscript what we need, without worrying much about "esthetic judgments." The model of an aircraft carrier is not a bad one, to give us an idea of what space colonies will be like. A few people in them will live like Admirals; most will live like enlisted men. The space colony, as O'Neill himself has suggested in a recent (Jan. 18, 1976) article in the New York Times Magazine, will be an ultimate company town, and history can tell us a good deal about company towns — particularly those on which the workers were not free to leave. If "national needs" (i.e. the interests of powerful pressure groups) dictate, we will find plenty of poor people to draft for the menial and dangerous work of colonizing space.

51. I do deny it — unless, of course, they have been told terrible lies about what life and work in space is really like. I expect that this will happen, and in fact is happening, and it is one of my ethical and moral reasons for opposing this project. 52. Oh, indeed they do!

To sum up: We are not claiming that we have found the solution to the energy crisis, or that we have discovered the future human destiny. We are not claiming the right to unlimited funding,[53] or for space colonization to be declared a major national priority.[54]

We cannot responsibly propose that it shall be national policy to undertake space colonization, at a cost of $100 billion.[55] But we are quite prepared to propose that it should be national policy to support the study of space colonization, at a cost of some $1 million.[56] Our proposals are in many ways new, and to some they may be disturbing. But if these further studies confirm present indications as to the feasibility and economic merit of space colonization, then on that basis we may lay this as an issue upon the national agenda.

---

53 - 55. This is disingenuous. O'Neill, in his articles in the mass media, has said over and over that we can have a space colony operating in fifteen years for 100 billion dollars, and that we should do so. There is nothing tentative or modest about his way of saying this; he is actively lobbying for such a project. Thus, from his article in the N.Y. Times Magazine of Sunday, Jan. 18, 1976, we have: *. . . it appears on the basis of technology being developed for the space shuttle that construction of a high-orbital facility could begin within 7 to 10 years and that it could be completed in 15 to 25 years. . . . The levels of technology required to do all of this have already been achieved.* (underscore mine) *. . . If the concept is realized as soon as is technically possible, something like the following "letter from a space colonist" might be written as early as the 1990's.* (There follows a fictional "letter" which is pure sales promotion, speaking as it does only of delights, saying nothing of difficulties, let alone dangers).

56. A case might be made for spending $1 million for such a study, as long as people representing my point of view — call it "pessimistic theology" if you like — are a part of the study. But we need adversary proceedings here. We need, not just people who say, "How can we figure out how to do this?", but people saying, "This is not worth doing even if we could do it." The almost certain danger is that the million dollar study will lead to another, and another; to a five-million dollar study, then a ten, and so on. Once this bandwagon, this permanent WPA for the Aerospace industry, gets rolling, it will be very difficult to stop it, and the further it roHs, the more difficult, as more and more people have a vested interest in keeping it going.

Let me close by referring once again to my statement that Earth's problems must be solved on Earth. This is not a statement of theology, or even philosophy, but a matter of hard politics and economics. To colonize space to the extent that O'Neill has proposed, to a point where it would make a difference to our population problem, will take, in all probability, fifty to a hundred years or more. These must be years of peace and relative economic stability and prosperity. It would take nothing larger than a minor war or major depression to put a stop to the project, and a major war would of course stop it indefinitely. Beyond that, it is hard to imagine a time in the next fifty or more years when a given amount of money, spent on dealing with Earth's problems, on Earth, would not bring vastly greater benefits than a similar amount spent in space. Thus it would have taken, and would now take, probably less than a billion dollars of research and development in the direct conversion of solar energy to electricity to make that form of power economically competitive over most parts of the world. We have the needed collectors, and need only find a way to mass produce them. In the same way, less than a billion dollars worth of research and development in wind power would probably be enough to make New England, and other comparably windy sections of the world, self-sufficient in energy. And it would take much less than a billion dollars worth of research and development of the kind the New Alchemists are doing at Woods Hole to give to large numbers of the world's poor the means, with very little capital and materials ready to hand, to double or triple their food production. Had we done such research earlier, we would now be energy self-sufficient, instead of at the mercy of the oil-producing countries; we would be prospering, instead of being in a deep and probably lasting depression; and we would be in a position, as we are not now, to make a direct, immediate, and important contribution toward dealing with the problems of the world's poor. All this we can still do, and in a short space of time, if we choose. To choose instead to spend billions of dollars in the way that O'Neill, Heppenheimer, and many others suggest, seems to me in the highest degree impractical, wasteful, and immoral. Someday, in a world where mankind has learned how to put a stop to war and to feed all the world's people, to try to colonize space might be a practicable, fitting, and even worthy enterprise. But not now. □

Note: Heppenheimer has done a book, Colonies in Space (1977, 224 pp.), $12.95 from Stackpole Books, Box 1831, Harrisburg, PA 17105. Holt also is doing a Space Colonies book. —SB

. . . energy for the Third World (the wrong end of the lollipop again). . .

Investment space

. . . there must be a proven private sector market for use of space. EARTH/SPACE is therefore dedicated to developing low-cost methods of using space, which are of value to individuals and commerical industry. It's not enough to say — we can get you there cheap. You also have to have a reason for wanting to get there in the first palce. We're giving space a reason.

Space can be used to increase the control of authority over the individual; or it can be used to free the individual from the strictures of earth. EARTH/SPACE is, quite simply, making space a place to free the individual by expanding his horizons. The real use and colonization of space will take place — not from the well-ordered designs of a few government planners, but from the random designs ot tens of thousands of pioneer individuals.

Paul L. Siegler, President
EARTH/SPACE
2319 Sierra
Palo Alto, California 94303

Space war

I just heard on the ABC network news that the Russians have been "blinding" our spy satellites with high powered laser beams. I took note of this development because of a book I've read recently called **Soviet Conquest From Space** by Peter N. James who is an intelligence expert with our government.

In his book is a highly documented comparison between our military/space programs and the Russians'. The book concludes that the Russians are pulling ahead of us in rocket development, space shuttle and space station deployment as well as militaristic use of laser beams. Mr. James said to watch for Soviet space developments over the next few years and we will see a better space shuttle system than ours which will enable them to embark on an extensive space station network. These could very easily be used to store and deploy nuclear weapons.

I've only touched on the most major points in the book and I mention it not only for its own significance but what it could mean to the future development of O'Neill's space colonies, oneupmanship being what it is between governments.

Douglas Nommisto
Wood Dale, Illinois

GURNEY NORMAN
Author of Divine Right's Trip
(*in* The Whole Earth Catalog.)

. . . I see the main use of space colonies as religious. They should be built, not as industrial enterprise, but in the spirit of the old Cathedrals, like Canterbury. We should take it all very slow and build in all the meaningful earth-stories and myths. Clearly space colonies have more to do with myth than science or industry. I want the connection between the Indian Coyote tales and the Space colonies to be very direct and clean. I want the building of the colonies to encourage folk life and country music and old time religion, not discourage it. I want the colonies to have a lot of winos and neer do wells hanging around the old computer consoles, singing and praying and spitting and telling lies. I want there to be places for Neal Cassady and Nimrod Workman, and Merle Haggard and Jeff Kizer and Ed McClanahan and me and Chloe. There's the real test. "Do I want to go, or not?" In my head I'm against all this space stuff. But in my heart, if they're goin' to build 'em, I want to be on one. I want to get to heaven, by hook or crook. I'd feel a whole lot better about it all, though, if that guy hadn't hit that golf ball on the moon. I sure do dread being locked up in outer space with ten thousand golfers.

GARY SNYDER
Poet, author of Turtle Island *(Pulitzer Prize, 1975); "Four Changes";* Riprap, *etc. Hero of Kerouac's* The Dharma Bums.

Thanks for the invitation to comment on O'Neill's space colony. I'm sure you already suspect that I consider such projects frivolous, in the all-purpose light of Occam's Razor my big question about such notions is 'why bother?' when there are so many things that can and should be done right here on earth. Like Confucius said, "Don't ask me about the life after death, I don't understand enough about life yet." Anyway, I'm hopelessly backwards, I'm stuck in the Pleistocene. That is, seriously, I'm trying to figure out what happened to man at the end of the last ice age, which seems to me to have been a major shaping transition, and so I'm still mucking around in paleo-ethno botany, which is a kind of *zazen.*

Sufficient error

. . . Space Colonies becomes a good idea only when error-correction becomes impossible here.

Are we there yet?
I don't know.
That's error enough. . . .

M. Phillips
Kokomo, Indiana

Brazil's space colony

The space colony idea is about as brilliant as making Brazilia the capitol of Brazil. When we figure out what to do with the allegedly rich interior of Brazil without massive disaster, there will be time enough to think about colonizing outer space.

It seems to me that it is violating the principles of the title of your publication. We (and all other living things) have adapted (coevolved if you will) to the Earth's environment. Success would undoubtedly depend on recreating the conditions of the surface of the earth as closely as possible, which is easier at home than in space. . . .

Albert Himoe
Houston, Texas

69

Gerard O'Neill at the standing-room-only Senate hearing on January 19, 1976

GERARD O'NEILL

*High energy physicist; deviser of Space Colonies;
inventor of particle storage rings.*

*The following is derived (with his help) from Dr. O'Neill's
remarks before the Senate Subcommittee on Aerospace
Technology and National Needs on Jan. 19, 1976, and
his keynote address at the annual national convention of
the American Institute of Aeronautics and Astronautics
in Washington, D.C. on Jan. 30, 1976. For complete
texts, write to Dr. O'Neill at Princeton University.*

In the long run it may well be that the people working at the
orbital manufacturing facilities may build very comfortable
and earthlike habitats. Much of the public interest in this
concept may be due to that possibility. In the early days,
though, it seems almost certain for economic reasons that
the orbital facilities will house a selected, highly-qualified,
highly motivated population nearly all of whom will be
working, and working hard. They will <u>not</u> be in a utopian
paradise or a laboratory for sociological experiments. The
orbital facility will be much more like a Texas-tower oil rig,
or a construction camp on the Alaska pipeline, or like
Virginia City, Nevada, in about the year 1875.

It is natural for most people, and particularly for reporters
and art directors, to become preoccupied with two features
of orbital manufacturing, both of which are non-essential.
One is "Where is it going to go?" and the other is "What is
it going to look like?" I think the proper answer to the first
question is "in an orbit high enough so that it almost never
gets eclipsed," and to the second, "It will be a rotating
pressure vessel, containing an atmosphere, with sunshine
brought inside with mirrors." Beyond that, any further
detail is almost certain to be wrong. For that reason, among
others, I think it's unwise to get personally identified with
particular designs. I'm for whatever works best, and it's too
soon yet to be sure what that will be.

As to resupply, first there's the question of atmosphere. In
some of the space habitat designs we're dealing with, the
skin thicknesses are 2 to 7 inches of solid aluminum, and we
ought to be able to make <u>that</u> gas tight. The problems could
come at windows and airlocks. There would be plenty of
power available for Roots blowers and compressors, so there
would not have to be loss of air each time a lock is opened.
Outside window areas it might be best to have a thin
secondary membrane with recuperation of any leakage. As a
guide, in high-energy physics for the past five years there has
been a single-stage ultrahigh vacuum system working with

several <u>kilometers</u> of length and thousands of joints, but with
a total <u>leakage</u> rate of less than one cubic centimeter per day.

Again, it will be necessary to be careful only if we have to
use a nitrogen-rich atmosphere. Oxygen will probably be
a waste-product at L5, because the chemical processing
plant needed for construction will be separating several
hundred tons per day of oxygen.

On the question of food supply, we presently think in terms
of fairly conventional but highly intensive agriculture. I doubt
that we need to provide cosmic-ray shielding for the growing
crops, because the 10 roentgen/year of radiation in free space
in a thin pressure vessel would be far below the level where
any effects on plants have been detected. It could be that the
<u>seed</u>-crops should be grown in shielded areas.

Water for agriculture would be totally recycled, as it is on
earth, with the initial stock consisting of 11% hydrogen
from the earth, and 89% oxygen from the processing of
lunar soils.

What if closed-cycle ecology turns out to be impossible, or
takes a very long time to perfect? The fallback position is
to bring dehydrated food from the earth, and to add water
at L5. A full diet for heavy construction work is about a
pound per day, dehydrated, so the budget would be less
than a quarter-ton per year per person. On the basis of
experience in the heavy construction industry on earth, the
productivity of manufactured goods should be from 10 to
30 tons per person-year, so even with the fall-back position
the economic leverage of having a man or woman working
at L5 would be a factor of 50 to 150.

If satellite power is to have real impact on our energy problems
here, it will be necessary to emplace about <u>15 or 20</u> stations
of 5000-megawatt size in geosynchronous orbit every year.
Therefore the lift requirement over a six-year period for
establishing a space manufacturing facility would be about
<u>1%</u> as large as for ground-launched satellite power over the
same period. Any adverse effects on the atmosphere due to
rocket flights would be correspondingly reduced, in the space
manufacturing approach, by the same factor of 100.

<u>Recommendations:</u>

I suggest the following as essential components of a balanced
program leading toward satellite power:

    1) The vigorous continuation and successful operation
of the space-shuttle.

    2) Continued development of microwave power
transmission, leading toward pilot-model demonstrations

here at ground level of phased-array power transmitters as well as planar receivers.

3) Detailed study of the electromagnetic mass-driver, not only as a launching device but for the easier role of high-thrust, high-velocity reaction engines for use outside the atmosphere.

4) Research on earth into long-term physiological effects of oxygen atmospheres and of rotation. Success in these studies could reduce substantially the cost of construction of a habitat for the workforce at an orbital manufacturing facility.

5) Study of continuous-flow chemical processing methods for minerals similar to those found on the moon.

6) Conceptual study of a human-rated version of the LDEF (Long-Duration Exposure Facility): a test laboratory capable of being put into orbit by the shuttle, in which the long-term effects of partial or zero gravity and of various rotation rates could be studied.

7) Studies on earth of high-yield agriculture, under conditions of controlled atmosphere and abundant solar energy, with human intervention (as is customary in agriculture on earth), as necessary, to maintain stability.

8) A balanced set of design studies of earth-to-low-orbit vehicle systems, emphasizing:

    a) Minimum development cost,

    b) Minimum cost per pound of payload,

    c) Minimum adverse effect on the biosphere,

but with less emphasis on massive single payloads. For example, shuttle-derivative freight rockets of moderate size, and single-stage-to-orbit fully-reusable vehicles of moderate size, would receive greater attention if this recommendation were followed.

9) A continued moratorium on the development of nuclear rockets. If our calculations are correct, the availability of liquid oxygen for refueling in high orbit, as a result of the processing of lunar materials, would give to ordinary chemical rockets a higher performance than could be achieved by nuclear stages, and without any risk of serious accidents affecting the environment.

10) Study of space-stations larger than a human-rated LDEF: facilities whose components could be launched by the shuttle or by a shuttle-derived freight rocket, and which when assembled would be suitable as construction and maintenance shops for larger objects. In my opinion the emphasis in these space-station studies should be on productive work, not on physiology, because I see no reason why the purely physiological questions could not be answered earlier and at much lower cost by a human-rated LDEF.

11) Design studies of large power satellites, emphasizing reliability, simplicity, ease of manufacture and conservative technology, with less emphasis on the achievement of minimum weight.

12) Study of an unmanned asteroidal probe, emphasizing the confirmation of carbon, nitrogen and hydrogen resources in the asteroids.

Of these recommendations, it should be noted that (1) and (2) constitute endorsement of programs already under way. (9), though negative in tone, reinforces a decision already made. (8), (10), (11) and (12) are recommendations for taking broader, less restricted viewpoints in conceptual designs already under way. Of the remaining five recommendations, four relate to research of a modest scale which could be carried out wholly at the surface of the earth, at a cost much lower than for operations in space. Only one, (6), is a recommendation for a study leading to a new operation in space. Even that is relatively inexpensive, because it would be entirely within the launch vehicle capacity of the space shuttle.

These are technical and economic considerations. But where if anywhere does this fit into a conceivable space program? The simplistic approach is shown as "direct goal logic."

Direct Goal Logic

This is the approach characteristic of wartime or a powerful sense of urgency. Robert Oppenheimer in the 1940's with the first atomic bomb project, Hyman Rickover in the 1950's with the Polaris submarines, James Webb and Tom Paine in the 1960's with Apollo, worked on that logic and they got results on time-scales like 4 to 8 years. My best guess is that the corresponding figure for high-orbital manufacturing is about 15 years.

We are certainly <u>not</u> ready as a nation to make that kind of commitment to a goal at this time. We have a deeply-troubled economy, and national wounds are only slowly healing after years of serious division.

In present circumstances we have to accomplish as much as possible <u>without</u> any charter to embark on a sizeable new commitment. In that climate agencies tend to be forced into the "queue logic":

Queue Logic

What can be accomplished under those circumstances? Still a fair amount, by working within the ground rules as they now exist, while preserving at all times an updated plan that would permit shifting to a fast, goal-oriented program if and when that becomes possible. In other words, "Combined Logic":

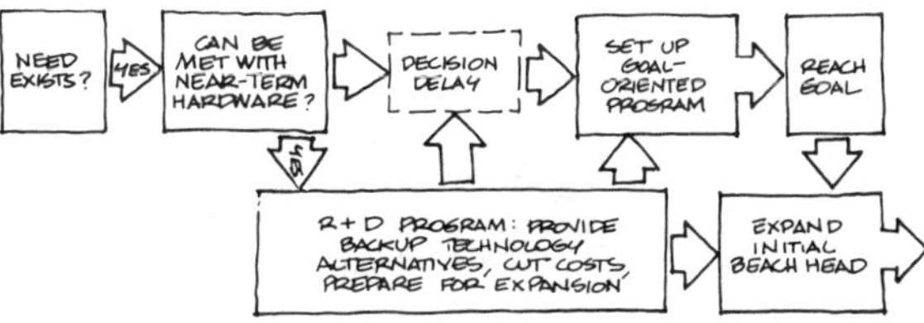

Combined Logic

There are a number of areas of technical development which fit within the ground rules of the queue logic, and which also make good sense in terms of combined logic. They include a physiological test facility within shuttle launch capability, large-structure assembly in space, new methods for automated metal-forming, a low-orbital space station, study of the mass-driver as a medium-thrust, high-specific-impulse propulsion motor for deep space, development and testing of microwave power transmission, and studies of advanced lift vehicles which could be of value to a number of programs.

There have been two unusual features in the development of the orbital-manufacturing concept: one is public response. It has been strong and mainly positive so far. I think there are four reasons:

1) Immediacy: the <u>possibility</u> of realization not a hundred years from now but within the next decades.

2) Personal participation: The prospect of taking part directly in the opening of an exciting new frontier.

3) The possibility that in space manufacturing there may be a real opportunity to solve the energy problems of the earth within the technology that we already have, and to do so in a non-polluting fashion.

4) The eventual earthlike character of the new territory being opened: that is, the possibility that the new options may involve not a machine-dominated robot-like future, but green grass and trees and flowers.

The second component is the <u>people</u> who have been drawn into this work from outside the traditional industry boundaries. They constitute a resource which is much like the citizen army that is at the nation's service in any threatening war. By the same token, they can be lost by mismanagement or by inaction. Here the time scale is critical.

Discount economics gives us the same message: once it's clear exactly how best to proceed, act fast, because a stretchout in time eats up the profits.

Clearly our task is very big, and specialists in every area can contribute to it. Those of us already working welcome all the help we can get. □

STEWART BRAND

Editor, The CoEvolution Quarterly

If built, the fact of Space Colonies will be as momentous as the atomic bomb. Each make statements that are equally fundamental. The one says, "We can destroy the Earth." The other says we can leave it, leave home. With that our perspective is suddenly cosmic, our Earth tiny and precious, and our motives properly suspect.

On the other hand, suppose that the Space Colonies don't work, that we do find some fatal flaw. It would be no less of an event. "We cannot leave the Earth" is a thought so foreign to the 20th Century that nothing would be unchanged by it.

Either way it goes the experiment should be made, because not knowing whether we can leave the planet begs all of our important questions. Either knowledge — knowing we can leave or knowing we can't — could make for more responsible habits. Either knowledge is a kind of growing up.

The same applies to the biology. If we can learn to successfully manage large complex ecosystems in the Space Colonies, that sophistication could help reverse our destructive practices on Earth. And if we fail, if our efforts to impersonate evolution in Space repeatedly run amok, then we will have learned something as basic as Darwin about our biosphere — that we cannot manage it, that it manages us, that we are in the care of wisdom beyond our knowing (true anyway).

Is balance really possible where even the gravity is manufactured? It would be nice to know.

The commonest complaint about Space Colonies is that they are merely more of the same — the same old technological whiz-bang and dreary imperialism. I'm arguing that the success of Space Colonies would bring needed whole-system sophistication, their failure would bring needed whole system humility, and only not trying them at all would bring more of the same.

Besides, Space Colonies are distinguished from other high tech mischief such as nuclear energy, the SST, and the Arms Race by a major difference. They take place outside the Earth's atmosphere. They are separate whole systems. The experiment of Space

Colonies endangers only the experimenters. When high tech goes wrong on Earth it is the innocent who get the consequences, down wind, downstream, and down the years.

What Space Colonists use from Space — energy, materials, and location — is taken from no one else. They are out of the Earthly "zero sum game" where one group's gain is another's loss.

"North America all over again," is everyone's first response. "Now we'll pillage Space." Could be. But most of that thinking has scant comprehension of the nature of the Space environment. There's somewhat more than a lot of it. There is in fact no perceptible or theoretical end to it. Space is not like a continent, or the Pacific Ocean, or anything else we've experienced except possibly death and rebirth. It's more like a Buddhist chant: "No-air-no-gravity-no-night-no-day-no-up-no-down-no-motion-past-no-standing-still-no-life-also-no-death-no-thing-only-waves-of-star-star-star-star-star."

Space may teach us much about the Void. Still it is not the same. It has both emptiness and form, both measurable.

Its emptiness makes it highly predictable. Things go where they're pushed and keep going. Pioneer 6 was designed for a six-month life in Space when it was launched into solar orbit in 1965. Ten years and twelve times around the Sun later it's still going strong collecting and relaying solar data, with no reason not to expect another ten years out of it.

In de-emphasizing the exotic qualities of life in Space O'Neill is making a mistake I think. People want to go not because it may be nicer than what they have on Earth but because it will be harder. The harshness of Space will oblige a life-and-death reliance on each other which is the sort of thing that people romanticize and think about endlessly but seldom get to do.

This is where I look for new cultural ideas to emerge. There's nothing like an impossible task to pare things down to essentials — from which comes originality. You can only start over from basics, and, once there, never quite in the same direction as before.

Earth and Moon, for once, drawn to scale. In this scale the nearest next item,
Venus, would be seventy-two yards away. Between, nothing. Except sunlight.

So much for all the wonderful benefits of Space Colonies. Is there any likelihood we can politically get from here to there?

The most political argument against is the trade-off one — couldn't we spend $100 billion elsewhere more beneficially? I'm not so sure. The Apollo Program cost $25 billion, and as Government projects go it probably did more good, did less harm, and made more friends for America than anything else we've done since the Marshall Plan.

In 1977 the U.S. is spending $3.7 billion on Space, $10.4 billion on energy (mostly nuclear), and $101.1 billion for the Department of Defense. At present the trend is for increased Defense spending and decreasing money for Space.

Now, I'm claiming that the prospect of Space Colonies gives us the best leverage on the Arms Race that we're ever likely to get this side of war. It employs the same nations, the same engineers, manufacturers, contractors, etc., and it's a more interesting story. The Arms Race is a big bore. Nothing ever happens.

Perhaps the leading U.S. and Russian Space planners should participate in the next round of SALT talks and turn them into Strategic Arms Conversion Talks. "We'll scrap the cruise missile and build a Model I Space Colony instead, if you'll do the equivalent." Conversion may be a great deal easier to monitor than Limitation — is the alternative project coming along or not? Look and see. It might even be that collaboration would have reason to replace competition — if only to check on each other's de-weaponization. Maybe Apollo-Soyuz <u>was</u> about something. America has better expertise in Space so far, but Russia has more imaginative long-range fantasies. And their astronauts sing better than ours.

I guess I expect there will be much more public participation in Space Colonies than there has been in the Space Program so far. Debate such as in this book. Public demand perhaps that new Space expenditures come out of the Pentagon's hide. Further international projects like Spacelab. And so forth. The difference being, after all, that each voter can consider using the program personally, can consider emigrating.

At least each young voter can, and that's where mass support seems to be building, among college age and younger. Most science fiction readers — there are estimated to be two million avid ones in the U.S. — are between the ages of 12 and 26. The first printing for a set of Star Trek blueprints and space cadet manual was 450,000. A Star Trek convention in Chicago drew 15,000 people, and a second one a few weeks later in New York drew 30,000. They invited NASA officials and jammed their lectures.

Now is the time for NASA to encourage people besides engineers to get into the act. The program needs administrators who are not afraid of excellent artists, novelists, poets, film-makers, historians, anthropologists, and such who can speak to the full vision of what's going on. And their voice needs to be a design voice, not just advisory. America (and Russia) were in Space for ten years before they bothered to get a photograph of the Earth. That's pretty arid thinking. There's no reason it has to continue.

To be sure, if the soft sciences and arts are let into the Space Program there will be constant argument and much silliness. Good. One of the best points that O'Neill makes for Space Colonies is that they lead to divergence — many visions travelling in many directions. That could become their best real function.

Ten thousand or one million or many millions of people in Space is somewhat about engineering, but it's mostly about people.

Returning to the question of Artificial vs. Natural, my friend Dick Baker has his doubts. Some years before he became a Zen abbot he worked in the merchant marine and observed that too long on board in a totally man-made environment tended to make the seamen a bit crazy. The same, he's noticed, goes for cities.

It's true, we make ourselves dishonest in worlds we have had too much of the making of. Still, "Natural" has a way of getting in through whatever barriers. As Baker said in another context, "From the Buddhist point of view <u>everything</u> is artificial." ■

Space colonies should keep away from the government for a while.

ASTRONAUT RUSSELL SCHWEICKART ON THE PHONE

Russell Schweickart: I haven't met O'Neill myself, but there are hardly any visionaries left in the system right now, and it's absolutely delightful to find somebody outside the system who's got that kind of vision. The kind of people who are being attracted to O'Neill is the most refreshing thing I've seen in a long while.

The problem is that NASA is part of the system, and new ideas, at least a lot of the better ones, come from outside, and then get co-opted. If I've got a fear about this thing, it is that the system will co-opt it too soon.

Stewart Brand: How might that work?

Schweickart: Well, it might work by NASA starting to seriously fund the whole project, which unfortunately a lot of people would see as a good thing. I would see it, depending upon when it happened, as either good or bad. The later it happens the better.

SB: Why?

Schweickart: Because, if the whole thing starts getting formalized and funded by the government, then the government puts its heavy hand in it in controlling further funding, and once people get used to being paid for their time and their energies, then they're no longer going to do without it, and then OMB and the Congress, and other people . . .

SB: What's OMB?

Apollo 9 Astronaut Rusty Schweickart (the first man to walk in space without an umbilical) was flying a desk in Washington for the applications end of NASA activities when I phoned him in 1976 and we fell into this conversation.

Most of us consumers who are demeaned by mechanical conveniences never see what it is to fine-tune oneself to the edges of possibility of a superb machine. When the Space Program is moving (it's been slow lately) the astronauts are athletes in a game that changes as fast as equipment design and their abilities can co-evolve. Help invent a bird and fly it and invent again. The trick is, making only mistakes they have a chance to learn from.

The adversary is appalling distances, vacuum, radiation, and cold just three degrees above absolute zero. The ocean of space has no other side, only other tiny islands.

We were wrong in perceiving the astronauts as crew-cut robots.

—SB

Schweickart: The Office of Managment and Budget, the old Bureau of the Budget. They're amazing. They have the lowest profile in government, and tremendous power.

They're not bad people, but their principal responsibility is to control the budget, which means mainly hold it down. The place where they cannot hold things down is where you have a constituency, where people's immediate well-being, livelihood, jobs, are affected by it. But any idea like O'Neill's, which is visionary, which is further out, which is long-term, is just a sitting duck for OMB, because it isn't going to lose anybody any votes. They can come out against it, and the President isn't going to be faced with a large constituency of people with a vested interest who are going to go after them.

As soon as something like this gets into the system formally in a full-blown way, then it goes through all of the necessities which it has to go through eventually, but if it goes through them too soon without the correct mental set in the country, without the momentum behind it and the emotional commitment to it, then things like cost-benefit analyses and the various trade-off studies and this, that, and the other thing that have to go on inevitably will kill it. The government does things when it has to, and if it picks up an idea like this too soon, it can kill it in a way which would make it difficult for somebody on the outside to pick up again.

Whereas, in its present state it was generated totally outside the government. . . . It's picking up a lot of momentum on its own outside the government, through things like CQ and others, and the government is having to respond to it because of the pressure from outside. So I'm not in any rush to see the concept come totally into the government. I would like it to be discussed in the media, at universities and college campuses and groups forming all over the country to support it, that kind of thing. If there's a little resistance by the government, it's going to pick up a lot more popular support.

SB: What do you think would be an optimum schedule?

Schweickart: I don't think it is so much time, Stewart, as it is circumstance. First of all I think we've got to reach the bottom of this cycle we're in now in the nation. We've got to see a more favorable economic attitude that can tolerate a little bit of luxury and long-range thinking. Nothing gets approved right now in the government unless you can show immediate utility.

There are only two things that get approved. Let me take NASA as an example. First are the kinds of things I'm working on right now, the earth resources satellites, the direct-payoff applications of space. You'll get support for something which shows that the nation will benefit ten dollars for every one that we invest. The other kind of project that NASA has that will get approved are those kinds of science programs which are moderate sized. Moderate sized for us

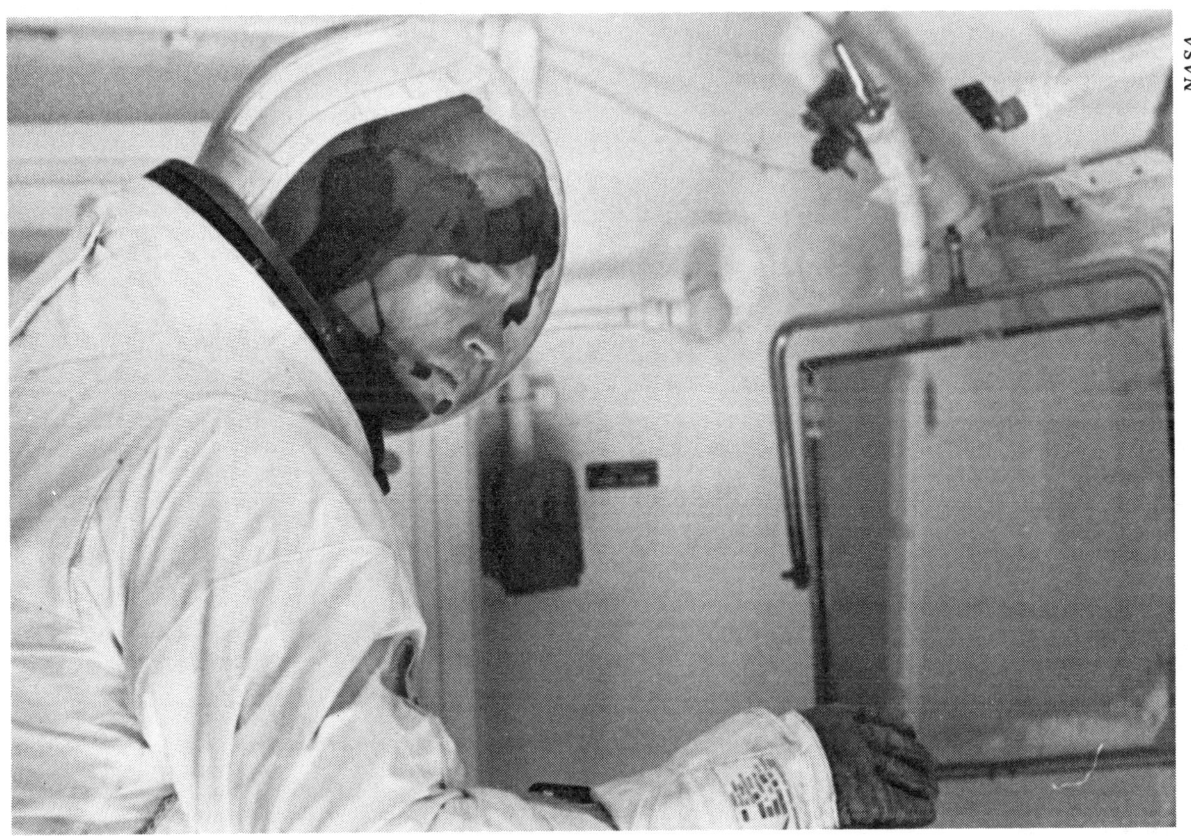

*June 1968, Houston. Schweickart at a simulated altitude of 200,000 feet testing
the Portable Life Support System which makes his suit an independent space vehicle.
With it he made the first spacewalk without an umbilical.*

is a few hundred million dollars. Programs like the Viking
mission to Mars for example, which doesn't have an
immediate practical value but which is legitimate science.
They'll get approved because they don't make any difference,
and they can be cut off immediately without the governors
of half of the states and half of the Congressmen coming
down on OMB or the President or somebody for cutting it
off. So we do science at a level which will keep us doing
science, not aggressively, but we will keep our hands in it,
and we will also do those kinds of things which show an
immediate benefit.

But a large-scale project like this which does not show
immediate benefits, something like the Apollo program, is
not going to be approved for a while yet. Now if the space
shuttle works, if we get that off the ground and it works
right, and the demand on its use begins picking up, there's
one other thing that's got to happen. I think we're talking
perhaps early-to-mid-80's before we're going to have a real
test of the space shuttle. People are going to have to look
at it and say, "Sure, of course we need it. How could we
have ever thought we wouldn't need it," before they are
going to start looking seriously at another large space project.

Now this trend towards utilitarianism I mentioned is even
evident in the work that NASA is doing on O'Neill's concept, in that NASA has pushed it quite a bit towards the idea
of power generation, to relay power back to earth. One of
the principal reasons for doing that is because the project
now takes on a practical value.

Well, that's ok, but you've got to be careful, because if you
once start playing that ball game, then you've got to come
out a winner within the rules of that ball game, or else you're
doomed. You're much better off refusing to play until you
know you can win.

The fact of the matter is that each step along the way to a
lunar base or O'Neill's concept or a trip to Saturn or whatever, each step along the way is going to justify itself on
practical grounds, but it will also, as a spin-off, permit the
next step. The big difference is that if you have that long-
term vision that O'Neill has inspired, it will generate the
enthusiasm for the intermediate steps, which will then each
and of themselves justify themselves on relatively
mundane grounds.

SB: What are some of the intermediate steps?

Schweickart: The space shuttle is one.

SB: Once the shuttle is up what does it do that makes
everybody glad it's there?

Schweickart: All kinds of things. It puts up useful satellites,
communications satellites, at reduced cost. You can put up
earth-orbiting earth-resources packages, which are going to
help in things like city-planning and forestry management,
and agriculture and all kinds of things. The shuttle is going
to open up space for low-cost utilization. It's really a transportation system into a useful environment which we haven't
been able to afford in a major way up to this point.

SB: Will there be any manufacturing facilities?

Schweickart: Oh yeah. Space manufacturing is a little
research-y right now compared with other applications, but
it's a comer. One of the nice things about space-processing

is that it's going to require man to actually be in space operating things in order to develop the technology, as opposed to things like remote sensing where you really do most of the development on the ground, put up a sensor, and let it go.

SB: Do you feel it's always better to have someone with the device?

Schweickart: It depends on what you're going to do. In the research phase, I think for complex operations the answer is yes. In the operational phase, where you already know exactly what measurements you want to make, you want to make them on a repetitive basis, you want to make them routinely, no, don't put a man up there. You can automate that. But in the research phase, it looks as though man will be very useful in space. The interesting question is whether man will be of real value in space applications in operational situations as opposed to research and development situations. That's a question people don't have anywhere near as clear a feeling on one way or the other.

The next step beyond the shuttle I think is going to be a space station in near-earth orbit, and that again I think will argue for itself partly as an outgrowth of things like space-processing. There are other reasons for going toward a space-station where you'd have a permanent crew, or at least a crew which you would cycle just occasionally.

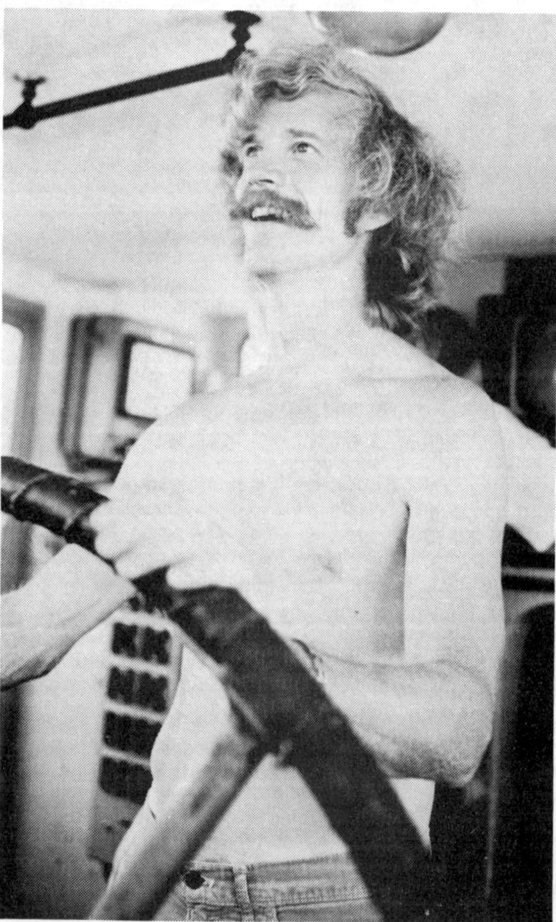

September 1975, Bahamas. Schweickart at the helm of Jacques Cousteau's research ship Calypso. *They were comparing satellite and surface oceanographic data — Landsat overhead, divers below.*

SB: A crew of what size, do you suppose?

Schweickart: Oh, it'll probably start out modular, and build. Ten people to start with, and then you'd probably build up to stations which would have hundreds I'm sure. But that's a long way off.

SB: How thick are the Russians in that stage of the fantasy?

Schweickart: Well, you never know. They seem to have a lot more fantasies than we do. That is, they seem to be more willing to speculate and wave their arms about things of that kind than we are. It's questionable whether they have the real competence or the real determination when it comes to putting the ruble down to do it. There's no question that our technology is way out in front of them in most of the things we're talking about, in anything that requires sophistication, in the practical use of space, but when it comes to speculating about it and talking about far-out things, the Russians are more willing to do it than we are. Perhaps because they are further away from having to commit real resources.

SB: Do you know of any interest by them in the O'Neill concept?

Schweickart: As a matter of fact I don't. I have no knowledge of whether they've been looking at it, or even whether they're familiar with it.

SB: I've heard speculation that the Russians and we have some kind of unstated mutual program, that we seem to be making vehicles like mad, and the Russians seem to be more into pay-load like mad. For a space station or whatever. Is that the case?

Schweickart: No. Well, I can't state positively that I know it's not the case, but I'm high enough in the organization that I know we're not that cooperative with the Russians yet that we would forsake one element to them and risk getting caught with vehicles and no payload, or vice versa. We're a long way from that. Even with the Europeans where we're in an arrangement for the shuttle/space lab, even there our shuttle can do a lot of things without their space lab, but the space lab can't do anything without the shuttle. So we're on the safe end of the deal.

SB: There's a European consortium?

Schweickart: Yeah, the European Space Agency — ESA — which is a combination of ESRO, the old European Space Research Organization, and ELDO, the European Launch Development Organization. It's a consortium of something like 12 European nations which collaborate on a project-by-project basis. They're all members but they can opt whether or not to participate in any given project. With the Spacelab I think all 12 of the members are partners, with the Germans being the leading interest.

SB: Is that a growing involvement by ESA?

Schweickart: It's a total commitment. They have sole responsibility for the design, development, manufacture and testing, and everything, of the spacelab which will fly in the shuttle. NASA is basically a customer. We actually will purchase our spacelabs from Europe, so we are really totally dependent on them for that concept. On the other hand, what they did in doing that was to give up an autonomous space program, because they scrapped all of their major development in terms of boosters for independent pay-loads. They really hung it out.

SB: What level of budget are they into with that?

Schweickart: I'd have to do a little research to tell you. It's something I'm not directly connected with, so I don't know those kinds of numbers. Off the top of my head it comes to something like 500 million dollars. It's big.

SB: What are prospects of a Concorde-type debacle out of that do you suppose?

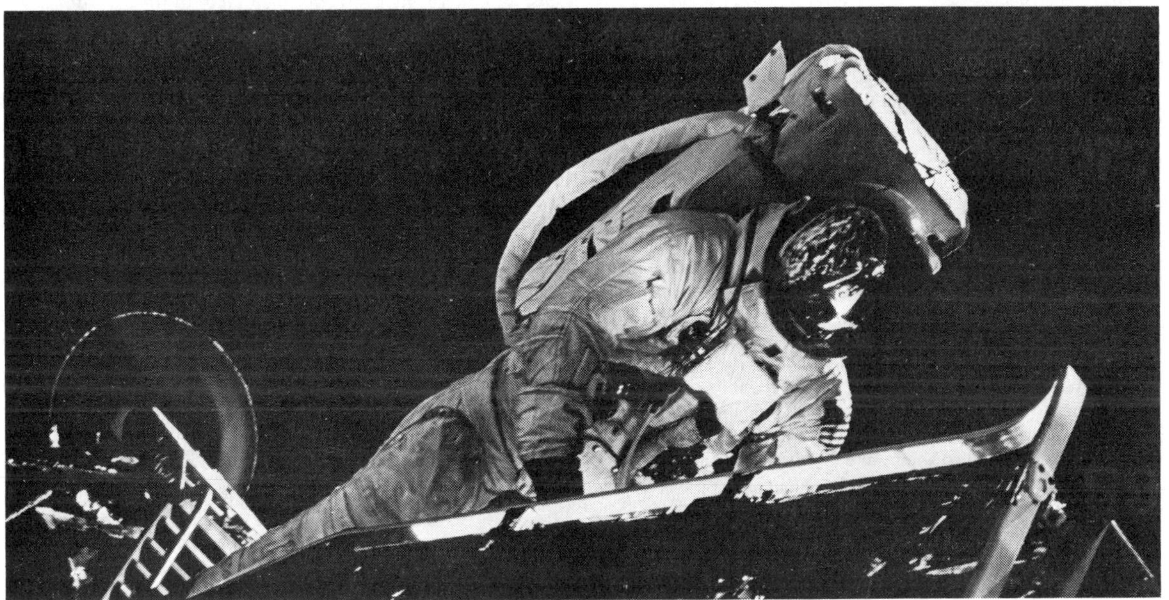

*March 1969, Earth-orbit, Apollo 9. Schweickart, pilot of the Lunar Module "Spider",
goes outside to get a thermal sample. More "outside" than anyone had ever been. With
him on the mission were James McDivitt and David Scott, who took the photograph.*

Schweickart: I think very very low. There's a mutual inter-dependence here which is an interesting thing to watch. Europe, because of its commitment now, has put a certain pressure on the United States to continue the program, and to stay on schedule, and not rock the boat, because of our partnership. At the same time Europe is somewhat the same way. Because the shuttle is going on, it's overcoming a lot of their political problems to keep the thing together. The inter-relationship obviously will have rough spots. When you marry somebody it's a little different from living with them. Things bother you that didn't bother you before, but at the same time you have a committed relationship, and the children will come.

SB: Do you know if ESA has any knowledge or interest in space colonies?

Schweickart: Yes, they do. But I don't know how formal it is, and I don't know what their specific reaction to O'Neill's thing is, but I do know that they have looked at long-term projects. In about '81 the first operational shuttle will carry the first spacelab. It'll be a cooperative European-American payload.

SB: Are there any Europeans in the astronaut program yet?

Schweickart: There are no Europeans in the astronaut program right now. The next selection however is being considered now, I think for a year from this summer if I remember right, and I'll be very surprised if we don't find European applications, and probably we'll be accepting some Europeans. Now, that's a delicate matter. NASA may ask ESA to recommend specific people rather than the United States trying to select among the member nations of ESA. Imagine the delicacy of that one. You can see Spain, with a 3% budgetary commitment, in the spacelab with an astronaut and Germany with 53% commitment having none, or something.

SB: Is the space shuttle very adaptable to moving from near-Earth orbit to lunar-distance orbit, for work on Space Colonies?

Schweickart: Not directly.

SB: How serious a change would be required?

Schweickart: It would require a total new vehicle. The shuttle is really designed to do one job and one job only, and that is to get into low-earth orbit from the surface and back down to the surface, and do it over and over again. That's where you realize the economies. But even if you severely penalize the payload and put in the maximum number of additional fuel modules and things, you're limited to something like a 600 mile high orbit, which is not very high. The normal orbit would be a hundred miles or so. Well, that's a long, long, long way from going out to lunar distance and returning.

SB: Is fuel the limitation?

Schweickart: Well, it's two things. It's fuel, but also there's no reason for taking wings to the moon. The only reason to have wings is to mess around in an atmosphere.

SB: Would you end up with a two-stage shuttle — the shuttle-shuttle to get to near-earth distance, and something else to get to lunar orbit?

Schweickart: Right, you would have a vehicle which would have to be designed probably to be carried up either whole in the shuttle, in the payload bay, or more likely be carried up in modular form and assembled in orbit perhaps near a space station, and then go from there to the moon, and back again. You end up with basically a stable of different vehicles for different purposes. You would have a vehicle which would provide transportation to lunar orbit from earth orbit. You would probably have another vehicle which would go from lunar orbit down to the surface.

SB: Would you use the same vehicle to get to O'Neill's colonies at L-4 or L-5 as you would to get to lunar orbit?

Schweickart: Yeah, no reason that you couldn't. Once it's been built, that vehicle is going to serve all its life in a weight-less state. It could take you anywhere. But if you're going to come down through an atmosphere, whether it's a Jovian atmosphere, or Earth atmosphere, or Martian atmosphere, or wherever there is a problem in aerodynamics, you need another vehicle.

SB: How's the space environment for maintenance on a piece of equipment like that? Is it going to be worn out in a number of years, or go on indefinitely because it's operating in a weightless vacuum?

Schweickart: Well, we don't understand all we might about that.

SB: Presumably the stresses are mainly in the propulsion system.

Schweickart: Not entirely. We don't understand rotating equipment, for example. Bearings. In some sense bearings ought to last indefinitely in space. You don't have any gravitational forces wearing them out. All they do is sit there with just occasional very very small stresses on them. In fact that's not the case. We had large rotating inertia wheels on Skylab which were the principle means by which we changed the attitude, or controlled the attitude of the laboratory. And one of them failed. I think it was at the end of the second mission, or maybe the beginning of the third mission, and a second one was having heart attacks and was threatening to fail as we approached the end of the third Skylab mission. Now, these momentum wheels are monstrous, they're just huge fly-wheels is what they are — not all that huge, only about a yard across, but they're pretty massive, and they spin at 10,000 rpm or something like that, very fast. You've got a lot of momentum in them, and you do generate side forces on the bearings, but we don't understand the failure mode. We don't understand why they failed or how they failed. There are different ideas about lubrication in weightlessness, and how the flow of lubricants and the behavior of surface tension effects in liquids is changed in weightlessness, so we don't really understand it fully. There are some pretty interesting engineering questions that have to be answered before you are going to be able to really design things in detail like Gerry's space colonies.

Skylab 3, 1973.

SB: Are there any problems like that with the centrifugal gravity that O'Neill is counting on?

Schweickart: Artificial gravity does do some nasty things. The only way to generate an artificial gravity, of course, is to rotate, and when you rotate you immediately complicate the process of looking outward from the spacecraft. If you want to look internally and work internally, it's no problem. However, if you want to communicate with the earth, or you want to point a radiotelescope, or an optical telescope, or you want to make observations externally, then the only direction you can look is along the axis. And even then unless you want to have things rotate as you look at them, you immediately have to de-rotate something. You've got to somehow get out of the rotating system, or counteract the rotation by having a counter-rotating hub, or perhaps some sort of separate equipment standing off to the side that doesn't rotate and a data link between you and it or something of that kind, so it complicates and therefore increases the cost of space operations to have an artificial gravity.

On the other hand there are obviously very nice things about having an artificial gravity, in that things behave "normally" and you don't have to take all the precautions with containing liquids. Convection works, as well as a lot of other things that you are used to here on earth. And if you're going to start growing food and vegetables and things, there's a lot less uncertainty about what you're trying to do if you have an artificial gravity. On the other hand, the work with Skylab was all in zero gravity. We had three guys up there on the last mission for about three months. After 84 days we saw no indication of anything that would stop us from going for significantly longer periods.

SB: How was it for them when they got back in 1-g?

Schweickart: Well, it was better for them than it was for the guys who stayed up a shorter time. They were in better condition than the crew who stayed up for 59 days, and they in turn were in better condition than the crew that stayed up for 28 days.

SB: Any explanation?

Schweickart: Yeah, there are two. One is that going from the shorter to the longer flights we increased the amount of exercise in each mission. We started out with a little over a half an hour per man per day. The second mission we went to about an hour per man per day. The third mission we went to about an hour and a half per man per day.

SB: This is just thrashing around in mid-air or playing with ropes or what?

Schweickart: Oh no. This is programmed physical exercise. Either riding a bicycle ergometer or running in place against a set of bungies on a teflon sheet, or arm exercises with a bungie, or various exercise devices we had. And putting out real work. Not just keeping limber but really putting out a hell of a lot of BTU's. As you do on earth, everybody felt subjectively much better after the exercise. They felt more invigorated, more energetic, more wide-awake, all the same things, except I guess you might say in spades compared with here on earth, because we put out a lot of work just sitting. In weightlessness, everything is floating, and you really power down. Your muscles are doing nothing. Here on earth just reacting against gravity, sitting up in a chair, walking around, getting the coffee, is a lot of work that you just take for granted.

The other reason is that we found that you reach in some sense a low point in the adaptation process somewhere around 21 days. If you look at all of the various physiological parameters — heart rate versus work, oxygen uptake, red blood cell count, heart rate and electrocardiogram under various stress conditions, they all have a different time history in their adaptation. But you could say that at some-

1970, some airfield or other. To maintain astronaut status Schweickart regularly flies a NASA T-38 high-performance jet. At the time of this interview, 1976, he was flying a desk at NASA Headquarters in Washington as Director of User Affairs, Office of Applications.

thing like 21 days you kind of reach a low point, and then begin returning to basically a pre-flight condition. Gerry Carr on the third mission actually came back at about the same weight he launched at. Body weight is a good indicator. It decreases and then goes back up. Subjectively you bottom out in about a day or two with space sickness, and then start coming back.

SB: Was that your experience?

Schweickart: I put my own reactions off by not moving around for about two days, so I had about a day to two day lag in there which I found out afterwards is not the right thing to do, but we didn't know it at the time. Basically you find that you physically feel the worst after about a day or two, and then from there on you feel a lot better. That's quite noticeable. But the things which you aren't aware of continue on this trend for several days, or in some cases weeks, and then start coming back up again.

SB: What's the effect on dreaming?

Schweickart: None that we know of. We measured the sleep in Skylab and the basic observations are that not only do you end up getting about the same amount of sleep but the same proportion of sleep is spent at the different stages and in REM sleep.

SB: How about dream content?

Schweickart: I don't think anybody ever consciously tried to do any analysis of dream content. Most of the guys aren't the kind of people who even recall their dreams on earth let alone up there, and nobody seriously proposed trying to train anyone to recollect their dreams or wake them up during REM sleep or something like that to recount dreams.

SB: Did you have anything?

Schweickart: No. I'm not even aware that I dreamt at all, but that's true here on earth too.

SB: Let me get back to grand strategy for a minute. I've found that a lot of people who don't like space colonies because it will take money from other things get real interested if they think that space colonies might take money from the Department of Defense. That's a trade-off they'll accept. Do you see any practicality in that?

Schweickart: I don't know. That's so wrapped up in the nature of man, Stewart. It's difficult. That's the dream of an awful lot of people. It's certainly mine. A lot of people have written statements or comments about that as a concept. I just don't see any . . . Angola is a great example. We just had one of the worst lessons in our life in Viet Nam, and yet here we are in Angola messing around, and it's only the Congress yelling and screaming which has forced the Administration to back out of it. This system is so oriented and conditioned to the idea of having to protect itself from other nation states that I'm not optimistic. I wish I could be.

SB: At least the Congress is resisting this time. They didn't before.

Schweickart: Yeah, I think they've gotten the message a little bit better maybe than the CIA and other people. I don't think that Congress is any smarter or any better intentioned than anybody else. The real difference is that Congress is accountable and the CIA is not accountable. The Administration is every four years, but even then it's not as accountable as the Congress is, and that's really the major difference. The people of the country have learned a lesson in Viet Nam, and they're gonna ensure that Congress knows it. The CIA and all the rest of DOD, the whole government (NASA included, NASA's no different, it's only that we're not responsible for national security) — if you don't have direct accountability you're going to keep going in the way that your momentum vector is pointed, and that is what we've been doing for a couple hundred years come July.

SB: NASA's not as subject to that as CIA or DOD I guess.

Schweickart: Not as much. But we're a bureaucracy too, there's no question of that. I'm not starry-eyed about that. I'm thankful that we've got a lot of bright people but I'm sorry that we don't have too many visionaries. I almost cried the day Von Braun left. Do you know Jesco von Puttkamer?

SB: No, who's that?

Schweickart: Jesco came from Germany but I think he's not old enough to have been associated with the original Peenemunde crowd. He went to Marshall Space Flight Center and worked for Von Braun, and he now has come to NASA headquarters in Washington. I forget what his title is, but he's basically responsible for advanced planning in the Office of Space Flight, which used to be the Office of Manned Space Flight. He's the guy in NASA who is charged with looking at O'Neill's stuff and other concepts like that. The long-range planning for NASA.

Anyway Jesco took his planning, which goes out through the year 2000, and some very nice graphs and slides, and went up to the Star-Trek conference in Chicago. He went to attend it, really. Well, they asked him if he'd show his presentation, and he said he'd be willing, and they said all right, we'll get a room and you can put it on. Five thousand people attended the damn thing, and he ended up putting it on two or three times because of the interest.

I heard his presentation back about two months ago, and it was the first thing I've seen in NASA in a long while that really lit my candle. Jesco's got a very nice way of saying that you don't justify a trip to Mars or a space station around Jupiter right off the bat. But looking at the logical stepping stones that allow you to do it you find that each one of them has real practical value and will justify itself on practical grounds. The way in which he brings that across is very nice, because it takes all the work that he is doing in long-range planning and it says it isn't just dreaming, it causes you to think about the process and where it's going and what it will permit you to do.

I'm so excited about O'Neill's project because it presents a challenge that's worthy of interest and time and energy on the part of young people, where so much of what we're doing is sort of the drudgery of space flight. We're trying to extract a practical value out of a communications satellite and those kinds of things, which are useful, which are good, which I have nothing at all against, but which are

1964, Moffett Field, California. Schweickart flies simulated Gemini re-entry in the centrifuge.

not the kind of a challenge which is going to cause somebody to climb Mount Everest.

SB: Well, we're hearing from quantities of those young people. They're in high school or they're in college, and they wonder what they're supposed to study to be of use. What do you tell them?

Schweickart: I generally point towards the sciences. The number of people who are going to be up there — it's something like airlines. You've got a lot of passengers in the back and you've got about three or four people up front who are doing the flying. After all, the glory days are gone in a sense. There's a lot of exploring left, and there's a lot of room for the individual, but the fact of the matter is that when we start operation of the shuttle we're going to have bus drivers and we're going to have passengers. That's a very unglamorous way to put it, but the flying is going to rapidly move toward routine, and the exciting thing is going to be what's done up there in that new environment. That's where the volume is going to be. You can transport a hell of a lot of people with a fleet of ten buses and 20 bus drivers.

SB: What are the passengers going to be doing?

Schweickart: They're going to be doing earth observing of all kinds — meteorology, oceanography, earth resources, forestry, agriculture (both in space and observing earth features for agricultural benefits). They're going to be doing all different types of astronomy — radio astronomy, optical astronomy, a whole new infra-red and ultra-violet astronomy, areas of the spectrum which aren't even open yet. They're going to be doing things like materials processing — operations where weightlessness is involved in the process, or very large capacity high-vacuum operations. Things that actually directly utilize the characteristics of the space environment. That whole realm of metallurgy and biological research into the effects of weightlessness on living organisms and materials. God, it just opens wide up. Anybody in almost any kind of physical science or engineering I think is going to be involved. The things in which I don't see direct involvement are the social sciences and the arts, at least in terms of government sponsorship. Now, when we get to taking passengers, I hope to hell we have a lot of artists and poets and other people going up.

SB: Here's a datum for you. As you know, our contributors are mostly environmentalists and are mostly against Space Colonies. But our graphic artists are almost universally for them — people like Arthur Okamura, Dean Fleming, Steve Durkee.

Schweickart: Where is Durkee now? Is he still in New Mexico at the Lama Foundation?

SB: Yeah. Do you know him?

Schweickart: Well, I've got a very good friend who's one of his best friends, who runs a thing for the University of New Mexico. It's a sort of a retreat near Taos at the D.H. Lawrence ranch. The Lama Foundation sits right down the road from them, and Al does a lot of sort of free labor for them, advice on mechanical things. Anyway, through Baba Ram Dass [Be Here Now] and Durkee's work with the Lama Foundation putting out the book, I developed an interest, but I've lost track the last few years.

SB: Do you know Ram Dass?

Schweickart: I don't know him personally, but I sure know his stuff. I've been following the guy since he was Richard Alpert at Harvard.

SB: We'll have to try to get you guys together some time.

Schweickart: I'd love to. We have a lot of common interests, I know that. You know, I never really wrote you back a letter and said anything to you, but it's on my tongue right now, and I think the kind of stuff you're doing with CQ is absolutely fantastic.

SB: Thank you!

Schweickart: I'm not trying to pull your head up, but I think very much of the kind of stuff that Bill Thompson's trying to do with Lindisfarne, the kind of thing that you're doing, the Steve Baers, the whole thing that you're wrapped up in there, and I sort of vicariously have one foot into it while the other one is still in the system.

SB: Vicariously, hell. You probably detected me going into interview mode a while ago.

Schweickart: Yeah, I was kind of chuckling about it.

SB: I put a tape on to be sure that I was getting a few things right that you were saying, and realized that I was hearing more good stuff than I've heard in months. What I could do if you're at all interested is work up a transcript of what we've been saying and let you look at it and see how you feel about it, and if it seems publishable, I'd love to do that, because what you're saying adds a whole other level of reality to this discussion. Mostly people are talking in this issue about space colonies as if it were an academic debating point but nothing that's actually going to happen. It's all sort of hypothetical. Some of your discussion I think would help bring it so much right down to cases for people that it would keep that pressure on them that it really is possible.

Schweickart: I have no problem with that. If you think you can make something worthwhile out of it, that's fine. The point you made is very interesting, and it's one that I've seen happen within the space program in the eleven years or so that I've been associated directly with it. Even in something like the Apollo program, or right now the shuttle program, people go along every day spending ten hours a day putting tremendous amounts of energy into the engineering, the design, all the hassles that go into making something like that happen, and they'll go that way for a couple of years and then one day they'll have a trip to California or somewhere where the actual hardware is that they've been messing with and are so familiar with that they can't stand it, and they'll finally get to the point where they actually see the hardware, and all of a sudden this reality sets in. They've been fighting the budget battles before Congress, and they've been doing all the hard engineering trade-offs, and all those things for years, but it's academic even to them until one day suddenly they realize, "By God it's real." People have got to recognize that the Apollo concept when we started out back in 1962-63, with the idea of landing a man on the moon and coming back to earth, that was at least as far out as O'Neill's space colonies. And seven years later in 1969, the 20th of July, a damned good friend of mine named Neil Armstrong put his damned foot on the moon and said here we are.

SB: What was the point of reality producing for you?

Schweickart: In Apollo?

SB: Yeah.

Schweickart: I think it was when I started traveling out to Rockwell in Downey and actually sitting in that spacecraft in the middle of the night. We'd test equipment and go through the development stage for 24 hours a day, seven days a week during a lot of it, and you would literally spend half a day in the spacecraft testing components as they went in and running the whole thing through different kinds of system and sub-system tests. Inevitably in something like that you have short periods of testing and long periods of trouble shooting and holding, and in those periods you basically sit there with a lot of time to think about it. That's useful time, but at 3 o'clock in the morning you tend to just sit there and realize where you are. "Here I am, a kid off a farm in New Jersey, sitting at three o'clock in the morning in the middle of a spaceship, a year and a half away from flying this thing off the earth." You can rap on it, knock knock, and it hurts your knuckles, and it's real.

August 1972, St. Johnsbury, Vermont. At a wedding reception Schweickart flies the rope hanging in the barn door.

The same thing happens to pilots, at least it happens to me when I'm flying. I fly a couple hundred hours a year in high-performance jet aircraft, and most of the time I'm busy getting somewhere, to a meeting, or a symposium or to a speech that I've got to make, or to something, and I'm thinking about the end point, the way you always do, but every once in a while it'll just be a spectacular night, and the milky way will be bright and the zodiacal light will be up, and the shooting stars are out, the whole place is a display. And you get to looking at it, and you say, "God, look at me, here I am flying! I'm a man, a creature with two legs, and here I am with an absolutely unexcelled view of the heavens flying along at 40,000 feet in a jet airplane, on those little bitty wings out there behind me. What a miracle!" It is an exciting exhilarating thing. People do it too seldom. You can do that when you're driving a car, when you're walking, when you're just breathing, but when you're in an unusual circumstance it's a little easier to do.

SB: Is there any chance you'd be flying out here sometimes in February? O'Neill will be in California, and it would be interesting to try and get you and he and Governor Brown together.

Schweickart: I'd like that. Let's see how schedules work out. There is one suggestion I'd like to make to Gerry O'Neill. I sure would like to have the inhabitants of those colonies have a relationship with the cosmos and not just be totally internal, inward looking. Their rate of rotation is considerably slower than the rotating restaurant at Los Angeles Airport, so if you wanted to look out you would see all right. You could design underground restaurants and meditation chambers where you could have a nice meal looking over a railing at the star scape, or meditate sitting on the stars. ■

THE DEBATE SHARPENS

Wendell Berry Angry

Wendell Berry

April 27, 1976

Dear Stewart,

Your promotion of the space colony idea is getting **more** and more irresponsible. Like O'Neill and Vajk, you begin with an air of critical reasonableness, and promptly resort to the glib logic of a salesman. None of you has yet foreseen a problem without at the same time foreseeing a more than adequate answer; indeed, as you represent it, a space colony will be nothing less than a magic machine that will automatically transmute little problems into big solutions. Like utopians before, you envision a clean break with all human precedent: history, heredity, character. Thanks to a grandiose technological scheme, nothing is going to happen from now on that is not going to improve <u>everything</u>; as you say, even if it fails, we will be much better off. You people are operating at about the cultural depth of an oil company public relations expert. All this prophetic-ethical computer-mysticism! What is wrong with it is that it is simply failing to make sense — unless, of course, one is looking at it as a sycophant of science, or from the point of view of a government agency or a corporation. That is exactly what worries me: that your coverage of this issue, whatever you mean it to do, will serve to recruit and train a company of intellectual yahoos to justify the next power-grab by the corporations and the government.

Your dismissal, out of hand, of so many people's objections and doubts — solicited by you — is an alarming display of smugness. It is also insulting. I thought I was being asked to take part in a debate on an issue that you felt to be debatable. I now sense, from the substance and tone of your various remarks in the spring issue, that I was asked to say something that you expected to be inconsiderable in support of what you had already determined would be the losing side.

As you might have expected, I hold this treatment in something less than esteem. Perhaps, I have said to myself, I should just leave my statement in the spring issue as my final words to **The CoEvolution Quarterly**, and say no more. The trouble is that those were not meant as parting words, and I think parting words should be offered and understood as such. Here, then, are some parting words. First, some objections to various statements in your editorial (p. 72):

1. "Either knowledge," you say, meaning either the success or the failure of space colonization, "is a kind of growing up." This assumes that all knowledge is good — which, of course, is not true. It is especially not true of knowledge that depends on practical proof or demonstration. Most people, one hopes, would not consider themselves improved by having killed someone, though, having done so, they would know more about it than before. There is no culture I know of that has not held that good people must refuse to know some things.

2. "If we can learn to successfully manage large complex ecosystems in the Space Colonies, that sophistication could help reverse our destructive practices on Earth." Sophistication, like knowledge, is a subject power, is good or bad according to the use that is made of it. Generally speaking, the more technological sophistication we have attained, the more destructive we have become. I do not think you recognize any of the doubts that now must surround the argument that still more of such sophistication will make us less destructive. It is not sophistication that makes people behave responsibly, but generous purpose and moral restraint. Peasants in Japan 4,000 years ago had these competences of character — and had a technology appropriate to them. This was a kind of sophistication, I think, though very different from the kind you are talking about. It was not inherent in their technical expertise, but in their willingness to live within strict moral-ecological limits. They did not waste anything. There have been other cultures that have had something of this sophistication. Some people in our own culture have something of it now. If such undestructiveness is so clearly possible on earth, by what logic shall we look for it in outer space? How can we expect to discover it by extravagance when its first principle is thrift?

3. You say that what the space colonists consume or destroy outside Earth's atmosphere will be "taken from no one else. They are out of the Earthly 'zero sum game' where one group's gain is another's loss." But do we not live in a universe? Is there no ecology of the heavens? You sound like Columbus taking "possession" of the Indies. I think you are only serving up again in space-jargon the ancient fallacy that we are somehow licensed to misbehave when we are away from home.

4. "The experiment of space colonies endangers only the experimenters." Not so. This continues your deliberate evasion of the fact that this project calls for more government and more government spending. Who do you think is going to pay the bill? And do you think people become more free by having their taxes increased? And what is the military potential of a space colony?

5. "People want to go not because it may be nicer than what they have on Earth but because it will be harder." This is essentially a warmed-over Marine Corps recruitment advertisement — the same irresponsible promise, appealing to the same sad fantasy: "If I could just get out of this nowhere place,

I could be a real man. I could show 'em." Let me point out
to you, Stewart, that we have not yet, in this country, faced
the hardship of the earth. As a people we lack the disciplines
either of character or method to live here without destruction. If some of your would-be space recruits want to be
sure-enough heroes, let them encapsulate themselves on a
strip mine bench and try to make it fit for life by 1990.
Let them "extend the biosphere" to the man-made deserts.

6. Space colonization, you say "employs the same nations,
the same engineers, manufacturers, contractors, etc." as the
arms race. Exactly. And this makes it certain that the worst
characteristics of this society will survive in space colonization. The assumption I am arguing from is that you cannot
escape character; you can only change it by changing its
understanding of its limits. The arms race mentality is
exactly the sort that would most like to re-enfranchise itself
by opening "infinite" sources of energy and materials in
space. That supposed infinity will be a perfect greenhouse
for bad character; look at what mere abundance has already
produced. Good character requires the disciplines of finitude.

7. "The Arms Race is a big bore." I have no understanding
at all of your willingness to be responsible for this statement.
Is space colonization, then, to be a kind of governmental
entertainment for those who are bored with war? "To us,"
said D. H. Lawrence, "heaven switches on daylight, or turns on
the showerbath. We little gods are gods of the machine only.
It is our highest. Our cosmos is a great engine. And we die of
ennui. A subtle dragon stings us in the midst of plenty."

8. "I guess I expect that there will be much more public
participation in Space Colonies than there has been in the
Space Program so far. Debate such as in this issue." Why
has there not been more public participation so far? What
reason is there to expect an increase? And what is the force
of public debate in the face of the economic and political
power of the engineers, manufacturers, and contractors you
were talking about? As you may remember, we carried on
a protracted debate with those people during the Vietnam
War. They stopped somewhat short of having us over
for supper.

9. "Now is the time for NASA to encourage people besides
engineers to get into the act. The program needs administrators who are not afraid of excellent artists . . ." The
fundamental totalitarian impulse is to officialize excellence.
We already have far too little free science because most
scientists are busy "applying" science for the corporations
or the government, which are therefore not afraid of excellent
scientists. Almost any conservation fight will reveal very
quickly whose hand most of the excellent scientists are eating
out of. Now you would like to see excellent artists applying
art for NASA. I hope the excellent artists will have the
decency to remain a little hungry. Excellence in art, and
in science too, requires enough independence so that one
can afford to tell the truth. Economic dependence makes
excellence servile. The government's prophet is always a liar.

I think you have wandered into a trap — one that is
crowded with so many glamorous captives that you think
it is some kind of escape. The trap is that a technological
"solution" on the scale of this one is bound to create a
whole set of new problems, ramifying ahead of foresight.
Here is a pertinent quote from Mr. Vajk, whose essay
you offer with evident approval:

"The popular wisdom currently holds that purely technological fixes are 'bad' because each technological 'solution'
creates five new and different problems. But the reverse
side of the coin is surely just as valid: purely societal fixes
are also 'bad' because each societal 'solution' creates five
new and different problems!

"What it is important to recognize here is that it is not
relevant whether a 'fix' is societal or technological; what

matters is whether or not the consequences of any proposed
program have been carefully thought out, and that steps be
taken to forestall or minimize any adverse effects on the
system and its parts."

Mr. Vajk apparently does not recognize that he is talking
about a condemnation peculiar to his kind of "scientific"
mind. It is a condemnation in two ways: (1) It would
commit us to a policy of technological "progress" as a
perpetual bargaining against "adverse effects." (2) It would
make us perpetually dependent on the "scientific" fore-
telling and control of such effects — something that never
has worked adequately, and that there is no good reason to
believe ever will work adequately. The fact is that when
you overthrow the healthful balance of the relationships
within a system — biological, political, or otherwise — you
start a ramifying sequence of problems (Mr. Vajk's "five" is
a figure of speech) that is not subject to prediction, and that
can be controlled only by the restoration of balance. If we
elect to live by such disruptions then we must resign ourselves
to a life of desperate (and risky) solutions: the alternation of
crisis and "breakthrough" described by E. F. Schumacher.

To say that we can only choose between purely technological
solutions and purely societal solutions is a gross oversimplification, and probably a gross deception as well. To begin with,
the distinction itself is counterfeit: it is impossible to
differentiate between a society and its technology; there is a
mutuality of causation and influence that I do not believe
can be convincingly picked apart. All that can be served by
this distinction is the self-esteem of a specialist who, for
moral convenience, wishes to ignore the social consequences
of mechanical "solutions."

Solutions that are only technological or only social or only
both are necessarily accompanied by adverse effects because
even both together fail to provide an adequate context or
standard of behavior. Mr. Vajk's false distinction between
technology and society rests upon another, implicit
distinction that is equally false: he supposes that the human
considerations of technology and society can somehow be
separated from all of creation that is not human: plants,
animals, soils, waters, climates, regions, continents, the
world, the universe. The universe of systems within systems
survives because it is healthy, it is in balance. Humans
survive within it because — only because — they are, so far,
more healthy than not. When the whole is considered, then
it becomes possible to conceive of solutions of which the
standard is not technological and/or social (wealth, power,
comfort) but ecological or organic (health). When health or
wholeness (not cure) is the standard, then solutions do
not create problems.

I doubt that either of those terms (ecological or organic) is
definitive; I use them for want of a better term. I can better
define what I mean by an example: The flush toilet is a
social-technological solution to a problem: How to get rid
of excrement inoffensively. This solution immediately creates
two problems (soil depletion and water pollution) which call
for solutions (agricultural chemicals and sewage treatment
plants) which create many other problems which call for
more solutions which create even more problems, and so
on and on. I doubt seriously that Mr. Vajk, if he had the
national budget at his disposal, could accurately trace out
and forestall or minimize the adverse effects of the flush
toilet, much less those of space colonization.

By contrast, we have here on our farm an outdoor toilet with
a concrete-block chamber underneath, in which, by the
addition of sawdust and some effort and care, we compost
the excrement of our household and make it fit to return to
the soil. We do not do this, the Lord knows, because we
want to be wealthy, powerful, and comfortable, but because
we want to be healthy, and we know that we cannot be
healthy if our soil is unhealthy. It is an ecological or organic

83

solution. It was not prescribed to us by technology or society, but by a need more comprehensive than both. It is less dependent upon a device than upon an understanding and a discipline. And it does not cause a ramifying series of problems — or of problems and solutions madly leapfrogging over the top of each other. It is a solution that causes a ramifying series of solutions. It withholds contaminants from the water, it enriches the ground, it calls for forethought, moral responsibility, physical exertion — and from those solutions other solutions follow. It begins a process of healing, and healing does not cause a problem; it only incidentally causes a "cure." Healing can properly end only when we are whole, when health joins us to the universe. The whole is a great concordance of solutions.

Such solutions, I think, come from the understanding and acceptance of limits. I do not think they can come from dependence on any kind of <u>quantity</u> to which there is "no perceptible or theoretical end."

But that is enough of picking at details. Let me see if I can give you a more concise statement of my objection. It is this: your thinking (and that of O'Neill, Vajk and Co.) on this matter is demonstrably superficial, and its superficiality slides over a political alignment that I find both morally repugnant and personally threatening. The fact is that you cannot advocate space colonization without implicitly advocating an enlargement of governmental power and the enlargement and enrichment of the corporations. As you are bound to be aware, this project will not be carried out in place of or at the expense of any other government project. It will be <u>added</u> to a budget that is already oppressive. The people will have to pay for it — which is to say that the people of moderate and low income will have to pay for it.

To point the issue more exactly, you are proposing to increase the tax burdens of those of your readers who are struggling to implement in craft shops, in communes, and on small farms ideas and hopes that you have supported. These are marginal enterprises, already threatened with being taxed and priced out of existence. In practical terms, your advocacy of space colonization amounts to a betrayal of these modest settlements of the earth. <u>That</u> is why I intend these to be parting words.

That, and disappointment. Since 1968 I have followed what you did with what, to me, has been a satisfying interest and friendliness — not to mention a steadily growing sense of indebtedness and gratitude. But now you have set yourself up as what I can only look on as a political enemy — not because we do not agree, but because you have now made it plain that you are willing to coerce my support of an undertaking with which I disagree: though you offer me room in your magazine to object, you are nevertheless willing to turn my tax money and my citizenship against me. I cannot be tolerant of that. I am not going to associate myself or my work with coercion.

Wendell

P.S. I shall consider this an open letter. You can publish it or not, as you wish. But I intend to send copies to interested persons. And I will probably publish it in some form — whole or in part — somewhere, sooner or later.

---

1 May, 1976

Dear Wendell,

You know of the admiration and affection I have for you. That is not diminished by your remarks about Space Colonies, but increased. Of all the 76 pages on the subject in the Spring '76 CQ *the statement I most often quote is your, "Humans are destructive in proportion to their supposition of abundance; if they are faced with an infinite abundance, then they will become infinitely destructive." I said it last week to the Deputy Administrator of NASA along with other misgivings having to do with satellite surveillance.*

I'm no astronaut, Wendell, I'm an editor. My function is to raise questions, and shortly after I lucked onto the subject of Space Colonies I realized that they raise a tangle of questions so fundamental as to leave no one uncleaved. They raise doubts, enthusiasms, and passions the like of which I have not seen since I gave $20,000 to an audience of 1500 people to argue over.

Irresponsible? I did not put money or Space technology into this world. They push us around to the exact degree that we ignore them (rise above them). For me responsibility requires that we wade in and see what we really want. Governor Brown has convinced me that there is such a thing as "political truth" and that it is arrived at only by a sustained (trusting) personal dialectic until everyone involved realizes they share a common humanity and participation with life, in whose perspective lives resolution.

I cannot anticipate that resolution, Wendell, ever. Evolution (especially co-evolution) defies prediction. Or control. Or even categorical understanding.

Recently I heard Space Colonies scolded by Elise Boulding as a typical male project. My notion of responsibility required that I mention to her Margaret Mead's 25-year support of Space Colonies, my mother's early and undiminished enthusiasm, and someone you know who probably had more influence on me than any one on this subject. That was Lois Jennings, Ottawa Indian and my wife for six years. When we were courting in the Nation's Capitol she expressed disappointment over one of the early Vanguard satellite misfirings. "I didn't know you were interested in that," I said. What she replied was, "Every one of those shots gets us closer to home."

Home to the tribe? To Washington, D.C.? To a certain star, or the stars? She never said.

To reply to your main concern (Gary Snyder shares it; so do I, less fervently): as near as I can tell, a large centralized government has been sufficient for a Space program, but it is not necessary. The next big project, Spacelab, is being designed and constructed wholly outside the U.S. by a consortium of European nations. Would you favor a world of proliferating smaller governments and occasional major collaborative enterprises such as ocean farming and Space exploration? I would.

I think that nuclear energy can be stopped, Wendell, and should be. I don't think Space exploration can be stopped, but it might be made wiser. I may be wrong about that. I agree with you that a society can loot or it can learn, but not both. I would like to see us learn in Space, as well as here. If we find they're mutually exclusive, then preferably here. Let us carefully watch the incremental steps into Space. It is a new natural history, and natural history, says Bateson, is the correction for piety — including yours, mine and NASA's.

I believe that **The CoEvolution Quarterly** *is useful to the extent that it is impious and acutely observing and a forum. That's why I value your voice in the magazine. Please don't withdraw it.*

I value you, in any case.

Stewart

P.S. For what it's worth, the mail and personal communication indicate that the Spring CQ *raised far more misgivings than faiths among the readership about Space Colonies. (I'm glad about that.) Don't quit while you're ahead. Besides, we've other fish to fry.*

May 8, 1976

Dear Stewart,

I have felt your friendship, and valued it, and returned it. For me, that greatly increases the burden of this argument. I fear, as I fear nothing else, these great projects — all of them — with their accompanying shifts of continental blocks of wealth and their millions of little coercions and oppressions. (It doesn't make any difference who does them; they can't be done without extremely dangerous concentrations of wealth and power.) I have feared them, I think, as long as I have been conscious of them, and in the last ten or twelve years I have spent a lot of energy opposing some of them. The Space Colony proposal, unlike the others, involves a very complicating personal difficulty: it comes wearing the face of a friend. My response to the public issues involved has been constantly troubled by the question of what such an undertaking will do to friendship. For what we are disagreeing about is not how we will run our own lives, but how much power some people may be permitted to have over other people.

This is not an argument that I can enjoy very much, because the personal stakes are too high. I don't want to be unfriendly to you or to be estranged from **The CoEvolution Quarterly**. But I called my previous letter "parting words" because I wanted you to make no mistake about the intensity of my opposition. This is no academic debate. It's not the sort of disagreement that the two sides can resolve in a handshake and "still be friends." I can see it as nothing less than the most audacious scheme so far in the struggle to abolish all small solutions.

But I'm grateful for your letter. I can't overlook its generosity, and I respect your willingness to speak directly to a critic. I therefore venture these further words:

1. I agree that your "function is to raise questions." My complaint is that you have not been raising enough of them, or respecting enough the questions raised by others. I think you have been functioning more as an advocate than as an editor. If you want to be convincing as a question-raiser you will have to quit giving so many one-sided answers.

2. For me, responsibility does not require "that we wade in and see." It requires that we see very clearly before we wade in.

3. To say that "Evolution (especially co-evolution) defies . . . control" is to abandon the whole issue of responsibility. It is to say that anything we can do we must do. I wonder if you really mean this. Again, you have bewildered me.

4. It has never occurred to me that I should be against space colonization only if it can be stopped. I shall be against it even if it can't be stopped.

If **The CoEvolution Quarterly** can be run as a forum for this debate rather than as a mouthpiece for the "winning" side, then this parting can be mended. I hope very much that you and I will have other fish to fry. But it's hard to have an appetite for fish when you've already got a bone stuck in your throat.

Wendell

P.S. If you want to reply to this, feel free.

---

11 May, 1976

Dear Wendell,

Thanks to your response I do feel free. And I invite you to help me inspect the question of freedom and control. I am feeling and thinking and learning increasingly two ways about it as I hear the messages of geographer Carl Sauer and even the most environmentally beneficial activities of NASA. The more we know the more we control. According to Sauer no acre of Earth is unaffected by "the agency of man" — especially the places we have tried to protect. Cultural now and increasingly dominates biological. Recently something has been added to cultural which is the product of intellect — technology in its full range of exquisite to brutal.

Maybe our function is to help maintain the best of biological wisdom and the best of cultural wisdom, and continually sort the exquisite from the brutal technology. The major criterion, I suppose, is balance. Where does freedom enter in all those controlling choices? Maybe through Heinz Von Foerster's guideline — choose the path that leaves the most choices open.

Does that include paths that open new choices — like Space?

Stewart ■

85

'Spaceship Earth' comes home to roost

The U.S. Space Program is rapidly headed for moral and ethical problems it may not be equipped to handle.

With the advent of the Space Shuttle in the early 1980's and the ability to construct very large antennas in near-Earth orbit, the resolution of visibility of the Earth from Space will jump several orders of magnitude. The traffic of communications via satellite will do the same. But what's important is, the kind of visibility and communications will change.

I think I'd best tell this story in the order it came to me from Jesco von Puttkamer, the person at the National Aeronautics and Space Administration responsible for long-range planning. I had just completed the interview with Jacques Cousteau and NASA Deputy Administrator George Low which appears on p. 98 when astronaut Rusty Schweickart said he had another appointment for me, with Jesco. The nature of planning at NASA was improving, he said, and Jesco was in the middle of that change.

Sure enough, in quick outline Von Puttkamer showed me how the customary "extrapolative" planning — figuring your future needs from your present and recent capabilities — was being expanded to include "normative" planning — where you envision distant future goals and then work backward to see what is needed to get there.

The problem with normative planning, he explained, is that you need consensus on which distant goals everybody wants. And that's usually impossible to get. So what Jesco did was put down a list of all the interesting distant visions and then see what intermediate steps they all have in common. That's a bright move, because now you can work toward the intermediate projects without having to decide yet about the long-range stuff. Posterity can work that out as it goes, its freedom to choose preserved.

Also you can see your way around some potential blind alleys that extrapolative planning (he also called it "evolutionary") can get you into. NASA administrators had been planning to build Space Tugs following the Shuttle. After a searching look at Jesco's "relevance tree" they agreed that Space Stations — orbiting construction camps — are the real primary need to get to any of the interesting projects, since they all have to be assembled in orbit.

The correct term for that next step, Jesco felt, is "space industrialization."

I asked how NASA's long-range planning relates to science fiction. Jesco said that he used to write science fiction to pay his way through school. So did Werner Von Braun. "Arthur Clarke was here just the other day. He drops in maybe twice a year. He gets new ideas from here, and we try out our ideas on him. After all he was the inventor of the communications satellite, the synchronous satellite.

"He was interested in some of the new communication initiatives we have in mind — large antennas from 150 feet up to a mile in length. It turns out that with an antenna of that size — taking all the complicated and expensive machinery and technology and putting it in space only once, and relegating the low-cost terminals to Earth use — you can build person-to-person communication. You end up with something like Dick Tracy wristwatches. Just one 150-foot antenna could have 25,000 channels and could serve millions of people.

"The same would apply to navigation. If you have a two-mile cross antenna in Space, you can have two wide beams going over the United States regularly. If you are down on the ground with a little wrist band receiver all you need to do is push a button and as the beams come over you, at the intersection the radio shows you exactly where you are. It can even tell you in what direction and how fast you're moving. You can measure your location to an accuracy of 100 feet, and the whole little system doesn't cost more than $10.

"We have even looked at building large structures to illuminate parts of the Earth with light from space. You can get ten times the intensity of the full moon by having maybe 50 mirrors floating out there. You can save electricity on the ground. You can reduce crime in certain areas by making it bright enough. You could facilitate around-the-clock construction. There are even now investigations going on whether crop-growth can be increased by having illumination from Space.

"If you want to find out how many people are getting across a frontier — maybe between the United States and Mexico, or some other country — you can put a million or so low-cost vibration-sensors into the ground. One big satellite receives the output from these sensors, and a central control station can then immediately see whether anyone has been moving around. So you can have better security.

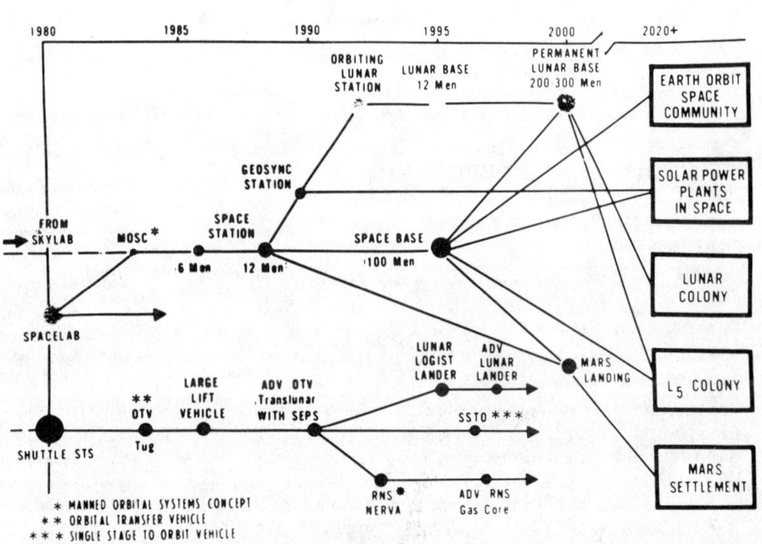

Jesco Von Puttkamer

Von Puttkamer's normative planning working backward from all the possible early 21st century space projects. For any of them a Space Station (construction camp) is needed soon.

NASA illustration of universal personal communication from wrist to wrist via satellite. A whole different <u>kind</u> of society is implied.

"Or package location. Using one two-mile cross antenna you can locate packages all across the United States on a central control board. Ten billion packages tracked, each located to 100 feet every hour."

He went on. Vastly improved air traffic control. Sophisticated radar imaging available to every boat near a coast. The complete replacement of the post office by instant hard-copy

FUTURE NEEDS REQUIRING LARGE STRUCTURES IN SPACE

BORDER SURVEILLANCE SYSTEM	5000 ft. x 10 ft. ANTENNA
COASTAL PASSIVE RADAR	10 N.MI. CROSSED ANTENNA
ASTRONOMICAL TELESCOPE	800 ft. CROSS
ATMOSPHERIC TEMP. PROFILE SOUNDER	10 ft. DIA. ANTENNA
VOTING/POLLING SYSTEM	150 ft. DIA. ANTENNA
NATIONAL INFORMATION SYSTEM	150 ft. DIA. ANTENNA
PERSONAL COMMUNICATIONS	150 ft. DIA. ANTENNA
NUCLEAR ENERGY PLANT IN SPACE	3600 ft. ANTENNA
ENERGY GENERATION PLANT (RTG)	(TO BE DETERMINED)
ENERGY GENERATION - SOLAR/MICROWAVE	7.3 x 2.6 N.MI.
AIRCRAFT BEAM POWERING	15 ft. DIA. MIRRORS (169 TOTAL)
CITY NIGHT ILLUMINATOR	300 ft. DIA. MIRRORS (100 TOTAL)
NIGHT SEARCHLIGHT - LASER	(TO BE DETERMINED)
VEHICLE SPEED CONTROL	120 ft. DIA. ANTENNA
ELECTRONIC MAIL TRANSMISSION	150 ft. DIA. ANTENNA
URBAN-POLICE WRIST RADIO (2-WAY)	150 ft. DIA. ANTENNA
etc.	

A NASA list of the uses of large near-Earth-orbit antennas.

transmission via satellite. The tagging and tracking of all nuclear fuel, to prevent theft and blackmail.

Listening to the lengthening list, I found myself dealing with a case of the horrors. My outlaw nomad was yelling from his hill — no, no fair, too much power to the cops! If there is no place to hide, there is no place to really invent. The Earth one big hydroponic garden? Let me out of here.

I had to flee anyway, to get back to a scheduled lunch with Low, Cousteau, and Schweickart. Over my potato chips and jello I voiced my fears to them. All of that satellite monitoring, I said. With that much information goes too much control. Who is worried about that, who is thinking about limiting it? Who is concerned about protecting privacy or freedom?

Jacques Cousteau said, "If you are doing something that will hurt your neighbor, it should be known. If you have syphilis I think you should wear a sign, 'I have syphilis'."

I said, "I used to smoke marijuana. It was against the law. Are you telling me I should have worn a sign? A lot of inventiveness would have been stopped by that. Too much information, too much control, too much order stops inventiveness. Anything that's newly right starts out by being 'wrong'."

While Schweickart took my side and Low tried to mediate, I privately tried on Cousteau's view. Maybe we do have to become that intimate. God knows the present system I was defending seems hell-bent for disaster. How would it feel

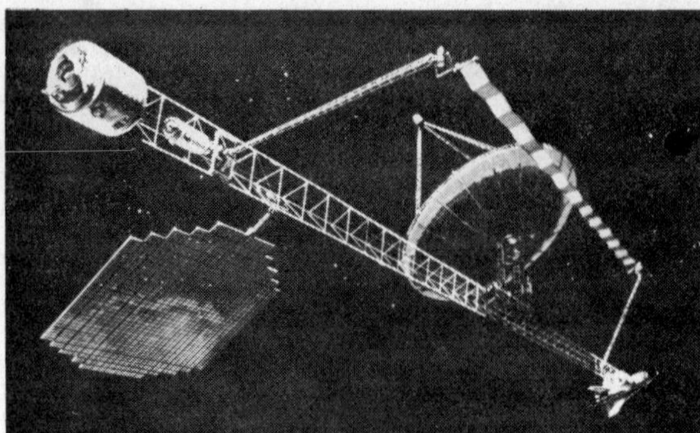

The Space Station. An orbiting construction site using the "strong-back" principle.

knowing everything going on and making new moves only by agreement? Might be a comfort and a wisdom.

The argument was not resolved then. Or yet for me. I continue to admire Cousteau and support NASA. I read with pleasure in National Fisherman *that the new 200-mile limit will be policed by satellite — at last better protection for the fish and fairplay on the seas.*

But who polices the policemen? If technology can have normative planning, how about ethical philosophy? We know what's coming. We don't have to wait for the excesses to occur, such as they already have in the trespasses of computer records. Retroactive regulation is harder.

Robots aren't here yet, but Isaac Asimov admirably anticipated them years ago with the now-famous Three Laws of Robotics:

> 1. A robot may not injure a human being, or, through inaction, allow a human being to come to harm.
> 2. A robot must obey the orders given it by human beings, except where such orders would conflict with the First Law.
> 3. A robot must protect its own existence, as long as such protection does not conflict with the First or Second Law.

Wait, robots are here. The damned "smart" bombs of Vietnam violate the First Law. It is time to legislate the devices of war.

The Congressional vote against the SST, the public vote in California about nuclear energy, the agitation about computer records, the self-curtailment of chromosome surgery — effective debate has come to technology to stay. NASA is no exception.

*But I'm not sure exactly what it is must be preserved from NASA, and what are the specific threats. Marshall McLuhan observed, "Since Sputnik there is no Nature. Nature is an item contained in a man-made environment of satellites and information," (*Culture is Our Business, *1970). And, as reader Ferguson Johnson reminded us, William Burroughs wrote in 1959 (*Naked Lunch*), "Americans have a special horror of giving up control, of letting things happen in their own way without interference. They would like to jump down into their stomachs and digest the food and shovel the shit out."*

Wild freedom, trees falling in the wilderness unheard, the unimpeded health of ecosystems — that's our banner. But I wonder if it's already long gone, like the Earth-centered solar system or the seven days of creation.

Maybe the image of wild freedom is true even if the fact no longer is, and wisdom can still be based upon it.

Nighttime illumination of crops from Space? Filter-proof policing of borders?

I thought at the end of our last issue on Space Colonies that our involvement in debate about NASA was over for a while. It's just begun. Address your comments to CQ, *Box 428, Sausalito, CA 94965.*

—SB

Driving or driven to Space?

Dear Stewart,

Glad to see your fair interview with me in the Summer CQ. I particularly value your concern about moral and ethical problems as may be introduced by the new initiatives offered by space, made possible by economic transportation to and from Earth orbit and our emerging capability to emplace and service large and complex facilities (tools) in the unique space environment. Those new "Public Service Systems" which I touched upon in the interview are no more than unconstrained ideas at this time; undoubtedly there would have to be regulatory control on what their capabilities are used for, but on the other hand their potential benefits to mankind appear truly immense.

I share your concern, and I welcome a debate in CQ, of the rather unprecedented humanistic aspects of what promises to become a most fascinating new frontier, a new era in space. You would contribute a good deal to what I feel is the most urgent need: the "humanizing" of the Space Program.

In reading Wendell Berry's anguished letter against space colonies, I cannot help getting the feeling that some misunderstanding regarding NASA's proposals for the next steps in space may be involved here. A Space Colony Program is not proposed. I would probably share at least some of his concerns if indeed the advocacy of space colonies were that solidly founded and unquestioned as some of the advocates want us to believe. There are snake-oil salesmen at this new frontier just as they were in the Old West. Space colonies may indeed be in our future — let coming generations decide on that — but so may many other "visions" people are dreaming about. Let us keep our options open at this time where we are but at the threshold of new frontiers, and let us plan our next steps to be responsive to mankind's near-term needs and wants while building a solid foundation of ethical responsibility and technological capability from which an open, choiceful long-term future (or futures) becomes accessible. That alone, in my opinion, provides validity to the Space Program.

By doing this one step at a time where each stepping-stone is of practical benefit to man's contemporary needs, the Space Program will pay its own way over and over again. It is a fact that the pursuit of space goals generates innovations in virtually all fields of science and technology, and therefore helps stimulate progress in areas not even remotely connected to the original program. Such high-technology endeavors are strongly anti-inflationary; they result in significant rates of social return out of all proportions to their cost. From secondary applications of space technology alone, each new dollar invested annually on space research and development would presently return $23 over a 10-year period, — that's more than 40% return-on-investment annually over 10 years. How much more can we gain by actually <u>designing</u> the "stepping-stones" to the long-range dreams primarily for near-term Earth-benefitting use! This is what, in your interview, I called "Space Industrialization," a progressive program to provide permanent, practical and commercial utilization/tools of space through products and services that create — in the long run — new values, jobs, and better quality of life for all mankind instead of just for a measly 10,000 living in some space colony. Its funding would gradually shift from the public to the private sector as risks are lowered enough to become acceptable to private industrial entrepreneurs. This does not preclude, but in fact validates, the option that Space Industrialization may grow by natural force into Space Colonization.

What do I mean with "natural force"? Let me speculate.

When some life form stands at the limit of its further development, it receives "instinctive" properties from nature that enable it to surpass these limits. Eternities ago certain types of fish gained the ability to go up on the land and turn into life forms capable of spreading out over the continents. Long long afterward man came and in time gained the ability to build ships and emigrate to new land areas. Life is something that grows constantly, expands in all directions, wants to fill voids and bring order to chaos. The pressure of life forces life forms further. We are experiencing this today

EVOLUTION OF SPACE INDUSTRIALIZATION
Example: Space Processing and Manufacturing of Goods

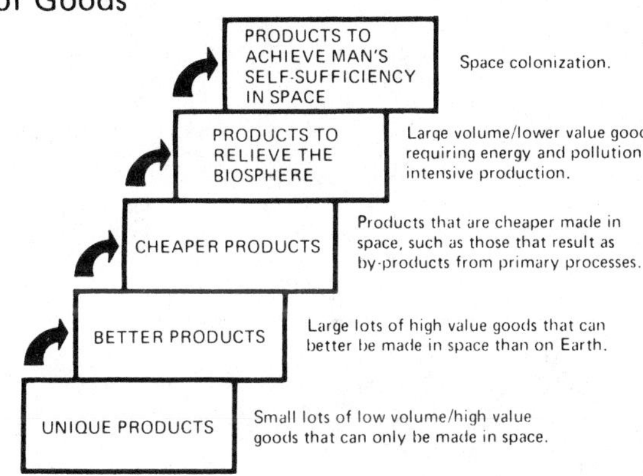

on a different scale. Today we are building space ships. And we have made the first journeys to another world.

Will there then be large-scale emigration from Earth in the distant future?

Yes, undoubtedly. Pressure in that direction will increase, and our capabilities for emigration will also increase. The only risk would be that nature's experiment with man might not succeed. Nature has made mistakes before, created forms that have been unable to survive. The dinosaurs may be an example. Man with his desire for knowledge, with his critical brain, with his self-awareness, is also an experiment that might be a dead end.

Things could go downhill if we begin to doubt ourselves, if we allow developments that are basically instinctive to come to a halt because we do not find any meaning in them. Our civilization could begin to rot from within. In that case nature will make new experiments with other forms, — but probably not on Earth, it may be too late for that. But we can hope that we will fulfill nature's demands.

Are we driving — or driven?

Many years ago, when I entered the space program with tremendous idealism, I used to see this power of nature to spread life as something of a religion — a matter of faith rather than provable evidence. Since it is a driving force upward toward higher goals and higher meaning, some people call this power God but that is an expression which came into existence because of the deficiencies of other explanations. I used to be unable to comprehend that any-one would question the validity of our thrust into the unknown. Today, I'm still no closer to a better understanding. All I see is the experts warning us of population explosion, resources depletion and "limits to growth," and at virtually the same instant in eternity, we have begun to venture into space . . . confound it! that can't be just coincidence!

I appreciate your concern about the questions of moral and ethics, Stewart, but I'm confident that our maturity will be commensurate to their challenges by the time we're there. I appreciate Wendell Berry's outcry, but I submit that the finitude he sees as basic requirements for a good character is a matter of individual viewpoint. It is relative. Man will always carry the disciplines of finitude with him/her, even into the far reaches of space, for it is precisely in the confrontation with the infinity of space that the realization of his own finitude will be thrust upon him.

With warmest regards,

Jesco von Puttkamer
Advanced Programs, Office of Space Flight
National Aeronautics and Space Administration

Satellite surveillance

The vast communications network that ticks off the passage of one person or coyote across a border, that can spot a field of marijuana in a Mexican mountain valley from 40 miles up, that can pick up a conversation whispered 10 blocks away — this network is a police construct. The information, after all, does not come back to you and me. It goes to Authority Central, where it is processed to make sure all movement is in conformance with The Law, which the police powers are authorized to uphold.

The question becomes: How much effort and money are we, as a society, willing to put into the perfection of the police function? Do we really want a border patrol that nabs every desperate farm hand, coke runner and laetrile dealer? Is such perfection desirable? Will it be worth the effort? . . .

If we assume that the earth is in the middle of a process of self-discovery, using the human consciousness (among others) to understand that it is a planet hanging in curved space, then this desire to know "every sparrow that falls" via electronic CNS extensions becomes natural. To take an evolutionary, deep-seated desire for knowing, which can deliver to the earth's total consciousness a sense of place and meaning, and give it over in its full power to police mentalities and authorities of whatever stripe, is to abuse it woefully. In other words, telemetry used to monitor manifestations of biological change (". . . emergence of daphnia in stagnant water in Quebec Province within the past two weeks has resulted in the largest sunfish fry count in history. . .") can lead to earth's improved self-understanding, or used to enforce some dimwit law can lead to increased authoritarianism. Humans are, finally, the bits of earth that leap up from the planet's surface, look around, tell what they see to each other, and die. The sum total of all this seeing and telling is the story of one planet waking up to itself. . . .

The voices of those who would give our best telemetry to the bug scientists and bio-dreamers are crushed under the magnitude of police communications. This is a cop world (as any evening of network television demonstrates so convincingly). But it won't be a cop world for long.

I personally am going to stop it. Cops are going back to loitering on streetcorners waiting to see a crime happen. Our ability to monitor everything is going to be given to scientists who have learned to use intuition, and so have developed a heart that feels for the earth. We will listen, not for the outlaw (who is only wresting himself or herself free) but for the cougar, so we will know how better to avoid it and leave it to its hunting. I personally will bring this about. I don't know how.

Jeff Cox
Organic Gardening
Emmaus, Pennsylvania

Other Voices

. . . As to cities in space — I recant. Earlier on, I sent you a note stating my enthusiasm for O'Neill's space colonies. Since then I've read just about everything on the subject, and have been on O'Neill's newsletter list. I no longer think they are such a great idea. . . . I feel that the colonies will come out looking like the image in Slater's **Earthwalk** — the equivalent of ". . . farting Annie Laurie through a keyhole." . . .

Bud Spurgeon
Austin, Texas

. . . On a space colony we can start nice and clean and fresh and engineer it right. How nice to "build in" recycling from the start, and not have room to accumulate crappy consumer junk. How nice to emphasize elegance and esthetics and bring about a resurgence in craftsmanship, personal creativity, fulfillment, and a true change in consciousness. I almost hate to point out that me and most of my neighbors are doing our damndest to live like that right now on this planet, and succeeding pretty well. Doesn't cost the taxpayers 100 billion either. . . .

Gerald E. Myers
Briceland, CA

... For me, Life's purpose is to become increasingly aware of itself and to evolve into increasingly hostile environments (sea→land→air→space). . . .

Roger C. Girouard
Tallahassee, Florida

... Imagine the artistic and intellectual implications of living in an architectural space where "down" can be seen to be in three different directions at once! . . .

Thomas S. Cooper
Albany, New York

... the whole idea of space colonies is making my brain itch . . .

Linda Sherwood
Brooklyn, New York

... The industries most likely to be used — metallurgy, electronics, etc. — are high capital, high technology, they don't employ many people. So why build a large expense colony? Cheaper to rotate your work crew every 3 - 6 months. That's what they do down here in places like Antartica. Nobody I know has proposed a self-contained colony there. . . .

B. Gold
Hope, British Columbia, Canada

... while Earth (Gaia) can still teach us much about meaning and purpose and rhythm, I think geocentric life ultimately will pale in comparison to the life among the stars. It's a gut feeling. And perhaps, there in the spaces between galaxies, will come the purest wisdom.

jon alexandr
Point Pleasant, Pennsylvania

Life Spreads Before the Sun

Dear Stewart,

Your Spring CQ has stirred up my feelings and random bits of information about space colonization. Thus I am inspired to write a letter when I should be packing for the summer.

A few notes on ecosystems, small is beautiful, and environmental ethics: Working on space colonization has given me perspective on many of these things. They have been great antidotes to the industrial mentality of ignoring side effects and the finiteness of our globe, but they fall down when proposed as absolute, universal ethical principles. Small may or may not be beautiful: good is beautiful, and good varies with the situation. Wendell Berry suggests that hydrogen bombs in space would be monstrous "in any context." In the Universe at large (yes, our universe, part of natural reality) there are places where all the hydrogen bombs on our planet would be like a candle in a forest fire! (None the less, nuclear explosives are not part of space plans, and appear unnecessary.)

I noticed two strong objections to stepping into space among the letters. The first is that space should not be meddled with, lest man hurt it. The second is that space is such a deadly environment that man cannot live there successfully, and was not meant to be there (if god had intended people to live in space, he would have given them the capacity to build habitats there . . .). Space, or at least our corner of it, is a place of dead rock, vacuum, and radiation. No ecosystems to disturb. No groundwater to pollute. No atmosphere to foul. Very few historic landmarks to deface. The worst we can do is disturb it with life. In its way it is hostile, but I wouldn't care to try a temperate zone winter without shelter either. Someone said that man is the only animal that fouls his own nest. The obvious response is that man is the only animal that is trying to grow up without leaving the nest.

On living in space colonies: what is possible is open space, sunlight, growing things, airy tension structures, and clean air. What is needed is imagination. There is little reason to put machinery or industry inside a colony, however much there may be in the vacuum outside. The only kind that comes to mind is farm equipment. If the mad bomber on page 28 tried anything, he might be disappointed. The result would be some broken panels, a stiff breeze a few yards away, and a temporary patch brigade on the scene a few minutes later.

On ecosystems in colonies: try going through everything that has been written on the difficulties of making colony ecosystems work, but replace words with the prefix "eco-" by phrases involving the tending of farms and parks. Two and a half acres per person under ideal farming conditions! If John Todd believes that figure, he must not believe in his own work. The New Alchemy Institute is in many ways taking on a bigger problem than the one we face. Growing fish and algae in the same tank is much like letting the cows in the corn and not pulling weeds; good stuff but all we need is something close to conventional farming.

Pictured here, left to right: 1) Pregnant Saanen dairy goat; 2) Pomegranate bush; 3) Eric Drexler; 4) White crested black polish rooster; 5) Elephant garlic, marjoram; 6) Scrap iron to be recycled. This was taken at the L-5 Society commune in Tucson, AZ.

About phosphorus: yes, the moon contains phosphorus, as do the asteroids. The reason the Earth is in phosphorus trouble is the same reason that goiters were common in the midwest and that much Wyoming rangeland didn't produce until it was scattered with selenium beads. That reason is that we don't have a closed ecosystem on Earth. Unlike the colonies, minerals on Earth end up in the seas; human phosphate supplies dissappear into our 400,000,000 tons of seawater per person, to be recovered through (largely) haphazard, non-ecologically controlled geological processes. Why do animals risk death to reach salt licks? Gaia's blood tends to pool in her feet, but then she is rather old . . .

Radiation (and two-headed calves): a couple of meters of dirt under foot will do the same thing for a colony as the air over our heads, and for the same reason.

To John Holt: I have worked on a dream for six years. I have done most of the calculations and considered most of the design points you mention. I know the people you are calling mad scientists. All I can say in response to your letter is please respect our ability to see the obvious, and please use a trace of common sense and creativity on a problem before proclaiming it serious. The most interesting questions are becoming economic, and are not simple. If you (or anyone else by the way) wish to visit our group at MIT to discuss anything, please feel free.

To those who have looked at the night sky and not seen a black wall, blessings. Talk to your friends. Life proceeds, and spreads before the Sun. Cambridge is turning towards our planet's dawn line, and I must pack.

Eric Drexler
Cambridge, Massachusetts

If the Sun Dies

Philip Morrison, the peerless book reviewer for Scientific American, *suggested this book by handing his copy to me. "Read it."*

He's right. It's far the best book on space exploration, and I suspect it is Fallaci's best work. She addresses herself throughout to her father, who loathes Space, and reaches him through the experiences they shared in the Italian Resistance during the war. Unlike most American reporters she has an unashamed perception of heroism, and she is abundantly dubious of the freeze-dried delights of American culture.

It is thorough journalism. A year on the project (early, before the moon landing), she talked to the scientists, the engineers, the flacks, the astronauts especially, and to Ray Bradbury, who gave her the title. It's clear that the ones who weren't scared of her loved her and said unsayable things. (When I read Wally Schirra's remarks aloud to Rusty Schweickart, his response was "I don't believe it.") Fallaci finds herself liking Von Braun in spite of politics, and adoring Deke Slayton who personally bombed her in Florence and is stricken by that. Violating her every European instinct, Pete Conrad sells her on fast-food culture.

It's the chronicle of a conversion, reluctant, hard-fought, richly perceptive, convincing.

—SB
[Suggested by Philip Morrison]

If the Sun Dies
Oriana Fallaci
1965; 400 pp.

from
Atheneum Publishers
Out of Print
Try your library

"I love the Earth, do you understand? I love the leaves and the birds, the fish and the sea, the snow and the wind! And I love green and blue and all the colors and the smells, and that's all there is, do you understand? That's all we have, and I don't want to lose it on account of your rockets, do you understand?"

You grew white with anger. And your every muscle warned me to be quiet, not to go on with my nonsense. But I couldn't keep quiet any longer: it was as if a war, a gulf, had opened up between us. And I told you, though I don't know if these were my words, that I love the Earth too, Father. It's my home and I love it. But a home you can never leave isn't a home at all, it's a prison, and you have always told me that man isn't made to stay in prison, he's made to escape from it and too bad if he risks getting killed escaping.

•

"Don't pay any attention to people who tell you they have such a wild look because of tension, of exhaustion, of joy at having made it. It's go nothing to do with these things. It's rage at having come back to Earth. As if up there they're not only freed from weight, from the force of gravity, but from desires, affections, passions, ambitions, from the body. Do you know that for months John and Wally and Scott went around looking at the sky? You could speak to them and they didn't answer, you could touch them on the shoulder and they didn't notice: their only contact with the world was a dazed, absent, happy smile. They smiled at everything and everybody, and they were always tripping over things. They kept tripping over things because they never had their eyes on the ground."

"You seem to know them well," I exclaimed.

"Sure I know them well!" she said.

"What did you say your name was?" I persisted.

"I didn't say," she replied, amused. "My name's O'Hara. Dee O'Hara. I am the astronauts' nurse."

•

"We need art as we need dreams," Wally Schirra concluded.

"Dreams? Did you say dreams?"

"Without our dreams we wouldn't be where we are: dreaming of going to other planets, to other solar systems, and finding other Earths, our Earth, among billions of stars."

"Our Earth? Did you say our Earth?"

"Certainly. Because it's our Earth, it'll always be our Earth that we're looking for, it'll always be our Earth that we discover. I don't dream about the Moon. I know enough about the Moon to know how unpleasant and inhospitable it is. There's not one bit of Moon that's worth the Earth or that we could bring back to Earth as a trace of civilization. I don't dream about Mars. I know enough about Mars to know that you can't live there, you can't settle it. Mars and the Moon are two ugly islands. So then, you say, what's the point of going to them? The point is to be able to say I've been there, I've set foot on them and I can go further, to look for beautiful islands. . . ."

•

Wally Schirra: Feeling weightless . . . I don't know, it's so many things together. A feeling of pride, of healthy solitude, of dignified freedom from everything that's dirty, sticky. You feel exquisitely comfortable, that's the word for it, exquisitely. . . You feel comfortable and you feel you have so much energy, such an urge to do things, such ability to do things. And you work well, yes, you think well, you move well, without sweat, without difficulty, as if the biblical curse *In the sweat of thy face and in sorrow* no longer exists. As if you've been born again."

•

We must be ready to meet an intelligence and a justice that are the fruits of different evolutions. For example, we must forget our principle of "Don't treat others as you wouldn't like to be treated yourself" or "Treat others as you would like to be treated" and establish instead a principle that says "Treat them as they would like to be treated." The first thing they want, seeing that they're alive, is to live; so we mustn't kill them, we mustn't land on their territory in such a way as to damage it, we mustn't go there at all if we aren't invited.

•

"Don't you pray?"

Pete Conrad scratched his anchor, revealed his wide-spaced teeth.

"Well, not in swimming pools. Nor in churches either. I mean, I don't go to church and all that. But as for believing, I believe just the same, we all believe, more or less, even those who say: I don't believe in a thing, neither Heaven nor Hell nor anything. But I'd like to see them when . . . Boy! At least three times I've nearly crashed and each time I commended myself to God like crazy. Those damn controls wouldn't be working one goddam bit and I'd pray to God, make them work, God! And you want to know something? I believe God helped me, that He made them work, those goddam controls that weren't working: because I was really going to crash. It's the same, you see, with the Moon. You fool around, you joke about it, but when you think about actually going to the Moon the first thing you do is ask God's help. Then the second thing is you thank Him."

"But what if He doesn't help?"

"Dammit! You thank Him anyway. It's good manners. If I ask you for a match and you don't give it to me, I thank you anyway, don't I? It's good manners. So I ask myself, why should I be polite to you and not to God?"

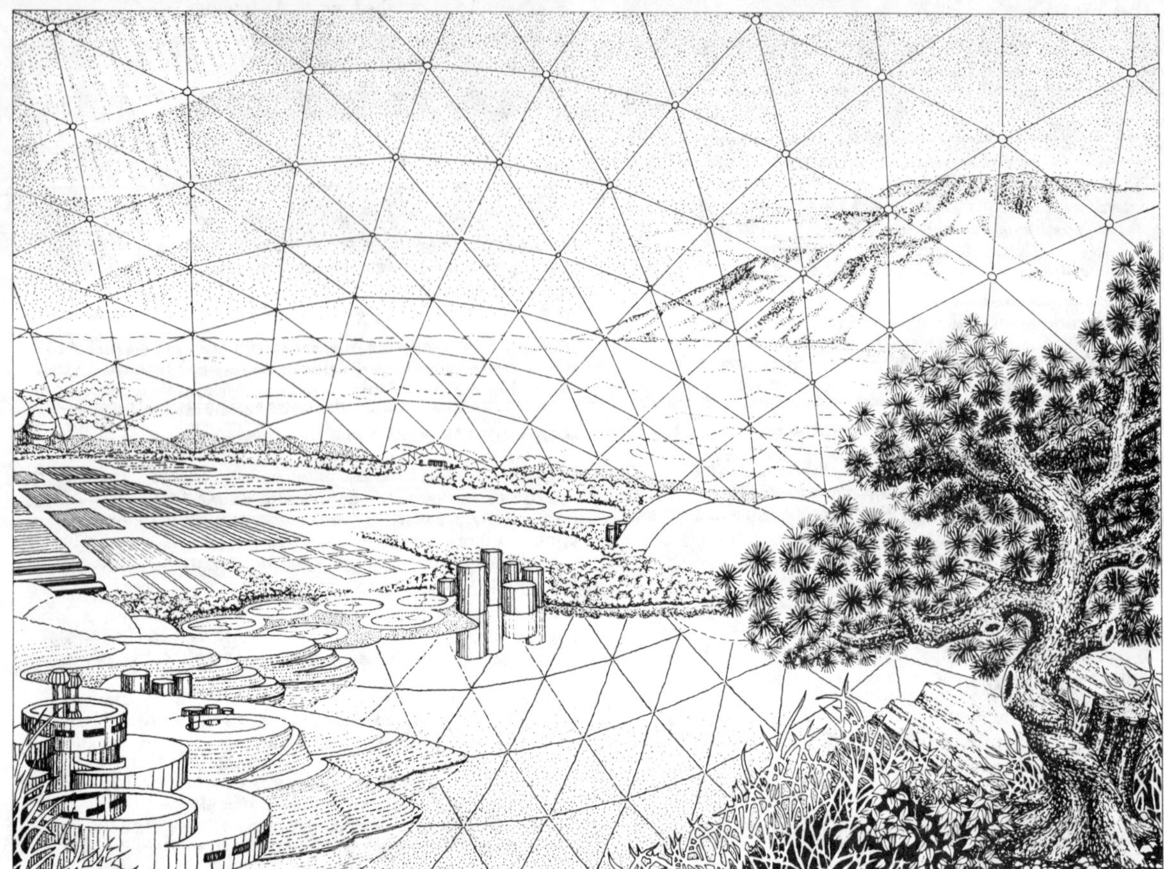

Ecological Considerations for Space Colonies

by
Antonio Ballester, Barcelona, Spain
E.S. Barghoorn, Cambridge, Massachusetts
Daniel B. Botkin, Woods Hole, Massachusetts
James Lovelock, Bowerchalke, Wilts., U.K.
Ramon Margalef, Barcelona, Spain
Lynn Margulis, Boston, Massachusetts
Juan Oro, Barcelona, Spain
Rusty Schweickart, Washington, D.C.
David Smith, Bristol, U.K.
T. Swain, Boston, Massachusetts
John Todd, Hatchville, Massachusetts
Nancy Todd, Hatchville, Massachusetts
George M. Woodwell, Woods Hole, Massachusetts

There appears to be growing interest in the possibility of establishing large space colonies capable of supporting hundreds or thousands of people in isolation from the earth for long periods (for example, see G. O'Neill, **Co-Evolution Quarterly**, Spring, 1976). Such colonies would present extremeley difficult biological and ecological problems. These should be addressed at the very outset if any serious effort toward designing satellites or colonies on celestial bodies other than the earth is to proceed. This statement is the product of a series of discussions held in Woods Hole on May 14 and subsequently with scientists who have had experience in the study or design of closed ecological systems and who express concern lest the problems of developing congenial livable conditions on artificial or natural satellites be considered engineering problems rather than basic humanistic and biological ones.

The proposal is considered a logical and exciting extension of our space exploration and pioneering. It is to build a new meta-stable ecosystem, complete with biotic resources and closed cycles for other essential resources, and capable of supporting man over long periods. No such system has ever been constructed on earth. The probability that such a system can be built and maintained indefinitely at present seems remote. It seems especially remote when we realize that we have no background in the analysis of the problem and no technical or scientific research programs underway at present to develop the background. One of the current theses of economists and technologists is that economic and industrial growth is necessary for support of contemporary

industrialized societies. As the earth becomes more crowded
and its resources depleted the options for growth become
progressively circumscribed. Yet problems of expansion into
space also seem nearly unsurmountable. People from industrialized societies might not accept the limitations on their
activities that life in such a colony would require: limitations
on ownership of private property, procreation, travel, and
psychological adjustments in interaction with others. Nonetheless, the question of space colonization should be explored.

Our experience leads us to believe that the greatest probability
of success in establishing a space colony that might remain
for as long as 25 years would be derived by adapting a large
natural ecosystem on earth to the support of man. By large
we mean in addition to surface area, a high diversity and
high degree of patchiness. The versatility of forested areas,
subtropical savannahs, and coastal environments and the
success of man in adapting such ecosystems to human
support leads us to the conclusion that the focus should be on
such areas. With this background we would advance a very
preliminary list of issues as worthy of immediate consideration before further commitments of funds or time are made.

1. Do all space colonizations intrinsically depend on
   exploitation of earth resources or, at least in
   principle, can independent and eventually productive
   ecosystems be established anywhere in the
   solar system?

2. A viable space station that has the possibility of being
   stable in its basic biotic structure over a period of 25
   years would probably require an area of several square
   miles. The total amount of primary productivity, its
   diversity, and its distribution between food and fiber
   and the stabilization of other resources and potenti-
   ality for recycling will obviously determine the upper
   limit of the number of people that might be sustained.
   Do we in principle have the technological capabilities
   to launch or establish such large colonies?

3. Decisions would have to be made as to the minimal
   volumes required to sustain segments of the ecosystem
   and minimal volumes to sustain individual populations
   within it. The knowledge of minimal volumes is in
   its infancy and the relationship between stability and
   minimal area has been established only in the most
   rudimentary ways. In general larger more diverse
   ecosystems are more stable; if stable and productive
   closed ecosystems could not be made to function on
   Earth they certainly would not function in orbit.

4. The distribution of major mineral nutrient elements is
   a major problem. All organisms on earth require six
   major elements in abundance: carbon, hydrogen,
   nitrogen, phosphorus, sulfur and oxygen. There are an
   additional six to ten that seem to be universally
   required and there are probably 30 more that are
   required in special circumstances by certain organisms.
   Almost any ecosystem would contain these elements,
   yet their proper delivery and removal, their cycling, is
   a serious challenge. Elements must be delivered in the
   optimal quantities and proportions and in appropriate
   chemical form. Often they are required as gases. We
   must be concerned with the biological and geological
   and man-made informational systems that distribute
   essential elements.

5. Most productive forests, mixed agricultural and coastal
   ecosystems are attuned to seasonality, which implies
   appropriate variations in precipitation and temperature.
   One must recognize at the outset that natural ecosystems will not remain in their initial condition but
   will change by both succession and evolution. The

design of the allowable maximum and minimum ranges
of variation of ambient conditions in an orbiting space
colony may be a serious problem.

6. Because the major source of energy to drive this basic
   unit of nature is solar, the question of the spectral
   distribution of incident energy, the quantity and
   periodicity of electromagnetic radiation at each wave-
   length, must be considered and controlled.

7. It will be essential that water be recirculated within the
   colony, delivered to the vegetation at appropriate
   intervals, collected, used in irrigation and for other
   purposes. We recognize the desirability of something
   similar to a coastal system with both terrestrial and
   aquatic components.

8. We must consider the choice of an agricultural base
   that will tend to maximize photosynthesis and provide
   a stable supply of food.

9. The stabilization of the basic ecosystem will require
   redundancy in its diverse systems. For example, often
   interaction between populations of organisms ranks
   more important for the stability of an ecosystem than
   the activity of single organisms. Very little is known
   about this. How many different types, for example, of
   nitrogen-fixing organisms should be present? What organic systems and mechanisms for chemical conversion
   of nitrogen would be required to allow for failures?

10. Eventually, the psychological and sociological
    challenge of confining people to small places for much
    of their lives must be considered. The potential for
    malfunction of small ecosystems seems large: the
    probability of disease microbes, fungal infections of
    plants and other disasters must be estimatable.

11. For a rational space strategy the order of approach to
    the problem is critical. The basic ecological problems
    must be attacked on earth first, then in near space
    orbit. Only after successful self-contained systems
    have proved feasible on earth for some time ought
    such systems be launched.

This list of issues is obviously incomplete. The challenges of
space are much greater than the same challenges on the
surface of the earth. Until there is evidence that we are
capable of meeting these challenges on earth, there is little
point in attempting to address them in space. If a serious
effort is to be made in the development of livable space
colonies, research should be started on the design and
construction of closed agro-industrial ecosystems on earth.
The experimental systems might be two to ten times larger
than those expected ultimately to be used in a space colony.
This extra size might ease the problem of stabilizing the
system over longer periods. Earth based and preliminary
orbital experiments should include both aquatic and terrestrial components and should be conducted in an atmosphere
simulating in its main features that of the earth. The
experiments might use technological approaches in partial
recirculation of gases and water. As experience is gained,
there is the hope that the techniques developed might be
used in the design of cities and in the modernization of
existing cities.

Although the probability of developing space colonies appears
to be extremely remote, we would suggest that the scientific
challenges and the magnificent opportunity for exploration
intrinsic in that objective are appropriate to the contemporary
world. We would agree with others that the time to explore
these issues is now and that the answers may prove to be not
only exciting but a great boon to earthbound man and to
earth-based problems in human ecology. ∎

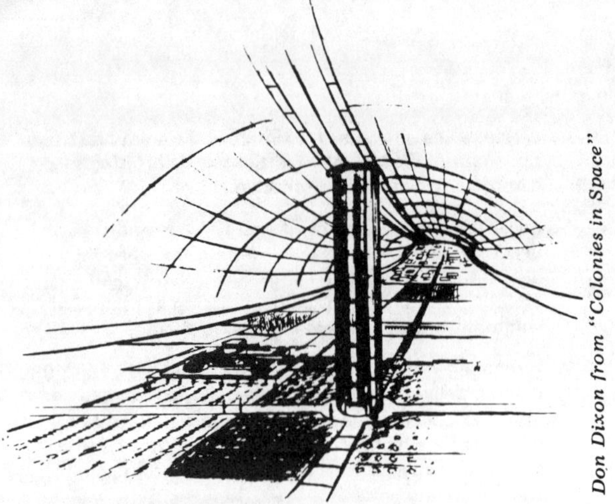

Space agriculture and Space cops

The correspondent is T.A. Heppenheimer, author of Colonies in Space. *(See p. 129 for review.)*

I think I can define the core of our disagreement, or our reason for debate. In the **CQ**, you are accustomed to thinking in terms of natural systems, such as those of the New Alchemists, as promoting healthier human futures. You are accustomed to dealing with the question, "how can we make agriculture more nearly natural and organic, while still preserving reasonable yields?" You want your agriculture to not be energy intensive, or to require the support of advanced technology.

But in space colonization, we take a rather different approach. We want to maximize the yields, period. If we can do it with no advanced technology, fine; if with some advanced technology, fine; if with a lot of advanced technology, also fine. We expect to have at hand plenty of energy and lots of advanced methods, and in practice, the agricultural concepts showing greatest promise appear to be a super-hothouse system which is quite different from the Alkies' work. Instead of a complex interdependent organic web, we appear to envision a sequence of growing plants or animals, tied together with industrial chemical systems which are controllable with a computer. The CO_2, the oxygen, the fertilizer, the industrial processing of wastes — all these key features of the problem involve industrial plants rather than green plants.

As one who loves his Granola and his fruit straight from the tree, I can imagine your feelings at such a prospect. But in a larger sense, it represents a response to the space colony situation which is no less valid than the work of the Alkies in the context of their situation. The space colony will be heavily dependent upon advanced technology, rich in energy, rich in the materials which can be obtained from the Moon. Any system of agriculture, to be valid, must reflect these features.

The real question, then, is whether the humanness of the colonists is served or defeated by such a system. I would argue that it will be served. In our modern society, few need labor merely to win the ordinary creature comforts of life; our industrial technique makes it possible for most people to be free from constant concern for life's necessities. In that sense, they have the opportunity (whether they use it or not) to pursue life-styles which will enrich their sense of life. In the space colony, agricultural technology can easily produce such an abundance of food and other necessities as to leave the colonists entirely free from fear that they will come to mishaps. They thus can live in freedom and concern themselves with the possibilities of the High Frontier, rather than ceaselessly worrying that someone, by breaking the rules, will bring disaster to all.

In all this, you can clearly see my unspoken assumptions and attitudes toward technology, industrialization, and the like. Whether or not you share these with me, this is my point of view; and I would welcome an opportunity to present more material (specifically, the chapters in my book on the colony's agriculture, architecture, and social organizations) in **CQ**.

•

Dammit, Tom, you continue to miss my point and the point of this whole magazine. I got in an uproar when you said, "Agriculture in Space will be no different than a steel mill" because it's such an input-output abiological statement. Aesthetics aside (but not far, because there's no surer sign when you're off the path), no organism will be as docile as iron ore — not algae or bacteria or hybrid corn. They are complex systems with ideas of their own and require complex systems around them to stay alive. The level of complexity required is still outside human knowledge much less computer regulatability. No one has successfully maintained a terrarium or other isolated organic system for very long. Hydroponic greenhouses, etc. are richly linked with the rest of the Earth's biosphere. Ecosystem management is still a speculational science, like exobiology. What we do know about monocrops is that when they go down, they crash. And you're back to Space Sticks for food.

Anyway, continue . . .   *—SB*

. . . We were talking on the phone about the problem of agriculture in space colonies, and the implications for the colony's internal policies. That is: would it be regarded as a robust system, difficult to damage, and hence compatible with human freedom and ordinary human cussedness? Or would it be regarded as being in a delicate balance, requiring close control over the populace to keep them from fucking it up?

On to my comments about "Apocalypse Juggernaut, hello."

The Aerospace Corporation is one of those military-industrial places around Los Angeles which grew up during the 1960's. It is distinguished by its display of full-size models of Air Force missiles, out in front; and these indicate its purpose for existence. Its main job is to provide advice to the military on new weapons systems. Once in a while, though, it does work for other agencies, such as NASA.

And so it was that some of the people at Aerospace recently carried out a study on the possible uses for large antennas in Earth orbit. Some of their suggestions were useful and valuable: wrist radios linked by satellite, electronic transmission of some types of mail, checking the location of packages of nuclear fuel. But quite a lot of their ideas gave Stewart Brand the unhappy feeling that 1984 might arrive right on schedule. These included: border surveillance; improved police communications; satellite monitoring of practically everything. So he wrote of his worries, in the article entitled "Apocalypse Juggernaut, hello."

Well, this is the kind of large-scale thinking that needs to be seen in perspective. True, the Aerospace proposals would envision a society very different than that of the U.S. today. But the significant thing is not that Aerospace has put forth its proposals. One can hardly expect them to have done otherwise. The significant thing is that even at this early stage, when the proposals are merely suggestions in a report, they are already being opened to the scrutiny and criticism of the public. It is no new thing that an Air Force contractor should propose to use space technology to turn the U.S. into a police state. It is a new thing that magazines like **CQ** should learn of these ideas, and criticize them, even before agencies of government can take the first steps to put them into practice.

After all, these suggestions are not official Federal policy, signed, sealed, and ready to roll. They are ideas in a report, and there is not a one of them which will ever fly without considerable support from the public. (Remember the SST.)

So let's look at Aerospace's laundry list and try to sort out the genuinely good ideas from the genuinely dumb ones. What uses for large satellites can genuinely help us?

The wrist-radio idea is an intriguing one. Even if it is literally possible to put a two-way radio in a wrist-size package, the tinny sound of the loudspeaker and the poor performance of its tiny microphone might well make it unpopular. But what is important here is the idea of making telephones portable, like CB radios. Imagine that your telephone, instead of being tied by wires into the Bell System, actually just runs off the wall current and communicates with a little antenna to a big satellite overhead. You could take it off the wall, power it from batteries, run it off the cigarette-lighter socket in your car.

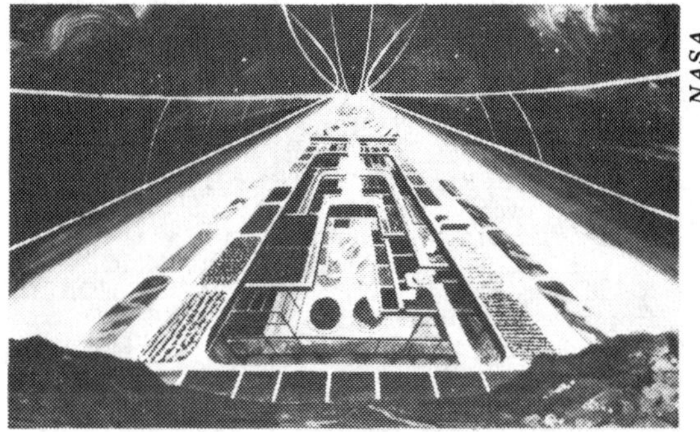

All phone calls would go through the satellite, so a call half-way round the world would cost you no more than one next door. With thousands of channels on the satellite, any caller would have the convenience of today's system; there would be no need to call "Breaker, breaker" as in CB, with everyone listening in. The portable telephone would be as major a change as the hand-held computer; and it would probably be something with lots of benefits and few disadvantages. Already much long-distance phoning is carried by microwaves, beamed from tower to tower; this new development would merely complete that process.

But the business of automatically polling the public, or having them vote on issues through their portable phones, does not exactly sound like a good idea. If you want to arrange for Congressmen to sit in their offices and take calls, for free, from their constituents — that's one thing. It would sure beat having to have them go home every weekend. But to have the population polled — "Shall the tarriff bill pass? Shall Title VI of the Pickle-Pepper bill be adopted" — could only lead to a dangerous oversimplification of the issues. With all its faults, as Edmund Burke pointed out two centuries ago, there is much to be said for a system in which the public laws are written by representatives who maintain a general awareness of the public will, yet who have a necessary degree of independence of judgment.

The idea of locating things by satellite could be useful. Nuclear fuel, railroad cars, trucks of a nationwide road outfit, packages in transit — all would be very useful to be able to track and locate. Homeowners could put little transmitters inside their cars, or other valuable items, and these would be easily tracked if stolen. The problem is that if you can track the car that is stolen, you can also track the car that belongs to the man whose politics you don't like. But there is a solution. The tracking would be keyed to a code, a special string of numbers broadcast by the little transmitter. So what you want is to let the code be chosen by the user, and known only to him, and that it can be changed at will. Then, the police might track a certain code but would have no way of knowing whose it is — until the owner comes and tells them that it identifies his car which has been stolen.

Finally, there are some ideas which are just plain silly. The idea of making our borders super-tight (an electronic Berlin Wall?) is one. Apart from the fact that such an idea isn't exactly sporting — wetbacks are human, too — there's the question of whether this isn't the same old idiocy from the same geniuses in the Pentagon who gave us the "people sniffers" for Vietnam. The Vietnamese people foiled these by hanging bags of urine or feces near them. If Aerospace is proposing a new type of detector to discover the sound of footfalls, then it's easy to predict the Border Patrol will spend a lot of time chasing jackrabbits and coyotes.

In his article, Stewart Brand raised the question, "All that satellite monitoring means too much control; who is worried about that, who is thinking about limiting it? Who is concerned about protecting privacy or freedom?"

His question gives its own answer. He is concerned; so are his readers; so, in turn, will be much of the public when such matters are brought forth into open debate. It is not new that people should propose to regiment or survey the public at very intimate levels of their lives. What is new is the unprecedented public discussion such proposals now receive, even at the earliest stages of their genesis.

T. A. Heppenheimer
Center for Space Science
Fountain Valley, California

Space agriculture retort

I enjoyed the Fall '76 CQ. I notice that Holt and Heppenheimer are at it again. At least Holt is beginning to develop a degree of humility about his expertise in the fields of physics and electrical engineering. (By the way, O'Neill's group is studying linear synchronous, not linear induction motors for the transport linear accelerator.) Planetary scientist Heppenheimer, however, is more certain than ever that he knows all about agriculture in space!

As the unwitting soul who turned Tom Heppenheimer on to the references he used for his chapter on space farming, it looks like it's my time to speak up. It may be true that a closed cycle agriculture could be set up using non-biological techniques to recycle liquid, solid and gaseous wastes. But I have not been able to find any reports of such a system having been put into operation for an extended period. Whether or not such a system will work is open to speculation. However, I also object to those who state *a priori* that living ecosystems could not possibly be held together with plumbing and chemicals. Why can't the debaters on both sides just admit to not knowing (yet)?

The University of Arizona has an interdisciplinary team, under under the direction of systems engineering professor Wayne Wymore which has tackled the "space farm" problem. In Wayne's eyes, the worst possible sin is to rush on to solutions before you know what the problems are. Our team is considering problems such as: what happens when there is a fire in the colony (the smoke won't just "go away"!) If someone dumps a lead battery in the trash, will people be eating it in the form of spinach 3 months later? What about the effects of spray deodorants? (The Navy banned aerosol cans from their nuclear submarines — which are semi-closed environments — long before anyone else noticed they were bad news!

Of course, no amount of library research and mouth flapping and typewriter key pounding will uncover all problems.

NASA's Phillip Quatrone, Chief of the Environmental Control Branch, has suggested that we let the astronauts of the mid-eighties grow vegetables in their space base. Those first lettuce and tomato plants can tell us how well they do or don't like their habitat far more eloquently than any assemblage of ecologists or chemical engineers.

After all, life didn't leap from the seas and blanket the land overnight. It was a gradual process. I imagine the "greening" of space will be similar, with the forests and fields of the O'Neill style settlements evolving from the house plants and vegetable gardens of the early space bases.

Now, to get back to the subject of Holt: I was not entirely surprised by his belief that O'Neill thinks space colonies will solve all problems. It reminds me of a conference this summer where O'Neill was being grilled by a panel of professors and NASA officials. One professor smugly announced he was going to be the bad guy and shoot O'Neill down. He read aloud an article he said was written by O'Neill. The article claimed that it had been proven that space colonies would solve the energy crisis, bring world peace, cure dandruff, etc. Finishing, he then demanded of O'Neill how he could have written such an atrocity.

"But I didn't write that article," Gerry replied mildly.

Gerard O'Neill has never claimed anything more outrageous than that we ought to consider whether space colonies might solve some of our problems. In spite of this rather unmessiah-like stance, there are those who believe Gerry is greater than Jesus and Moses combined and are desperate to touch the hem of his robe. When O'Neill catches them at it he banishes them to Siberia. . . .

—Carolyn Meinel Henson
L5 Society
Tucson, Arizona

95

The smile of Timothy Leary

Timothy Leary - I have always been an admirer of him - although bored by most of his books except Psychedelic Prayers and Neurologic - a racy quality about him, a recklessness, his breathless LIFE IN THE FAST LANE attitude, the rascal cum scoundrel image has a certain romantic attraction if you don't go into the details (and romance never does) - he once said to Baba Ram Dass, "it seems to me as the years go by that you are getting gooder and gooder and I'm getting badder and badder" -

ANYWAY, this time around Leary, the HOUDINI of the HALLUCINOGENIC set and setting, the I AM NOT A CROOK of the counter-culture, has escapologised himself into yet another day-glo corner - FRONT MAN for NASA and AMERIKAN INDUSTRIAL HARDWARE, SALESMAN for SPACE-CARS and SPACED-OUT ORBITING BIONIC PLANETOIDS where Leary (talking faster, slowing down) envisages a continuance of the ALL-AMERICAN ALL OUT CONSUMPTION AS THE ROAD TO HAPPINESS DREAM-MARE - SMALL is not BEAUTIFUL in Leary's eyes, the ECOLOGY movement is JUST a piece of shit, limitation on consumption is a limitation on vision and expansion of consciousness, and seeing the EARTH is now too small for its population, it is time to MIGRATE to space - MIGRATION is MUTATION, the EARTH is merely the COCOON, we are CATERPILLERS who now are ready to METAMORPHOSE into BUTTERFLIES and fly off into SPACE - although SPACE to Leary now seems only to mean these orbiting colonies, perfect Yosemites up there beyond the constricting laws of gravity, using current technology. No STAR TREK DREAMS, NO TIME TRAVEL, NO MORE JOURNEYS TO THE CENTER OF THE HEAD, just EASY LIVING with the ELITE in the ULTIMATE IN SUBURBAN LIVING, the SWIMMING POOL in SPACE, and he PROMISED we Marin Mutants, successful survivors of the sixties, if not immortality, at least RESORT-VISITS for gravity-free golf within the next twenty five years. Leary is a SMARTIST (smart was his favourite word for the evening) a man in a HURRY, particularly for someone who is into IMMORALITY - you would think he had ALL THE TIME IN THE WORLD, and advocate of the POLITICS OF ABUNDANCE, NO FOOL (just slightly ridiculous), a man with a PATTER and a PLETHORA OF POSES, including lots of COLLEGE JOCK HANDWAVING (they looked like stop-signals) and CLASSY GURU GRINNING (slightly crooked these days) and PAUSING FOR APPLAUSE, a STAND-UP COMEDIAN and with the sort of lecture he gave on Friday he would be PERFECT at the CIRCUS-CIRCUS or CAESARS PALACE in LAS VEGAS. He certainly was ENTERTAINING, skittering along on the surface of the ultimate topics, hinting at DEPTH and SECRETS but not delivering anything of any substance, a GAD-FLY with the gift of the GAB... and I could go on and on I'm afraid but I will stop and mention that I managed to struggle downtown last night to see MIRACLE IN MILAN which is a real HEART and FOOT WARMER, made by Vittorio de Sica in 1950, a fantasy about the PEOPLE gettting the better of the PROPERTY DEVELOPERS - at the end, incidentally, the PEOPLE all go off into outer space, leaving the earth and Milan to the cigar smoking top hatted industrialists - Maybe there is something in SPACE MIGRATION after all.

Jim Anderson   *Nameless Hearsay News*

Space just means bigger weapons

...Clausewitz once said something to the effect that war was too important to leave to the generals. The point I am trying to make in my communications to you and in the book which I will write is that science is too important to leave to the scientists.

One reason why science is too important to leave to the scientists is that the scientists have shown that they are not very smart, far-seeing, or imaginative in guessing the kinds of uses that the military will make of their Magic Toys. An example is the laser, which, as you may recall, when it first appeared was called the Maser. As soon as I read the first descriptions of this device, I thought to myself, "At last we have the death ray that the science fiction people have been talking about for so many years." But not for at least five years afterward did I ever read or hear of any people in the scientific community itself suggesting that this new invention might be used for such a purpose. It is only in the last year, and perfectly predictably, that military research on lasers has begun to come out of the closet. Both in this country and in Russia military scientists are working round the clock to get this thing perfected....

...In addition to anything else that might be said against it, most of so-called peaceful space research is simply military research wearing a smile button. It is a hawk in dove's clothing.

I can just hear the L-5 boys (and girls too, I suppose) saying, sometime in the future, as nuclear powered laser carrying space ships start cruising around out there, "But we never thought anything like this would happen."

I am not much impressed with the rhapsodizing and poetic thoughts of the first astronauts. The first people who flew above the Earth in airplanes used to talk very much the same way, about a newer and purer world, about being nearer to God, about realizing how petty and unimportant were those quarrels and so forth going on below. The fact is, in a world that has not learned to make peace, whatever has war-making possibilities will be used for war making purposes. To suppose that space is going to reform, all by itself, the character of human beings reminds me of nothing so much as those fatuous political commentators who thought that the White House would reform the character of Richard Nixon. If there is military advantage to be had in space, and there clearly is, then nations seeking for military advantage will try to get it there. If there is wealth to be had in space, which for the moment I doubt, then nations which are battling each other for the Earth's wealth will battle for wealth out there.

John Holt   Boston, Massachusetts

John, one out of every five space shuttle flights is expected to be for military use.   —SB

8 am of the world

Keith Hensen of the L-5 Society sent us a talk by Peter Vajk that has some interesting points on the Space question. The talk was given at Southern Illinois University in April, 1976.

Let's consider again those remarkable Apollo photographs of the Earth. I have already remarked on the new perception of the Earth and the biosphere which these pictures have produced, but I want to call your attention to another new perception which is latent in these pictures. Once you see this, you can never again fail to see it.

The unity of Earth is prominent in this picture. But what's in the OTHER half of the picture? IT'S FULL OF SUNLIGHT! Space is not a pitch black void — every cubic meter is filled with solar energy. The next time you look at the Moon on a dark night, you will SEE that all the surrounding space is likewise illuminated. . . .

The old analogy in which the 5 billion years of the Earth's history to date is compressed into a single year, with all of recorded history happening in the last five minutes before midnight on December 31st, is perhaps deceptive, in that it suggests that we are rapidly approaching some imminent endpoint. If instead we take the past 5 billion years of the Earth's history together with the 10 billion years or so in the future that the Sun is expected to last, permitting survival of the Earth as a planet, and if we compress that 15 billion years into a single day, then it is now 8 o'clock in the morning, the biosphere has just woken up, and the workday is only beginning! . . .

The statement "We have only one Earth" is radically different in its implications from the assertion "We have only the Earth." The former recognizes the uniqueness and sacredness of the Earth and its biosphere among the other heavenly bodies; the latter denies the existence, for any practical purpose whatsoever, of the rest of the universe, and it seems, has now become as obsolete and outdated as the pre-Copernican view of the Earth as the Center of the Universe.

Limits to Growth - wronger than ever

Peter Vajk is one of the more thorough, original, and enthusiastic analysts of Space Colony effects. In the Spring '76 CQ he wrote a piece called "Space Colonies, Ethics and People" which applied Limits-to-Growth computer modelling to the subject. Here is his 1977 update of that.
—SB

At one time, it seemed that exhaustion of non-renewable resources, especially metals, would result in the collapse of the human socioeconomic system within a few decades. If that were true, then we would have only a brief "window" in history during which we could take the extraterrestrial option O'Neill showed us was possible. Industrial, technological society would have to reach a large enough, sufficiently sophisticated state to build a necessary beachhead in space, but could not long defer that option or our descendants would forever after be trapped on this single planet, deprived by our inaction of a choice they might make differently than we had. If the principal assumptions incorporated in such computer models as had been used in the study **The Limits of Growth** were true, it would be irresponsible not to start in immediately on a large-scale program to establish civilization in space.

In my original article in the CQ, I expressed serious misgivings about the validity of such computer models and about their basic assumptions. My own computer modeling was this, in a sense, a parody of Forrester and of Meadows *et al.* which serves to show that such models can be used to "prove" any conclusion you like. (In fact, I had to restrain myself from making the answers come out even a hundred times better yet.)

The basic assumption of the M.I.T. models which produces collapse is that non-renewable resources are in imminent danger of depletion. In the last year and a half, this entire question has been carefully reconsidered by many different groups, and it is now clear that industrial, technologically advanced civilization here on Earth could be sustained, at a North American standard of living, for a population of 10 to 20 billion people, for as much as several million years. (H.E. Goeller and Alvin M. Weinberg spell this out in some detail in an article "The Age of Substitutability", **Science 191, 683,** February 20, 1976). In the age of technological substitutability of virtually inexhaustible materials for scarce materials, society would be based on wood, glass, stone, cement, plastics, iron, aluminum, and magnesium. The energy costs associated with going to lower grade ores would not exceed a two-fold increase per ton of finished metal above present costs.

The space options have, then, three long-term motivations: (1) global "insurance" against a major calamity, whether of natural or artificial origins; (2) removal of adverse environmental impacts of industrial activities from the biosphere; and (3) satisfaction of the very human need to explore. These motivations will not justify the expenditure of tens of billions of dollars for the establishment of space manufacturing facilities and lunar or asteroidal mining equipment.

But a number of short-term justifications will, I believe, bring us to the same point within two or three decades all the same. These include (1) short-term profitability of increasingly large communications satellite systems; (2) international development programs based on Earth-observation satellites such as Landsat; (3) high profitability of new, unique, or improved products possible in the space environment during the 1980's with the Space Shuttle; or (4) political decisions to shift, with the Soviets, military spending into space programs. The combined effects of such programs could very well reduce the *incremental* cost to build O'Neill's Island One to a scale well within the reach of corporate investments by the late 1980's or early 1990's.

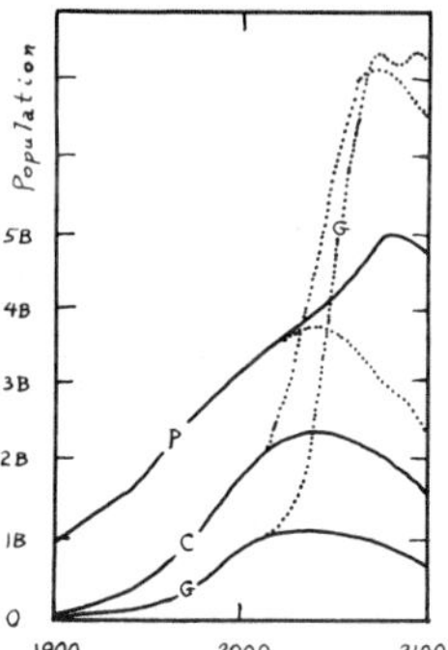
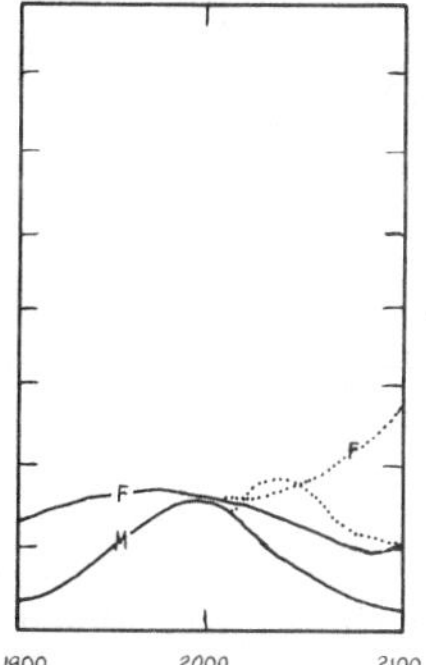

Results of world dynamics calculations for the underdeveloped world. The dotted lines show the effect a space colonization program and solar power satellites might have if the program began in 1982. Population is in billions; all other scales are in relative units. P—population; C—capital investment; G—pollution; M—material standard of living; F—food per capita. No satellite energy is delivered to the underdeveloped nations until 2007, when the project has paid for itself from energy sales in the developed world.

Whether or not solar power satellites will be necessary to meet terrestrial energy needs is now less clear than a year and a half ago, but the economic and environmental features of these satellites still appear very attractive. In referring to the firewood crisis in the Third World, I rashly stated, "No other energy technology presently foreseen [can substitute for excessive use of wood and dung]." I have since been educated on such alternatives as carefully managed village woodlots and more efficient stoves, but I still believe that solar power satellites could be a valuable tool in ecodevelopment of the Third World.

—J. Peter Vajk
Science Applications, Inc.
Pleasanton, California

Grandiose failure

When I read Dr. O'Neill's description of the space colonies, I was reminded of how I first felt about the plans for the Bay Area Rapid Transit.

BART was supposed to relieve traffic congestion in the Bay Area, offer transbay service by 1969, cost about $990 million, and operate at a profit. It was to be state-of-the-art technology, as they say. It was obvious to me that anyone who did not endorse BART had to be very shortsighted. In 1962, a $790 million obligation bond was narrowly approved by district voters for the construction of the system.

In 1974, a taxpayers' suit and a supplier's strike later, we finally got our transbay service. At last count, the cost of BART had risen to $1,620 million; it will always operate at a deficit. State-of-the-art technology means that when the tracks are damp, the computers go out for coffee.

But the biggest disappointment is that BART did not ease congestion. While we were busy solving the technological problems of BART, the Bay Area just kept on growing (especially around the suburban BART stations). The Metropolitan Transportation Commission estimated that

BART's contribution to the reduction of commuter traffic over the Bay Bridge was on the order of one year's recent traffic growth. This small initial decrease was promptly wiped out because it encouraged transbay trips which otherwise would not have been made.

Despite its problems, BART was probably the best technological solution available for relieving congestion in the Bay Area. Unfortunately, technology is not enough.

If we are willing to spend enough time and money, we probably can put colonies into space. The effort might stimulate the economy and keep some engineers out of mischief. That's fine with me. But I don't feel that these space colonies are going to save us. We are going to have to make some hard decisions concerning growth down here on the ground —and the sooner we get around to it, the better.

But to you space people —bon voyage. Thanks for giving up your places. Appreciate it.

Graham Holmboe
Berkeley, California

Jacques Cousteau at NASA Headquarters

Stewart Brand: Jacques, I'm interested to know why you're interested in the Space program.

Jacques Cousteau: At the very beginning of the space program, there was among all the ocean specialists a kind of scandalized reaction, saying, "How is it that so much money is going to be spent to explore distant things that have very little to do with our life, while the ocean seems so little known, and needs so much funding?" I don't know why, but at that time I had the vision of what Space would bring to oceanography. Since the beginning, I was convinced that our duty was to explore outer space as much as or more than inner space. The reasons that I had were purely philosophical, and since then they have become very practical. Everything that happens is demonstrating the need for Space technology applied to the ocean. The only thing that worries me is that for the moment there is still no coherent program for systematic monitoring of the ocean from Space.

SB: What would be an ideal monitoring of the ocean from your standpoint?

Cousteau: Well, I think that it cannot be done unilaterally by any nation, but I think it has to be organized, at least between the nations who have space technology, to make it a worldwide survey in order to constantly have the pulse of the ocean. The health of the ocean has to be checked all the time.

SB: What are the indicators of health that you can get from Space?

Jacques Cousteau is not the only environmentalist to have discovered the uses of satellite monitoring of habitat, but his subject, the oceans, simply overwhelms any other method of measurement. As a result there is a growing alliance between Cousteau and NASA. In charge of that liaison is Astronaut Russell Schweikart who also set up this interview, which took place in 1976 in the office of NASA's No. 2 man, Deputy Administrator George Low. Low, who had been with NASA since the beginning, was about to leave to become president of Rensselaer Polytechnic Institute. Also on hand for the interview was Cousteau's son Phillipe, who remarked after the tape ended that the American naval commanders in the Revolutionary War always had the crew break up the lifeboats before a battle.

Cousteau's comments read most easily if you mentally restore their sexy French accent. He is a powerful Gallic presence as well as a mythic hero (Odysseus, Captain Nemo). You may want to join the Cousteau Society at 777 Third Ave., New York, NY 10017 or 8150 Beverly Blvd. Los Angeles, CA 90048. $20/year.   *—SB*

Cousteau: Well, you can not have an indicator of health without having a thorough knowledge of the ocean, same as we can not predict the wind if we do not have a full knowledge of the atmosphere. So, to my opinion, a complete monitoring of the ocean involves two different techniques — remote sensing and tele-measurements made by interrogating sensors that are deeply anchored in the ocean. It's two different techniques, but both pass through a satellite.

SB: How rich would the data have to be for it to be useful?

Cousteau: This is a gigantic program. It's a gigantic program that at least has to be started one way or another, you just can't go full scale on a program like this, because it will cost probably billions of dollars. But it has to at least be experimented in all these phases systematically before a vast program is launched. Seasat was not a systematic oceanographic satellite.

SB: Well, has anything come out of the space program so far that has been useful in ocean monitoring?

Cousteau: Oh, yes. Lots of things. Bits and pieces.

SB: Like what?

Cousteau: Well, our knowledge of swell, currents, temperature differences. Day to day there are quick changes in currents, for example, that information could be forwarded to ships.

George Low: Jeek, you had some ideas about looking for sea farms from Skylab. Did anything come out of that?

Cousteau: For the moment, not much that I have heard about.

Russell Schweickart: That's an interesting little thing. During the joint mission with the Russians, the ASTP mission, Jeek and I were down in Houston, I was showing him the Control Center, the tremendous accumulation of nerve endings there in Houston, and the way in which the mission is controlled. We went back through some of the support rooms, and as we got to one of the rooms which was assisting in what we call the visual observations program — which was the astronauts and cosmonauts looking at particular Earth features to gain information and to take photographs — Jeek went up to this big world map and immediately started circling about six or eight different areas of the world which are sort of bays, as I recall . . .

Cousteau: Areas which seemed to me to be favorable for mariculture. But of course they have to be checked very carefully.

Low: How do you define mariculture? What are you looking for?

Cousteau: Well, as we were discussing yesterday, in exploiting the Earth agriculture was the break through. Before, it was nomadic collecting, and when the first settlers began to plow the earth, they did this only because they had

George Low, Deputy Administrator of NASA, flanked by Astronaut Russell Schweickart and Commander Cousteau.

had ancestral knowledge and observation from thousands of years of father-to-son transmission of understanding of all the plants and the animals in the forest, and they had selected a number of things like wheat, like rice, like corn, a very few, that were favorable for culture. And also they selected among all the animals those that were favorable for livestock because they were <u>directly</u> transforming vegetables into protein. They did not choose the eagle, the falcon, the tiger, et cetera; they chose the cow, and so on. This was a formidable knowledge of nature that was accumulated for thousands and thousands of years, probably hundreds of thousands of years.

Now, we suddenly understand that in the sea we are not doing that at all. Fishing is hunting. Scientific fishing is scientific hunting. When we see those big factory ships we can compare them with the people who shoot wolves from an airplane or from a helicopter. It's high technology in the service of just hunting. It means raping the sea for what it has. That's all it is in fishing. So fishing must be eliminated completely and replaced by farming, if we are to be civilized in the sea as we are on land. What we call civilization originated in farming. We are still barbarians in the sea. When we are farmers we are all going to get many times the yield per acre that we are getting now, because we are skimming the production of the ocean at the top level, which means at the level of the eagle. The tuna is the eagle of the sea. The marlin is the tiger of the sea. But none of these is eating directly vegetables. For 1 pound of cow meat, we need about 10 pounds of grass. For 1 pound of tuna you need 10,000 pounds of grass or vegetables.

SB: We have an article in this issue of the magazine that is espousing krill as a good source for human food.

Cousteau: Well, that's a very shameless substitute for the whale. The whale was a very good way to exploit the sea, if we had kept it at the maximum yield. The maximum yield is something like 70 - 75,000 whales. Now the yield is 32,000 a year. A few years ago we were killing several hundred thousand. If we had kept the number of whales at

the maximum yield, we would be exploiting the krill in the best possible way for mankind, because this is one of the most favorable chains that there is in the ocean. The krills are eating directly phyto-plankton, and 1 pound krill needs only 10 pounds of phyto-plankton. The whales are eating the krill directly, and there the transformation factor is fantastic. You need 4 pounds of krill to make 1 pound of whale. Which means that you have a 40-to-1 ratio instead of the 10,000-to-1 for the tuna. And besides, the krill is not so easy to digest for mankind. It's very good for whales, not so good for man. So, if you are just speaking of efficiency, the whale was the best transformer there was. The ratio was offering us a beautiful way to exploit the krill, and we've slaughtered them, eliminated the beautiful way. But you know, in any case I don't approve of using whales for food.

SB: Would it be possible to restore the whale to that function?

Cousteau: Yes. In 2027 there is a chance that there will be whales again. If we start now.

SB: What's good maricultural land? I mean . . .

Cousteau: You have good lands and bad lands in the sea. You have all the places where there is upwelling currents bringing in the nutrients from bottom — good lands. You have also the places that are not enriched.

Low: Can Space help you find the good ones and the bad ones?

Cousteau: Oh yes, because the upwelling currents are automatically near the surface, enriched in phyto-plankton, which means that by luminescence, by chlorophyll detection via satellite you can detect it.

Schweickart: In fact we talked earlier about getting together today with Dave Chalener who is the assistant secretary of the Smithsonian for Science. It happens that he ended up working with the Panamanian government on trying to predict the cold-water upwellings which then start the whole food chain through the indigenous shrimp there, and which are presently

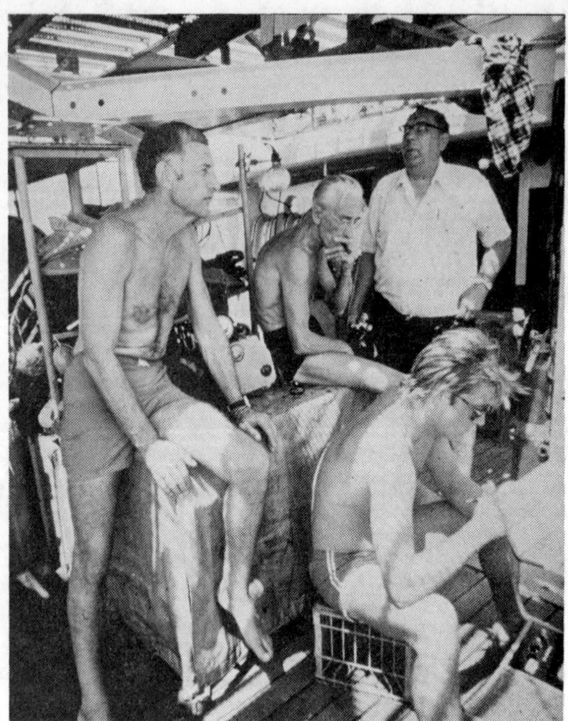

Aboard Calypso *off Eluthera Island in the Bahamas for the joint NASA/Cousteau Society research on use of the Landsat satellite for shallow sea bathymetry (water depth measurement). Left to right: George Low, Jacques Cousteau, Fabian Polcyon (from Environmental Research Institute of Michigan), and Jack Ford (President's son).*

in serious trouble. They have to preclude shrimping during the critical time when the shrimp first emerge until they reach a certain size. But that is depending on when the up-welling starts, and it's unpredictable, so the government forbids the shrimping at the wrong time.

Cousteau: But you see, here again, you are confusing fishing and mariculture. What the Panamanian government is trying to do with NASA is improve the destruction of the sea — the elimination of the shrimp once and for all.

Schweickart: Nope, let me disagree with you, Jeek. I can't allow that to pass, because in fact what the Panamanian government is interested in doing by monitoring the up-welling, is to preclude shrimping during that critical phase when the shrimp are first emerging. Right now they do that on a relatively arbitrary basis, and they miss the right time, and as a result they wipe out the baby shrimp.

Philippe Cousteau: That would be efficient if they had a limit to the amount caught every year. But that is not true. I just came back from that area, and there is no limit to the catch. So really, what you're talking about is an improvement in the amounts that are caught.

Cousteau: Improving the fishing is always bad. It's a dead end. It's like improving hunting with machine guns and toxic gases. I know that administrations such as NASA, and mankind in general, will probably be obliged to make the switch progressively. The switch cannot be made overnight. Scientifically we are not at all in the same position as the primitive man of 12,000 or 15,000 years ago, because we do not have the accumulated knowledge of the behavior of the marine ecosystem that our ancestors had on land. So we are

unable today to chose which species is good in what area. Marine culture is for the moment almost a dream that is not about to come true before we increase tremendously our knowledge. That is why fishing is still as a kind of intermediate evil which is necessary. But improving fishing is in the long run a disaster.

SB: Dr. Low, why is Space interested in the Cousteaus?

Low: I spent a very very enlightening week with Commander Cousteau on *Calypso*. It was Christmas a year ago, off a small island in Mexico. We started talking about what we can jointly do — the Space program and the Cousteau Society — in learning how to better understand the oceans and to begin to get on top of some of the problems that Jeek has been discussing here. The one thing that impressed me in those few days more than anything else is the great similarity of how science must be done in the oceans and how science is being done in space. When we went to the Moon we had men who were visible to the world in a hostile dangerous environment. The analog of that is the men under the sea. What our astronauts do on the Moon, or Cousteau's men do under the sea, is well-planned and has a detailed scientific purpose. When we sent astronauts to the Moon they had with them scientific instruments that they left there, or samples that they took, planned and programmed and detailed to get more knowledge of the Moon, but planned by people on the ground who were supporting them. They had the aid of communications systems, of navigation systems, of telemetry, of all sorts of instruments. Watching the kind of work being done on *Calypso*, it had great similarities. In fact, we did a joint project last summer in the Caribbean to try to see whether satellites can help determine the shallow ocean depths, the depths of the ocean where ships tend to run aground and spew oil all over. It was an experiment to decide whether a spacecraft called Landsat, or a variation thereof, could make those measurements on a global basis, because if you do it without a satellite, it is going to take you forever to make them.

SB: Most experiments don't work. How did this one turn out?

Low: Well, I think it was partially successful. We determined that down to a certain depth it can work. But the intriguing point was that in this experiment, thirty different satellites were involved, because they are the modern tools of science, the modern tools of exploration. There were six navigation satellites, I think, three or four weather satellites, the Landsats — two of them that actually made the measurements — communications satellites, direct communications back with NASA's Space Flight Center in Washington to help direct *Calypso*, to send weather maps back and forth, and to send photos of their area taken by a satellite that had just gone over. The *Calypso* divers were making measurements in the water of all sorts of quantities, of how the light was hitting the water, how turbid it was, things that others here can explain much better than I can. Instruments on *Calypso* were verifying how deep the ocean was while instruments in the satellites were making measurements and telling *Calypso* exactly where she was with a precision of what, feet? It was a tremendous example of how modern tools of scientists can be put together to get a better understanding of this globe we live on. I think that's really the key to our total future. We've got to use all of these tools, whether they're on the ocean, whether they're under the ocean, whether they're in Space, to help understand the fundamental factors that govern our environment, our life, and everything else, before we can really help fix it.

SB: This is not widely perceived among environmentalists. In my experience most of them regard NASA as irrelevant.

Cousteau: May I add to what George has said? Three things. One is that we are using satellites already for the study of another environment than outer space. This is the land surface and weather. The second thing is, the very elaborate instrumentation in Space that we were using to support a very primitive instrumentation in the water demonstrates the fact that we have not yet started harmonizing the tools to work together. There is a big gap between the quality and the sophistication on both sides of the surface.

A third reason why we are trying to work together was illustrated to me (I'm not going to say this for my own glory, it's just because it is a matter of fact) when I was at the Jet Propulsion Laboratory discussing with all these scientists and looking at a program of Seasat, I had great hopes when I heard the word Seasat, "Ah, at last, an oceanographic satellite." Then I went to Jet Propulsion Lab, and it was a demonstration of the fact that oceanography is a very complicated, unrefined science. Most of the specialists, beautiful scientists in oceanography, they look at their own little alley, and they don't have a general global view, which I by definition have because I do not go very far in each one of these alleys. I'm more or less a sponsor of scientists than a scientist myself. Speaking their language, I can collect and try to synthesize what they're doing. To establish a program by putting together all the scientists into a committee will result in very little, because each one will pull on his side. There is a need for someone who understands the global problems of the ocean.

Schweickart: Jeek, you were talking about determining a "vitality" index for the oceans . . .

Cousteau: We are still working on that and we are not very far because of the hostility of some specialists. It's very very funny. What Rusty is saying is that we need to understand realistically what's happening in the ocean. Our divers look to the place where they have been twenty years before, and they see that it is going down the drain. They are unable to quantify, it is not a measurement, it is a subjective impression, and in science subjective impressions are laughed at. They are not considered as serious. I don't think it's right, because subjective impressions have their importance but nevertheless you cannot say, "I have observed a reduction in the general vitality of the ocean," without having scientists say, "Ha ha ha, how much, how do you measure it?" So we have to measure it. We have to determine an index, a coefficient that is not measuring really anything, but that will be comparable to itself over the years. It's like IQ. IQ doesn't mean anything. But it's comparable. And this is the type of coefficient that I want, using only measurements that can be done almost by anyone, not too difficult.

SB: This would be like the number of species in an area, the amount of bio-mass in an area, or what?

Cousteau: That's too complicated. We have to find a more synthesizing thing. For the moment I'm orienting my research on the hormones that are excreted by animals in the water.

SB: It would be a straight chemical test then?

Cousteau: Yes.

Low: But aren't you talking about very very slowly changing quantities?

Cousteau: The trouble is that these hormones disintegrate pretty rapidly. It's on the contrary very instantaneous measurement.

Low: Yes, but over time the changes are relatively slow, so you have to be in a position to make very very fine measurements.

Cousteau: I have the impression that if we had started this 20 years ago, we would already see at least 40 percent decrease.

Schweickart: In certain areas like the Mediterranean?

Cousteau: In most areas.

Low: We're facing a problem in Space now like that — the ozone layer. What are we doing down here on the surface of the Earth that may be destroying the ozone layer and destroying life on Earth ultimately.

Cousteau: It's very minute changes.

Low: Very very minute changes over any period of time over which you can make these measurements. In fact, the noise that changes daily because of other things may be much larger than the signal you're trying to measure. So the real question is how you make these measurements over a long enough period of time to really determine what's happening. I think what you're doing in the ocean leads very much to those kinds of measurements. We're worrying about things that may be perceived only over a period of five or ten years. Yet they become terribly important to what's going to be alive on the earth within 50 years.

Cousteau: George, I suppose you agree that even if you need five or ten years to understand it, you have to start.

Low: You have to. In fact, there's an interesting side-line to this ozone problem. That is, that this might be a problem was identified many years earlier than it otherwise would have been because we were looking at the planet Venus.

SB: Because of comparative data, or because of the sensing apparatus that you developed for Venus being used here?

Low: A little bit of both. It's really an analytical tie. Looking at Venus, the atmospheric scientists started asking themselves the question, "Why do we see in the upper atmosphere of Venus only carbon dioxide and no carbon monoxide? Why doesn't the Sun dissociate the carbon dioxide and make it into oxygen and carbon monoxide?" The answer was that there had to be another gas present that immediately recombined it. They then applied that same theory to the upper amosphere of the Earth, and said, "What would happen if there were chlorides present in the stratosphere of the Earth?" The answer was: they would surely dissociate the ozone, the ozone layer which protects us in keeping the ultra-violet light from hitting the surface of the Earth. It would dissociate the ozone into oxygen — O and O_2. Having

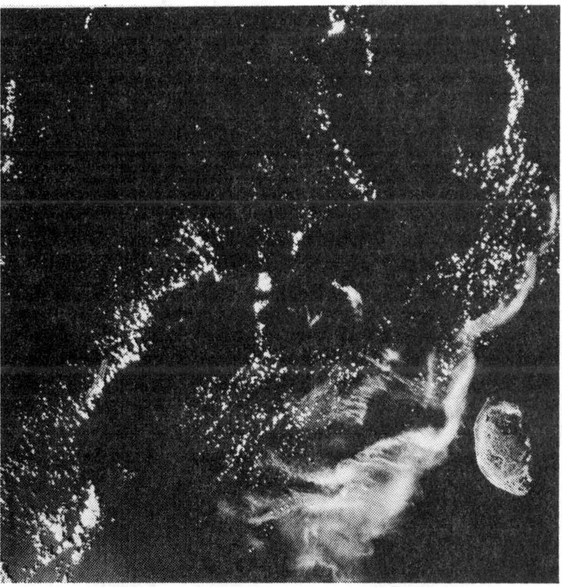

Photograph from Skylab III. The cloud formations cupped around areas of clear ocean may indicate cold water upwellings, nutrient rich. The feature on the right is an island.

asked that question, they said, "Well, are there any chemicals being released on the surface of the Earth that might destroy the ozone layer?" And the answer was: it could well be the freons that are used in all the spray bottles. That's how this problem was discovered, probably a number of years earlier than it would otherwise have been discovered, and that's why it's being attacked so vigorously now.

Schweickart: Although we still in fact don't know what the effect is of the lower atmospheric release of these things on the ozone layer. One of the more interesting pieces of data that we've seen in the analysis of ozone data we've collected over the last five years, (we're still not done with that analysis, it's a very laborious process) but in the two and a half years or so of data which we have analyzed, the largest identifiable change in a cause-and-effect relationship was, as I recall, a solar flare, which released a great number of particles, which then had their effect in the upper atmosphere. There was a noticeable drop in the ozone concentration, which took something between 30 and 90 days to come back up to normal. So can you say that one medium-sized solar flare equals 10^6 sweet-smelling armpits, or some equivalent measure? We don't really know that yet.

SB: What kind of an index would you call that? Well, Jacques, are you interested in being an astronaut?

Cousteau: Well, it would be my dream, yes. But of course it's a matter of age. I should have started earlier.

Schweickart: It's a matter of vitality, Jeek, not age.

Low: Von Braun made a speech about five years ago which said, in effect, "Five years ago I was too old to fly in space. Ten years from now, I will again be young enough, because then we'll have the space shuttle."

Cousteau: It's obvious that for me to have a visual contact with the ocean I have devoted my entire life to from outer Space would be the ultimate. Especially if I have the proper instruments. As I have said to NASA several times, I'm ready to go any time, but of course obstacles are formidable. Money, program, instrumentation, all those things. I'm keeping in as good a shape as possible in order to eventually be able to go.

SB: What are the stresses on someone with the shuttle? Would that be reasonable?

Low: Well, since I intend to be there when he flies, and fly with him, the answer is yes. Seriously, the environment in the shuttle ought to be one that an average human being in reasonably good physical shape ought to be able to fly in. We still don't understand some of the problems that we faced back in Apollo and Skylab. Why does everybody tend to get motion sick for a day or two? That applies to the best trained astronaut as well as to somebody who's never been in space.

SB: Do you think that'll happen to you, Jacques? Do you get seasick at sea?

Cousteau: I did, in my youth. No more at the moment, but it may come back. You never know. But very rarely.

SB: Did you, Rusty?

Schweickart: Yeah. On Apollo 9 I avoided the issue for two days by not moving around, thinking that was the right thing to do. We know a little better now. On the third day, when I had to move around, I went through the Space sickness thing for a day, and then after that was okay again.

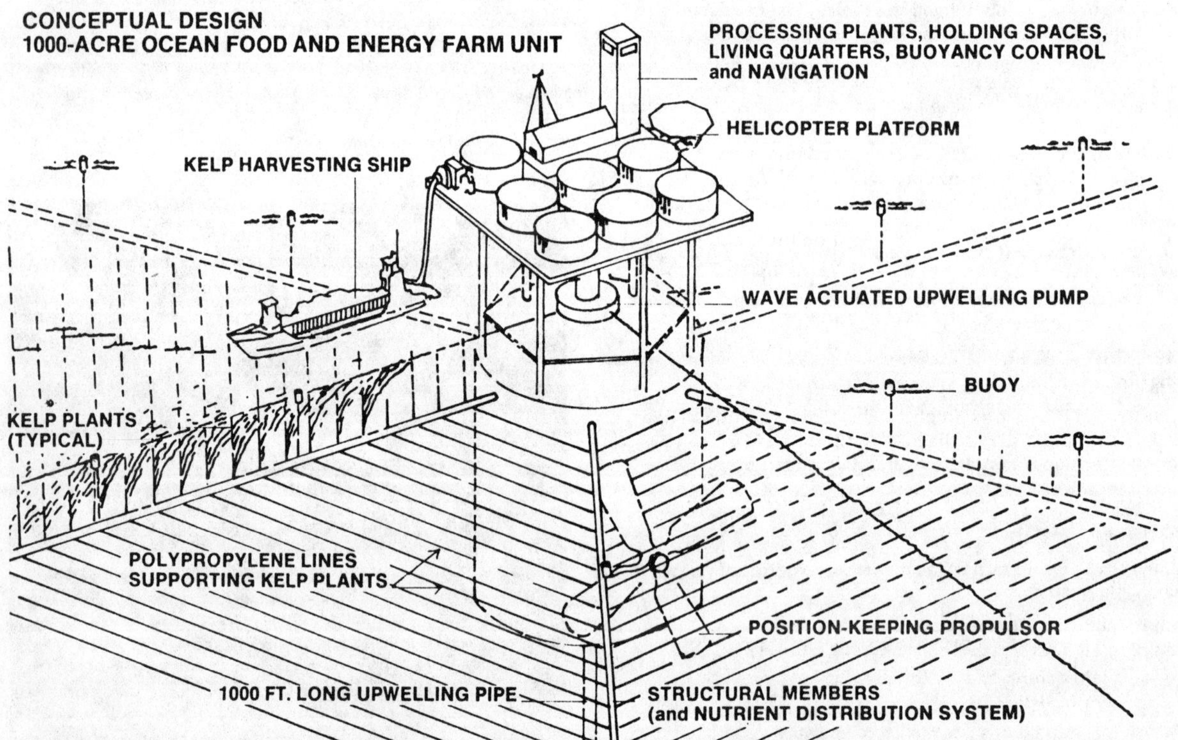

Illustration from "The Ocean Food and Energy Farm Project" by Howard A. Wilcox, distributed recently to the 150,000 members of the Cousteau Society. Dr. Wilcox presents a persuasive argument for open ocean kelp farming for food and energy use. "The marine farm concept is 'conservatively' projected to yield enough food to feed 3,000 to 5,000 persons per square mile of ocean area cultivated, and at the same time it will yield enough energy and other products to support more than 300 persons at today's U.S. per capita consumption levels, or more than 1,000 to 2,000 persons at today's world average per capita consumption levels." [Sent by Bill Steinberg]

SB: Once the shuttle gets really rolling, how many people is this apt to mean in space at any time?

Low: You know, we are terribly short-sighted when we project those kinds of things, because things happen much quicker than we expect, but we're talking about something like 40 to 60 flights a year.

SB: How many people on the bus each time?

Low: Say an average of 5. That's 200 people in space per year.

SB: Do you think this will change the public idea of the usefulness of Space, having that much traffic?

Low: Last Sunday I was preparing a speech I'm going to give in Salt Lake City this Friday night. It's a Bi-centennial event, and I thought one way of doing this is to go to the Tri-centennial and look back. I was trying to estimate how many babies will have been born in space by the year 2076. I came up with a number, I have no idea whether it's believable or not. I said it will be the event of the 100,000th birth in space.

Cousteau: 100,000th?

Low: I had a much smaller number at first, and then I thought, it's probably going to be larger.

Schweickart: That would assume Space colonies.

Low: It assumes Space colonies, and I figured by that time half a million people living in Space.

SB: Would these still be mostly Americans at that point, do you think? Or what?

Low: I'm not going to predict that. I stole a line from Carl Sagan's piece in your last magazine — "It's the 21st-century equivalent of 19th-century America." That it'll be limited to Americans I doubt very much. I just don't see how it can be.

SB: Did you see this quantity of people being mostly still in the vicinity of Earth, or scattered rather thinly in the Solar System by then?

Low: I guess I look at it as still being relatively close to the Earth-Moon system, and still counting on some resources from the Earth or the Moon or the nearby planets, but I may be wrong on that. You know, I was deeply involved in Apollo and the lunar landing, yet if you'd asked me in 1957, before Sputnik, twelve years before the lunar landing, when did I think a man would first set foot on the Moon, I don't think I would have said 1969. I doubt whether I would have said 1979, I might have said 2009.

SB: Is that rate defined more by technical matters or by political matters? That it was '69 was largely a political decision by Kennedy.

Low: It was a technical capability and a political decision.

SB: Gerard O'Neill claims that his Space colonies in most of the configurations he's thought of so far are technically feasible. Does that seem realistic to you?

Low: Yes. Now, I'm not sure whether they will look exactly like his Space colony, but I've seen nothing yet in the debate on those that says that it's not possible to do it technically.

SB: How about the short-term benefits, the cost-effectiveness, the political sexiness of the thing? He indicated a schedule of fifteen or twenty years to get a serious colony going. Does that seem realistic?

Low: Not to me today. But again, I'm an arch conservatist. I see other needs first.

SB: Such as?

Low: Solar power from Space. I don't know whether that's going to come or not, but it's intriguing, it looks possible, that we might be able to collect solar energy in Space, and beam it

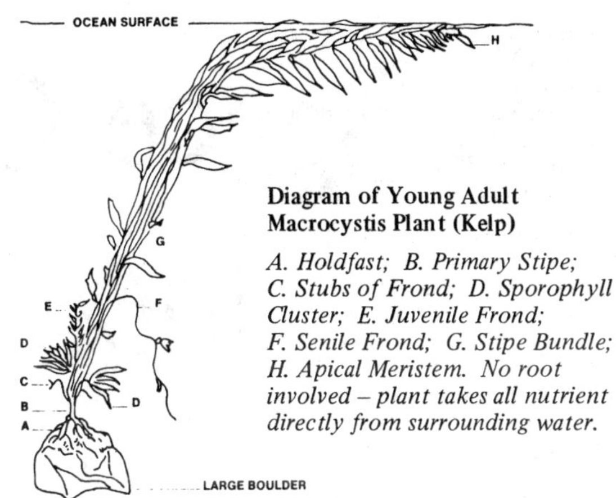

Diagram of Young Adult Macrocystis Plant (Kelp)

A. Holdfast; B. Primary Stipe; C. Stubs of Frond; D. Sporophyll Cluster; E. Juvenile Frond; F. Senile Frond; G. Stipe Bundle; H. Apical Meristem. No root involved – plant takes all nutrient directly from surrounding water.

down to the surface of the Earth to use it without doing all the bad things that we do when we generate power in any other way. I think that will probably take up our financial resources in the years where O'Neill first visualizes space colonies. But also they will help teach us how to build those colonies. There's an awful lot of things that we need to do in generating power in Space that will lead toward the colonies.

SB: Jacques, I'm curious about your longterm vision of what's going to happen? You said last night that you expect a major calamity in the next 50 years — or string of calamities — that would leave only 1 in 10 people alive and only 1 in 10 species alive. And that then a much more harmonious far-reaching ingenious civilization would develop. How does any of that relate to Space?

Schweickart: Do you see us moving out, Jeek, before the major calamities?

Cousteau: I think the calamities that we are talking about are inner calamities, which means that when they spread there will be no moving out possible, because there will be chaos as well in NASA as in the street. When we are projecting that far I'm not serious. I think that we have our duty today to work in the direction that might reduce the chances for such a calamity and in any event reduce the size of the calamity.

Schweickart: Do you put some of your resource into the generation of life boats, even though you put most of your resource into the good navigation of the ship?

Cousteau: I think we certainly will have communities in outer Space some day, and probably we will find uses for them that we have not anticipated, but I do not think that it is a way of life to live off the planet, same as I do not think that it is a way of life to live in underwater cities. I feel very much attached to my planet, and I think this is going to be true for generations to come.

One anecdote that happened to me just four days ago. I was with two eminent archeologists on a small island across from the ancient harbor at Knossos, Crete. We spotted from the ship some not ruins really, but a very old construction site. There was a slope which obviously had been polished by pre-historic man to pull a boat out on the rock. We landed, and while the archeologists were looking in the neighborhood for artifacts — and they found not much — I was picking up wild flowers. The very simple little flowers, I associated with them on this hot day. While I was doing so I discovered the hidden entrance to a cave which they had been passing by without noticing. We entered this cave and it was a primitive neolithic temple. So it's just by looking at the flowers that you may have the discovery. The specialists look at progress and science and see nothing. You have to remain deeply attached to the Earth if you want to understand what's going on. ■

The Space Colonies Idea 1969–1977

BY ERIC DREXLER

Space colonization and industrialization have increasingly filled my life since I began work on them seven years ago. In that time they have gone from an ancient germ of an idea to tons of research paper and increasing public attention. If the colonization of space happens, it will be both as a by-product of early, small scale space activity and as a result of a deeply felt yearning of most of the human race: a yearning for an open future for themselves and their descendants.

As a vague idea, space colonization is as old as myths of a better life beyond the sky. As an idea to toy with, it is as old as workable rockets and aircraft. As something to look at seriously, it is as old as the 1960's, when the future began to cloud and the Moon, with footprints, junk, and dust, flickered onto world television. As an evolving, public proposal, it was born in 1974 with Gerard O'Neill's article in **Physics Today**.

At this point I could wade off through the history of the idea, from O'Neill's conception of the idea in a freshman physics class that he taught in 1969 (a true story by the way, I recently indexed his working notes from 1969 to the present) to the first conference and publication in 1974 to the NASA interest and concrete proposals of today. However, if that story isn't told somewhere else in this book I would be most surprised. What may be lost in that story is a good impression of the range of people and of ideas that have participated in it.

O'Neill started with a basic idea: use materials from the asteroids to build attractive places for people to live in space. While not a new idea, this was at least unusual. Most thought in the 1960's suggested that the first thing to do _in_ space was to get _out_ of space, to land on another planet. Once there, under an unbreathable atmosphere, with night to ruin solar energy prospects, with enough gravity to make building difficult but not enough to match Earth conditions, and with other planetary unpleasantnesses, the intrepid expedition would dig in, and eke out some kind of more-or-less self-supporting life. The general, and justified, reaction to this kind of "space colonization" was, "No, thank you . . . perhaps a scientific base might be nice, once the world's problems are solved?" The alternative that O'Neill (and myself, independently about a year later) began looking at was to stay in the clean, sunny environment of space as much as possible, to find materials convenient there, to build, and to live.

Ideas evolved. O'Neill settled on a cylindrical shape for a colony, then on a pair of counter-spun cylinders that could easily be kept pointed at the sun, and on three mirrors, three windows, and three land areas in each cylinder. He shifted from favoring the asteroids as materials sources to favoring our own, nearby, Moon.

He considered getting materials off the Moon with a rotating device with an arm that literally threw pellets into space, then settled on an electromagnetic catapult (mass driver) at about the time that the entire idea, after five years gestation, finally went public.

In the year that followed, designs became firmer, more of the problems were considered and solved, the justification for the project shifted from population pressure relief around 2050 to energy supply around 1995, and cylindrical colonies became spheres, toruses, and dumbbells of various sizes, proportions, and proposed construction dates. Sociologists, anthropologists, engineers, biologists, artists, and celestial mechanicians became involved and contributed often conflicting ideas. This may seem like confusion, but, to an engineer looking at a young concept, competing approaches and solutions are a strong sign of basic health and vitality.

What has emerged after yet another year of refinement, criticism, and invention is a gradual program involving 100 - 200 flights of the space shuttle to set up a system capable of getting rock off the Moon and

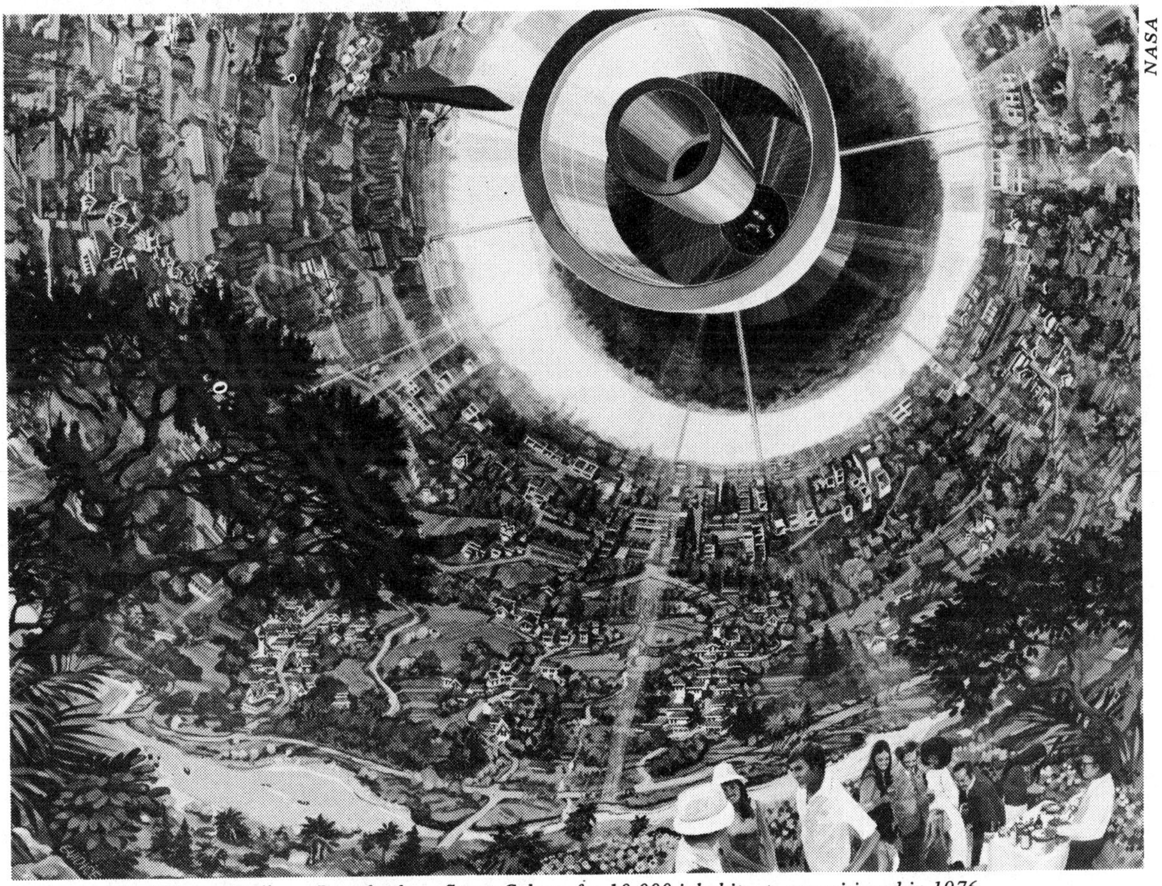

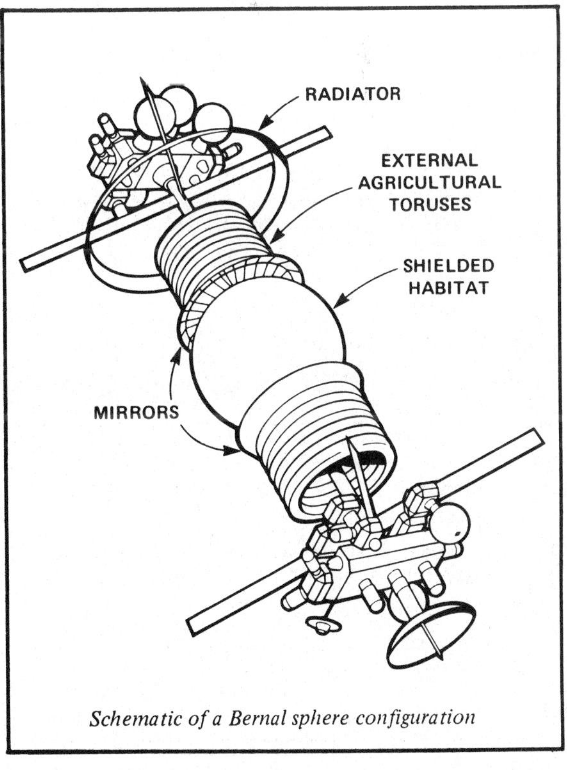

Schematic of a Bernal sphere configuration

"Island One" – a Bernal sphere Space Colony for 10,000 inhabitants as envisioned in 1976.

refining it into metals. After that, power satellites would be built to supply energy to the Earth, and small colonies would be built to house the people building the power satellites. It seems likely that the cost of doing all this can be repaid, with interest, by selling the electrical energy produced at competitive prices.

A point that seems to have been missed by many critics of this porposal is that no one is saying, "Let's spend billions of dollars on space colonies!! . . . maybe we can pay it back," but rather, "Let us find out if some proposal like this can supply us with economical, environmentally safe energy and other benefits, and, if the benefits outweigh the costs, and if the risks seem small enough, let's consider doing it." For anyone interested in cheaper, cleaner sources of energy than coal or nuclear power plants, the second suggestion is hard to argue with, much less become indignant over.

So, as you can see, the whole idea has finally been wrapped up in a fairly neat package. Engineering analysis has devised a system for putting people in carefully designed cylinders in space, and providing them with the means to pay for the adventure by selling power to the Earth. After a time there may be many carefully designed cylinders with many people leading carefully designed lives all over space.

Isn't that nice? No? I agree. Will it happen? No, because people are people and the world is packed solid with tricks and twistynesses yet undreamed of. Some are already dreamed of.

Here's what will really happen: work now in progress at Avco-Everett Research Laboratories, on stealing military laser technology and using it to drive rockets into space, will pan out. Anyone will be able to go into space in 1985 for $10 per pound in a vehicle the size of a Volkswagen using water in its jets. They do so. They go as corporate expeditions seeking steel in the asteroids and constructing power stations in Earth orbit. They go as political groups seeking a place to practice. They go as outlaw bands; the United Nations space treaty breaks down and is forgotten.

Here's what will really happen: we will mine the Moon, build power satellites, and small colonies to build and service them. In space, people find there are things to be done. Other industrial activities increasingly pick up, first with unique space processes, then with things that are cheaper in space, and finally with things done in space simply because that is where most of the people and industry are now. In a century the Earth has large sections of continents set off as wilderness area, other sections as space-non-intercourse zones, where ancient planetary lifestyles are practiced and refined, and still others are covered with disgusting urban sprawl, complete with rats and tenements. Space, meanwhile, has settled into a typical stagnant bureaucratic state (the kind the United States is trying desperately to become). Except for one small difference: space is a little too big for fences and national rule. At the open edges, the human race escapes forever the constraints of the central state, and begins its evolutionary migrational wave across the galaxy. Ten million years later, our trans-human descendants reach the edges of the galaxy, and snicker at their new limits to growth.

Here's what will really happen: power satellites will be launched from the ground by big boosters, built by Boeing. Solar sails will be built, and used to return steel from the asteroids. First it will be used for the power satellites, then for the Earth, and finally to satisfy the demands of space workers, now grown to like space, for improved housing. People stay in space, and . . .

Here's what will really happen: the Soviets continue their buildup of space armaments. The U.S., realizing the vulnerability of its domestic, foreign, and military communications satellites, builds a counter capability. Satellites become massive and armed for defense. Military space stations orbit the Earth. A by-product is decreased militarization on the ground, and cheap space transportation, developed by the Air Force. Space industry begins; the U.S. effort, at least, is kept free of military entanglements as much as possible, people stay longer in space, and . . .

Here's what will really happen: it is discovered that low gravity prolongs lifespan, and . . .

Damned if I know what will happen. I do suspect, however, that most of those who express confidence in a particular future (including those who are sure

USEFUL ATTRIBUTES OF SPACE
by Jesco von Puttkamer

- Easy gravity control from ambient zero-g (or micro-g) to any desired rotationally induced multi-g level

- Absence of atmosphere —
  - unhampered viewing of space for astronomy, astrophysics, etc.
  - perfect vacuum and freedom from seismic, acoustic, and convection disturbances

- Comprehensive overview of Earth surface and atmosphere

- Isolation from Earth's biosphere (for hazardous processes)

- Freely available light, heat, and power

- Infinite natural reservoir for
  - unlimited disposal of waste products
  - safe storage of radioactive products

- Super-cold temperatures (heat sink)

- Large, three-dimensional volumes (storage, structures)

- Variety of non-diffuse (directed) radiation

- Magnetic field

- Extraterrestrial raw materials

space colonies won't be undertaken or will be miserable outpourings of the worst aspects of the human race, etc.) are simply venting gas.

At this point a few words may be appropriate on the question, "BUT CAN IT BE DONE?" (as opposed to "Can it be done profitably?" or "Should it be done?"). First, however, I'd like to give some idea of how I feel, in my stomach, about most of the technical objections that have been raised in **The CQ.** Suppose you were in a garage, preparing a car for a trip across a desert, to move to a house in a pleasant valley. You've gone driving in other deserts before, and can make as many trips back and forth as you please, if you forget anything. When you stick your head out the garage door, the neighbors are gathered around. Some ask with concern if you have packed enough food or water for emergencies; you smile and nod. They ask if you've packed enough games for the kids to play on the trip, and you say hey! good idea! Then someone asks if you have any tires on the car. Someone announces that only murderers and villains like Hitler would consider moving across the desert, which can't be done anyway. Someone else announces that you can't go because it would be far too expensive. "Why," you ask, annoyed. He responds that the cost of bringing the corner store, the laundromat, and the sod around your present house (all of which you have used from time to time) is more than a year's salary, and that your lifestyle proves you can't do without them, so . . . At this point you are tempted to shut the door and check the oil.

Meanwhile, back in the real world, work has been done. John Holt helpfully pointed out that the Europeans have been working on magnetic propulsion (as in the mass driver) and that they have not

106

exceeded 12% of a gravity's acceleration, while O'Neill
has been talking 25 to 75 gravities, 200 to 600 times
more . . . Nonetheless, the car _does_ have wheels:
I personally helped build a model mass driver section,
out of ordinary wire (the real one would be superconducting, and hence have higher performance) that
reached 34 gravities about three weeks ago. Students
are upgrading it to 100 gravities. O'Neill is now talking 1,000, and taking bets on a number in that range.
If anyone had troubled to notice, the mass driver is
not a linear _induction_ motor (as Holt thought), but
a linear _synchronous_ motor (as virtually everything
published on the subject mentions specifically). The
difference is like paddlewheels and hydrofoils.

I could go on, picking away at technical objections,
one by one, but it would be tedious. Rather, two
common-sense observations on much-discussed
subjects: "closed ecosystems in space," and "social
systems in space." First, "closed ecosystem" in this
case equals parks plus farms. Given some sod, saplings,
jungle vines (or whatever), light, water, a soil analysis
kit, and an agricultural extension agent, we should be
able to make a decent park. If not, add some imagination to the mix. Given practically anything from
tundra to equatorial jungle, people have always
managed to farm in the past. Given light, water,
minerals, and some attention, it is an observed, tested
fact that corn, soybeans and, indeed, many plants will
actually _grow_. What is more, animals and people will
eat them, and a decent team of sanitary engineer and
agronomist can probably figure out how to get the
water, minerals, and so on back to the plants. As for
crop failures, which are rare enough _here_, give me a
designed agricultural environment any day. Ecologists: please remember this is an energy-intensive
farm, not a coastal wetland!

With regard to social systems, I must point out that the
earliest space colonies are like submarines or Antarctic
bases, where people have done useful work (and even
come back for more), except with the minor benefits
of sunlight, open space, more people, more regular
activity, and open communications to Earth. Taken
together, later colonies are like a world; endless room,
people, variety, bickering, and so on. The human race
has some partially successful experience with this sort
of thing. As for the "fragile, hence totalitarian colony"
idea, as an engineer I see no reason why a colony
should be considered more fragile than San Francisco.
They can certainly be made less fragile than parts of
(dreadful, totalitarian) Holland.

In short, yes, we need more study. Yes, neighborly
suggestions can be helpful. No, we do _not_ need to
pack the corner laundromat, and yes, the car has tires.
Successful new enterprises are often difficult, but
this is usually recognized ahead of time:

> Others again, out of their fears, objected against it and
> sought to divert from it; alleging many things, and those
> neither unreasonable nor improbable; as that it was a
> great design and subject to many inconceivable perils
> and dangers; as, besides the casualties of the sea (which
> none can be freed from), the length of the voyage was
> such as the weak bodies of women and other persons
> worn out with age and travail (as many of them were)
> could never be able to endure. And yet if they should,
> the miseries of the land which they should be exposed
> unto, would be too hard to be borne and likely, some
> or all of them together, to consumer and utterly ruinate
> them. For there they should be liable to famine and
> nakedness and the want, in a manner, of all things.
> The change of air, diet, and drinking of water would
> infect their bodies with sore sicknesses and grievous
> diseases . . .

The date is 1620, the writer William Bradford, one
of the Pilgrim Fathers, and later a governor of the
Plymouth colony.

Why go into space? Let's assume that space constructed power satellites prove economical (if not, a
number of ideas will slink quietly back into corners).
How about a steady, weather independent, day-night
independent source of solar-electric power? That can
be exported to the Third World? Whose apparently
minimal environmental effects disappear when you
turn it off? That doesn't gobble land like a ground
solar plant must? That could serve 20 independent
power systems in an average state? That is inexhaustible? Assuming the above is true, and, as always, that
it pays for itself (most energy proposals miss on the
latter point, by the way), most people would feel that
an adequate reason has been found to go into space
in a big way, probably leading to colonies. Some
people, I have found, remain ideologically opposed.

At this point we come to a topic that has me irritated
at my fellow inhabitants of this speck of a planet, or
at least some of them. Because I would like to make
very clear what attitudes and ideologies irritate me
before presenting the arguments, some explanation
is in order. Most of those ideologically opposed to
expansion into space are oriented in what might
loosely be called a limits-to-growth, alternative technologies, decentralized systems direction. Equally,
many supporters of expansion into space have a
similar orientation, but have seen the planetary/space
enterprise in a yin/yang, complementary, co-evolution-
ary light. They have recognized that limits-to-growth
is not a universal dogma or a universal good, that
different environments may have different appropriate
technologies, and that space may be the ultimate de-
centralizer. For them I have nothing but warm good
feelings. What follows is addressed to the positions (I
think) of the ideological critics. Call them Theocrats.

Theocrats have many denominations, but the religion
has a common belief: that the will of the gods is
manifest, and humans must be made to obey. They
seldom invoke gods by that name, but agree that the
conflicting desires and habits of today's human race
are wrong, and that major changes in human purpose
must be made in accordance with universal principles,
as revealed to _them_.

One sect, having been told that resources are finite,
seized the principle that growth is bad. When dragged
outside on a starry night and asked to point to the
limits to growth, hardcore members will explain that
material limits were never the issue (read Forrester's
more recent papers), but social limits are. This is
backed up with no sociology, and with a disregard of
the deep human desires of the Third World for a
decent life. It is by no means proven that world

poverty, even equitably distributed world poverty, will ennoble the human race and lead to peace. Experience seems to show the opposite, and redistribution of limited wealth in a world armed with hydrogen bombs may prove interesting.

Another sect, having learned that energy <u>has</u> value, seized the principle that energy <u>defines</u> value. Wasting energy is therefore a sin, and the government must enforce morality (as always, in theocracies) by forcing people to conserve, switch to solar power (whatever the cost), and so on. As yet there has been no suggestion from this sect that the government take action against the biggest energy waster of all: the sun (which wastes power 66,000,000,000,000 times as fast as the human race). Perhaps people are what is of value, rather than energy? And perhaps direct human concerns are more important than energy ideologies?

Another sect, realizing that the Earth, where the human race evolved, is a complex and delicate thing, seized the principle that the proper future for the human race is careful stewardship of our planet. The image that comes through this sect's writings is of the Earth as an evolved organism, and of the human race as a temporarily cancerous part of that organism. This I do not object to; it is a useful image, merely incomplete. I object, however, when the lure of the vision of an attractive, decentralized world is supplemented by blinders: "Do not offer people a future containing technology and economic growth lest they be seduced away from the paradise we have planned!" Somehow I just can't shake the conviction that a paradise should be able to stand on its own merits before the public. When I voice this conviction to sectarians, and find it rejected, I begin to suspect something a little dark. This rejection of freedom, the freedom of the people of the world to hear of and choose between options for the future springs from the same black, authoritarian place in human nature as does the controlled press and suppression of rights in Russia, Chile, Korea, China. . . .

These same people sometimes say that to plan expansion into space is to play god. Is it playing god to permit yet another of a long chain of expanding, evolutionary steps of this planet's life? Or is it perhaps playing god to stop that flow, to close a cage on the Earth, to cry, "Stop! Our <u>true</u> goal is in sight and lies <u>here</u>!"?

Yet another sect says that the answer lies within, rather than without. This is fine, but many in this sect are convinced that reaching their paradise of the mind demands denying people all hope of a materially satisfying life in the future, apparently because people are too stupid to make the correct choice, or to synthesize the best of both worlds. See the remarks above regarding authoritarianism.

In the long run, many people accept the thesis "one world or none." Space holds out the hope (culturally, politically, etc.) of <u>many</u> worlds. It has room for the sects, as long as they practice their theocracy on the willing and persuade by example. Those who feel threatened by freedom should perhaps examine their motives.

Those who are convinced that the human race should be refined and improved (by cooking in a sardine-can Earth) before a move to space should consider the chances of a pleasant outcome to the refining. On this planet there are a multitude of dangers to the survival of attractive societies and to the survival of civilization itself. Space may not save us, but it seems to offer a greater hope. Realities:

This planet is filled with ambitious powers armed with nuclear weapons. Russia, for one, would like to rule the world and shows little sympathy for the values of most of those reading this. As long as independent nations exist, as long as advanced armaments exist, military activity, industry, centralization, government, and so on will have to be maintained to ensure national survival. The possibility will remain, in spite of this, that nuclear war will destroy the world as we know it.

The obvious solution to a wonderful, terrestrially confined future is to work for world government and the dissolution of present national structures. Great. One powerful world government, unopposed by outside powers. Will it be like the U.S. government today? Like the U.S. government may be if it continues to run more and more of the country? Like the Soviet government? Or perhaps like Uganda's government? How do you plan to make sure? And make your plans stick for the next 200 years? Even de-centralised, agrarian societies have supported despotism in the past.

Other realities: "One World" can make one decision, as a society. Possible long range decisions include many dead-ends: police states made stable by technology, genetically engineered ant-heap societies, drug-controlled societies, wire-in-the-pleasure-center societies, etc., etc. I am not sure that having the <u>whole</u> human race pursuing one of these courses would properly fulfill human destiny. "Many Worlds" permits many decisions. "Many Worlds" permits mistakes.

A question for those who hold the idea of Earth as a stable mother, to be trusted indefinitely: did you know that 90% of the last two million years has seen the Earth in the grips of ice ages? That our civilized world sprang up soon after the ice sheets last retreated from today's temperate zone? That they are due back in a time short compared to recorded history and will stay for 90,000 years? Has anyone ever tried to write an environmental impact statement for an <u>ice age</u>?

Space waits for us, barren rock and sunlight like the barren rock and sunlight of Earth's continents a billion years ago. If there is a purpose to evolution, that purpose says <u>go</u>! Gather sunlight and barren rock and make life! Take the abilities of a thousand species, the minds of those who wish, and go! And stay! And scatter to the winds!

And, through the crusty turning of the wheels of government, through engineering, through economic analysis, through crass motives, bold vision, international agreement and conflict, public opinion, and perhaps the life force itself. . . it just might happen. ∎

In 1979 the first launch of the Space Shuttle will begin to settle arguments and reveal optimum paths as a new level of activity in Space sorts out the practicalities and impracticalities of Space industrialization and colonization.

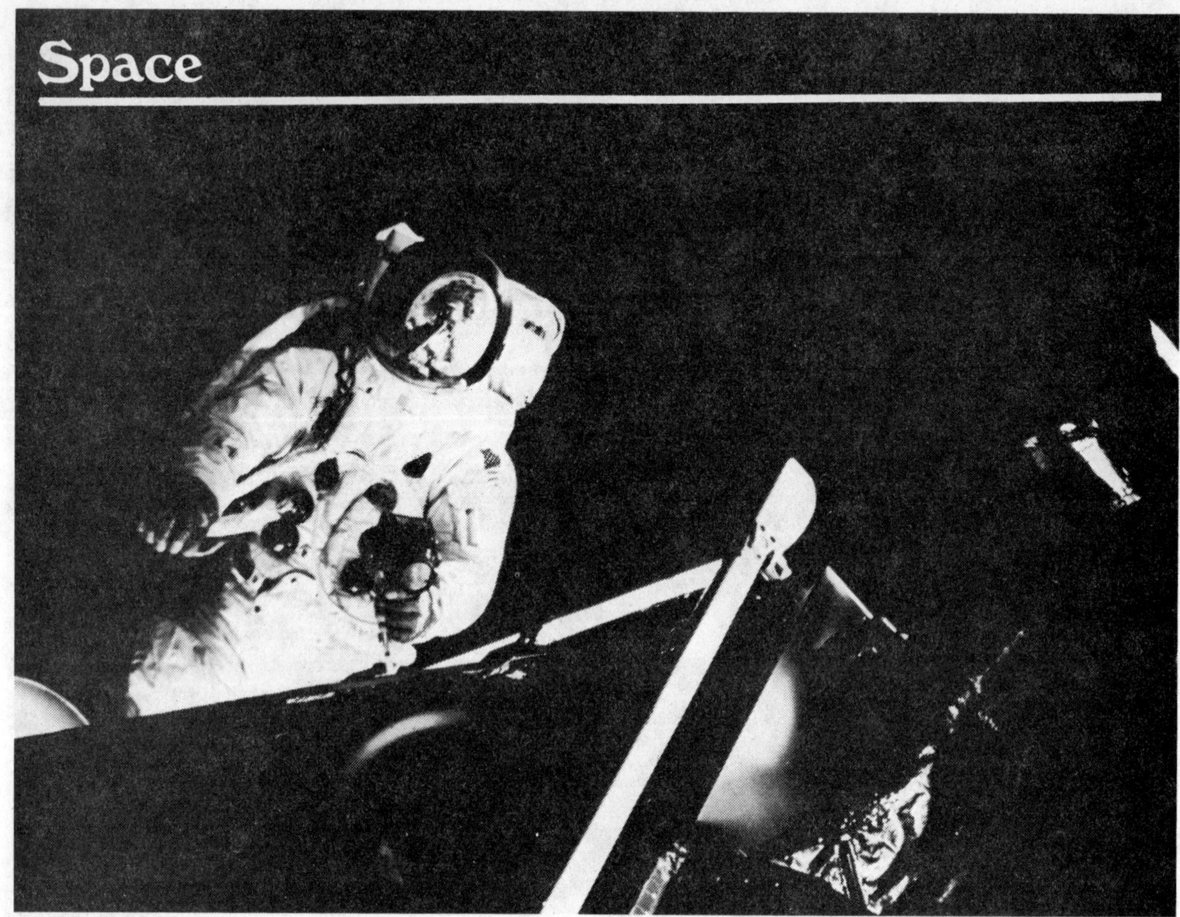

Russell Schweickart exits the Apollo 9 Lunar Module in Earth orbit, March 1969. *"Now you're no longer inside something with a window looking out at a picture, but now you're out there and what you've got around your head is a goldfish bowl and there are no limits here."*

Who's Earth

BY RUSSELL
SCHWEICKART

Photos by NASA

Up there you go around every hour and a half, time after time after time. You wake up usually in the mornings. And just the way that the track of your orbits go, you wake up over the Mid-East, over North Africa. As you eat breakfast you look out the window as you're going past and there's the Mediterranean area, and Greece, and Rome, and North Africa, and the Sinai, the whole area. And you realize that in one glance that what you're seeing is what was the whole history of man for years — the cradle of civilization. And you think of all that history that you can imagine, looking at that scene.

And you go around down across North Africa and out over the Indian Ocean, and look up at that great subcontinent of India pointed down toward you as you go past it. And Ceylon off to the side, Burma, Southeast Asia, out over the Philippines, and up across that monstrous Pacific Ocean, vast body of water — you've never realized how big that is before.

And you finally come up across the coast of California and look for those friendly things: Los Angeles, and

Like a long, pauseless prayer, astronaut Russell Schweickart spoke these words in the summer of '74 before a brainy group meeting on "Planetary Culture" at the spiritual community of Lindisfarne, Long Island.

Schweickart himself seemed amazed at what he was saying, amazed at the gathering he was attending, amazed — still — at the events which led him to drift bodily free between Earth and Universe. Remember the starchild at the end of "2001"? Like that.

This is just the conclusion of his tape — $6.50 for the cassette from Lindisfarne Association, Box 1395, Southampton, NY 11968.

—SB

Phoenix, and on across El Paso and there's Houston, there's home, and you look and sure enough there's the Astrodome. And you identify with that, you know — it's an attachment.

And down across New Orleans and then looking down to the south and there's the whole peninsula of Florida laid out. And all the hundreds of hours you spent flying across that route, down in the atmosphere, all that is friendly again. And you go out across the Atlantic Ocean and back across Africa.

And you do it again and again and again.

And that identity — that you identify with Houston, and then you identify with Los Angeles, and Phoenix and New Orleans and everything. And the next thing you recognize in yourself, is you're identifying with North Africa. You look forward to that, you antici-pate it. And there it is. That whole process begins to shift of what it is you identify with. When you go around it in an hour and a half you begin to recognize that your identity is with that whole thing. And that makes a change.

You look down there and you can't imagine how many borders and boundaries you crossed again and again and again. And you don't even see 'em. At that wake-up scene — the Mid-East — you know there are hundreds of people killing each other over some imaginary line that you can't see. From where you see it, the thing is a whole, and it's so beautiful. And you wish you could take one from each side in hand and say, "Look at it from this perspective. Look at that. What's important?"

And so a little later on, your friend, again those same neighbors, another astronaut, the person next to you goes out to the Moon. And now he looks back and he sees the Earth not as something big, where he can see the beautiful details, but he sees the Earth as a small thing out there. And now that contrast between that bright blue and white Christmas tree ornament and that black sky, that infinite universe, really comes through.

The Mid-East. *"You realize that in one glance that what you're seeing was the whole history of Man for years — the cradle of civilization . . .   You look down there and you can't imagine how many borders and boundaries you've crossed again and again and again. And you don't even see 'em."*

"Then looking down to the south and there's the whole peninsula of Florida laid out. And all the hundreds of hours you spent flying across that route, down in that atmosphere, all that is friendly again."

The size of it, the significance of it — it becomes both things, it becomes so small and so fragile, and such a precious little spot in that universe, that you can block it out with your thumb, and you realize that on that small spot, that little blue and white thing is everything that means anything to you. All of history and music and poetry and art and war and death and birth and love, tears, joy, games, all of it is on that little spot out there that you can cover with your thumb.

And you realize that that perspective . . . that you've changed, that there's something new there. That relationship is no longer what it was. And then you look back on the time when you were outside on that EVA and those few moments that you had the time because the camera malfunctioned, that you had the time to think about what was happening. And you recall staring out there at the spectacle that went before your eyes. Because now you're no longer inside something with a window looking out at a picture, but now you're out there and what you've got around your head is a goldfish bowl and there are no limits

here. There are no frames, there are no boundaries. You're really out there, over it, floating, going 25,000 mph, ripping through space, a vacuum, and there's not a sound. There's a silence the depth of which you've never experienced before, and that silence contrasts so markedly with the scenery, with what you're seeing, and the speed with which you know you're going. That contrast, the mix of those two things, really comes through.

And you think about what you're experiencing and why. Do you deserve this? This fantastic experience? Have you earned this in some way? Are you separated out to be touched by God to have some special experience here that other men cannot have? You know the answer to that is No. There's nothing that you've done that deserves that, that earned that. It's not a special thing for you. You know very well at that moment, and it comes through to you so powerfully, that you're the sensing element for man.

You look down and see the surface of that globe that you've lived on all this time and you know all those

Photo by Russell Schweickart of David Scott standing out of the Apollo 9 Command Vehicle. "There are no frames, there are no boundaries. You're really out there, over it, floating, going 25,000 mph, ripping through space, a vacuum, and there's not a sound."

people down there. They are like you, they are you, and somehow you represent them when you are up there — a sensing element, that point out on the end, and that's a humbling feeling. It's a feeling that says you have a responsibility. It's not for yourself.

The eye that doesn't see does not do justice to the body. That's why it's there, that's why you're out there. And somehow you recognize that you're a piece of this total life. You're out on that forefront and you have to bring that back, somehow. And that becomes a rather special responsibility. It tells you something about your relationship with this thing we call life. And so that's a change, that's something new.

And when you come back, there's a difference in that world now, there's a difference in that relationship between you and that planet, and you and all those other forms of life on that planet, because you've had that kind of experience. It's a difference, and it's so precious. And all through this I've used the word *you* because it's not me, it's not Dave Scott, it's not Dick Gordon, Pete Conrad, John Glenn, it's you, it's us, it's we, it's life. It's had that experience. And it's not just *my* problem to integrate, it's not my challenge to integrate, my joy to integrate — it's yours, it's everybody's.

I guess that's really about all I'd like to say, except that — and I don't even know why, but to me it means a lot — I'd like to close this with a poem by e. e. cummings that has just become a part of me, somehow out of all this, and I'm not really sure how.

He says, that

> i thank You God for most this amazing
> day: for the leaping greenly spirits of trees
> and a blue true dream of sky; and for everything
> which is natural which is infinite which is yes.

Thank you. ∎

There is a place with four suns in the sky—— red, white, blue, and yellow; two of them are so close together that they touch, and star-stuff flows between them.

I know of a world with a million moons.

I know of a sun the size of the Earth—— and made of diamond.

There are atomic nuclei a few miles across which rotate thirty times a second.

There are tiny grains between the stars, with the size and atomic composition of bacteria.

There are stars leaving the Milky Way, and immense gas clouds falling into it.

There are turbulent plasmas writhing with X- and gamma-rays and mighty stellar explosions.

There are, perhaps, places which are outside our universe.

—CARL SAGAN

"WHERE!?" 

We ran this provocative quote on the cover of the *Fall 1970 Whole Earth Catalog*, and Sagan re-used it in his recent Cosmic Connection *(EPILOG, p. 457)*. In 1973 we phoned him at his home in Ithaca, NY, for details. Come on, where? *Verbatim:*

The place with four suns in the sky. . . four different colors. . . and in touch. . . Well, first of all, most of the stars in the sky are not lone stars like the Sun but are binary or multiple star systems. A fair fraction of binary stars are called "contact binaries", in which the gravitational attraction of the more master star pulls matter out of the less master star—— it flows from the donor to the receiver. Now, there are many cases where two binaries orbit each other. Two stars are revolving around a common center of mass. Another two stars are revolving around <u>their</u> center of mass, and the two centers of mass revolve around each other.

Now, as far as color goes, the Sun is a yellow dwarf. A highly evolved star, like the Sun will be in another five billion years or so, is called a red giant. A red giant usually winds up as a white dwarf. And a very hot star but still in middle age like the Sun is called a blue dwarf.

A world with a million moons. . . is Saturn. The Rings of Saturn are composed of snowballs which are certainly less than a meter across, perhaps ten centimeters across. There are millions of such snowballs making up the rings of Saturn.

A sun the size of the Earth and made of diamond . . . Many white dwarfs fit that description. Where hydrogen has been substantially lost they are crystals, stars which are crystals, and they're cold and cooling still more. So, for example, Sirius has a white dwarf companion. It was the first one discovered, but there are enormous numbers of such white dwarfs, many of which are made largely of carbon in crystal form. Therefore diamond is the correct description.

An atomic nucleus a mile across that rotates thirty times a second . . . is a neutron star, which is the end product of the evolution of a star more massive that the Sun. It becomes, not a white dwarf, but a neutron star. That is, it's composed entirely of nucleons—— the elementary particles which make up the nucleus of atoms. Therefore they are atomic nuclei. And a mile across is how dense the thing shrinks to before the nuclear forces between particles pull the thing up against subsequent gravitational collapse. And they're rotating thirty times a second because of the conservation of angular momentum.

A star like the Sun spins once a month. When it contracts down to a mile across it's spinning something like thirty times a second. A specific example——the one that rotates thirty times a second—— is the pulsar in the Crab Nebula, which is a neutron star.

114

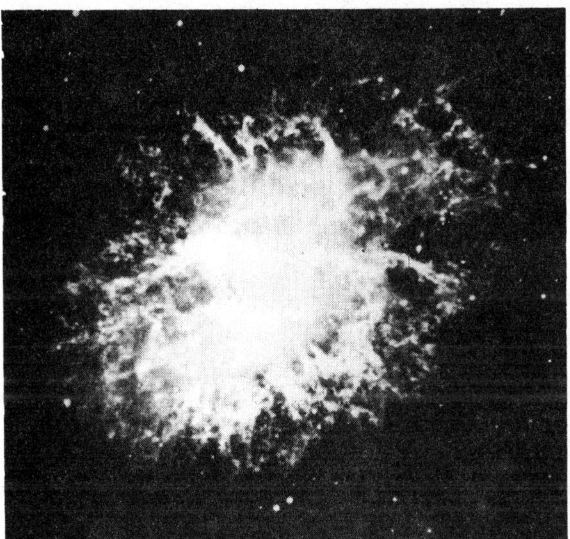

The Crab Nebula

OK, *Tiny grains between the stars with the size and atomic composition of bacteria . . .* Well, there's some absolutely tremendous number of them. If you take a look at a typical dark nebula, like say the Horsehead Nebula, the dark stuff is the kind of grains I'm talking about.

Hm, *Stars leaving the Milky Way . . . Gas clouds falling into the Milky Way . . .* Well, again it's quite common. We are a star which is in the plane, one of the spiral arms, of the Milky Way. But there are, for example, stars of a sort called "M dwarfs" which are oscillating out of the plane of the Milky Way—— they spend most of their time out of it.

OK, *Turbulent plasmas writhing with X- and gamma-rays and mighty stellar explosions. . .* Again, the Crab Nebula. Not the star in the center of it, but the nebula itself, is a good example of this.

Places outside our universe . . . Is a black hole. The nearest object which is thought by many astronomers to be a black hole is Cygnus X 1. I like to think of a black hole as a place where the gravity is so great that the fabric of space has become puckered—— isolated from the rest of space so that light can't get out of it.

"Does that do it?" he asked.

"Perfectly," said we. "Another question: What have you read that electrified you lately?" Sagan pondered, *"Well I'll tell you, the detective fiction of John D. MacDonald I find very interesting in terms of perception of character. Just for kicks it's far above Agatha Christie. . . Um, I've just reread* **The Odyssey.** *It's a good book. A lot of it takes place in Ithaca." "Which translation?" "Samuel Butler."* ∎

Steel asteroids

One can quite possibly turn a profit in supplying the surface of the earth with steel from the asteroid belt. The asteroids contain steel in chunks ranging from 100 kilometers in diameter down to dust. It is, in its native form, a strong, tough, ductile, and corrosion resistant material, and for engineering purposes, it is superior to most of the steel produced on the earth, because it contains about 5% nickel.

— Eric Drexler

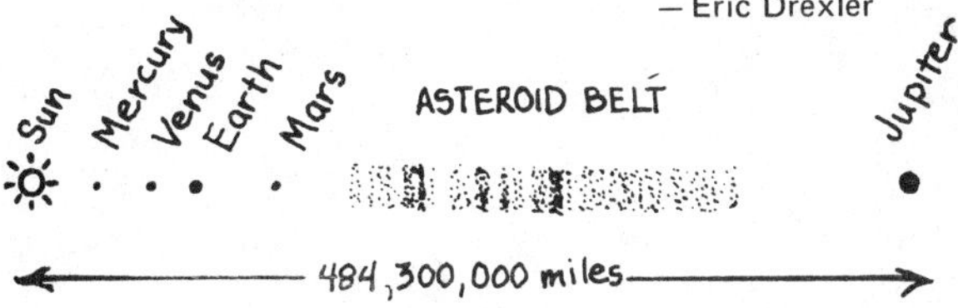

A House in Space

No book, including the ones by astronauts, has given so compelling an account of life in Orbit. Henry Cooper, on assignment from **The New Yorker,** *talked to all the participants in the three Skylab missions which accumulated a total of 171 days — nearly six months — in constant weightlessness. What they found there were the kind of amazing occurrences that you will find yourself starting conversation with. The astronauts spent hours searching for lost objects that wandered off. Whenever they opened a drawer the stuff inside exploded out at them in slow motion. Whichever way they stood in a room was "local down." The third crew went on strike. Fascinating problems, ingenious solutions in a definitively exotic environment.*

—SB

A House in Space
Henry S.F. Cooper, Jr.
1976; 184 pp.

$8.95 postpaid
from:
Holt, Rinehart and Winston, Inc.
383 Madison Ave.
New York, NY 10017
or Whole Earth

An astronaut could almost select, with his eyes, which vertical he wanted to follow, the room's or his own private one. "All one has to do is to rotate one's body to [a new] orientation and whammo! What one thinks is up *is* up," said Kerwin, the first crew's science pilot, who had discovered the phenomenon. "It's a feeling as though one could take this whole room and, by pushing a button, just rotate it around so that the ceiling up here would be the floor. It's a marvelous feeling of power over space — over the space around one. Closing one's eyes, of course, makes everything go away. And now one's body is like a planet all to itself, and one really doesn't know where the outside world is."

•

The astronauts were continually surprised at how much time they spent looking not only at oceans and deserts but also at snowfields and mountainous areas, in none of which could they see any sign of life. In contrast to the Apollo astronauts who had looked back from the moon and described it as an oasis in space, the Skylab astronauts thought the earth a barren place. The toughest part of the earth to survive in that they passed over, the third crewmen thought, was the area from Tibet across Outer Mongolia. "There is nothing but a great big nothing out here now," Gibson said during the third mission. "Northern China, Outer Mongolia, and all that gold stuff: the Gobi Desert." Carr, especially, thought man had a tenuous foothold on his own planet, where the checkerboards of his cultivation seemed to be packed into the few temperate areas, or fringed the deserts and oceans like a green mold struggling for existence. "Not much of the earth is hospitable to man," he radioed down one day — as though Mission Control's presence there had somehow made it seem an alien place. "We don't occupy much of our world. We're crowded into small areas."

•

Lousma had said once, "A guy like me, who likes both sunsets and sunrises, mostly gets to see sunsets. But here, in space, every day we get sixteen of each." Whatever they were doing, the astronauts frequently crowded around the window at such times.

There Ain't No Graceful Way

URINATION AND DEFECATION IN ZERO-G

ASTRONAUT RUSSELL SCHWEICKART TALKING TO PETER WARSHALL

*This is a spillover from Peter Warshall's "Watershed Issue" (Winter 76-77) of **The CQ**. Peter, who is one of CQ's Land Use editors, was quizzing many of the participants in the Space Colony debate on how much they knew about their local watersheds. Schweickart and O'Neill did pretty well. After the discussion of Rusty's Virginia watershed, they got to talking about, er, going to the bathroom in Space, where water does not shed.*

—SB

Peter Warshall: The thing that most people are really interested in, of course, is how you did it without gravity. That's what everyone asks: "Does it just float up?"

Russell Schweickart: Yeah, well it's kind of interesting Peter, because I just came back from a thing at Purdue University. I spent about two and a half days with a bunch of kids out there. It was a really nice program and I had four kids who were sort of my personal hosts and hostesses, and we really got into that line of material. You're right, everybody wants to know. That'd be kind of fun sometime, to just sit down and put together a whole article on how you do it in zero-g. Satisfy everybody's scatological curiosity.

It's something everybody's afraid to touch. Well, look if you've got a couple of minutes, I can describe a little bit of it to you. It's the end of the day, and why not.

Warshall: I'd love to hear it.

Schweickart: Well, of course there are basically two regimes. One is in the space suit and the other is out of the space suit. While you're in the space suit — which people always really mistake as a long period of time — we really wear the space suits relatively little. In Apollo and Skylab we wore the suits during launch and took them off shortly after getting into orbit, and put them on again only for EVA's (extra vehicle activities). So you can figure out how much time that is. And then again, depending on the mission, sometimes we wore them for entry and other times we did not. So you're basically talking about hours, maybe, depending again on the mission; anywhere from say, 4 to 20 hours, out of anything from 8 to 80 days.

Warshall: It's not really that much.

Schweickart: Yeah, it's not that much. But it's a fairly critical time, you know. When you're in there you don't have much choice, so you've got to design for it. Okay. So in the suit, for urine you use like a motorman's bag, which is basically composed of a bladder that holds about — boy, my numbers are really slipping Peter — but something between a liter and two liters, if I remember. A rubber bladder type of thing that sort of fits around your hips, and a roll-on cuff which is essentially a condom with the end cut out that's rolled over a flapper-type valve, you know, just a rubber flapper valve. It forms a one-way check valve.

Warshall: Oh, I see, so you don't have to do anything.

Schweickart: No, you don't do anything. You just roll it on as part of the suit-donning procedure, and then urinate into it through the one-way valve. There are lots of little cute problems and uncertainties. Unless you're an extremely unusual person, since the time you were about a year and a half old or so, you probably have not taken a leak laying flat on your back. And if you think that's easy, let me tell you, you've got some built-in psychological or survival programs, or something which you've got to overcome. So that's a tricky little thing. And then there's always the possibility that in maneuvering around in a suit you can end up pulling off the condom, and there's always — we have three sizes you know, small, medium and large — in diameter, and there's always this little ego thing about which one you do pick. Of course the smart guy picks the right size, because it's very important. But what happens is, if you get too small a size it effectively pinches off the flow and you just turn yellow because you can't go; and if, on the other hand you've got an ego problem and you decide on a large when you should have a medium, what happens is you take your first leak and you end up with half of the urine outside the bag on you. And that's the last time you make that mistake. So it's a cute little trick there.

In terms of defecation inside the suit, there ain't no graceful way to do it. So what we do is, we wear what's affectionately called a fecal containment system. The good old FCS is essentially like a pair of bermuda shorts with a hole for your penis to stick out of to roll on this other thing, but fairly well sealed around there. It's a tight fitting elastic type garment, and it fits especially tight around the thighs and around the

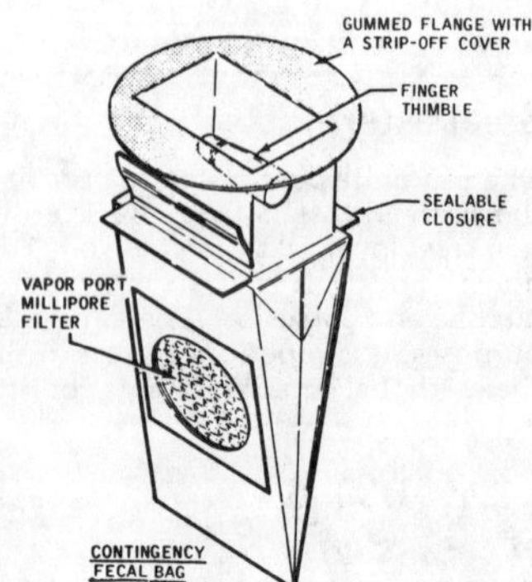

waist. And it's just like a pair of diapers is what it is . . . made of material which obviously is non-permeable but still breathes and all it does is contain it. Now, to my knowledge, nobody's ever had to use that. But you wear it, because if you don't wear it, the consequences are rather drastic. Okay. So that sort of takes care of the in-the-suit situation.

Warshall: Then you would take off that bermuda short type thing when you got back into the spacecraft. . .

Schweickart: Yes, and you take off the transfer system, and if you'd used it, you transfer the urine into, well depending upon the policies on the particular mission, you either take a sample of it, for a scientific investigation, or you just dump it, one or the other.

In terms of not in the suit, and in the spacecraft, again that's varied. In Apollo, for feces you just stuck a plastic bag on your butt, which was 6 inches in diameter, something like that, maybe a little bit less, 12 inches or so long and the mouth of it had a flange at the top with an adhesive on it, and you'd peel the coating off the adhesive and literally stick it to your butt. Hopefully centrally located. And if you think you know where your rear end is, you really find out, because you'd paste it on very carefully! So, you stick that to your butt, and then you go ahead and take a crap. But then the problem comes, because there's no particular reason whatsoever for the feces to separate from your rear end. So as a result, the problem is left as an exercise to the student to peel the bag off and make sure everything stays within the bag, and get all wiped off. It's basically a one hour procedure.

Warshall: For each time?

Schweickart: Yeah, from the time you start to peel down to stick the bag on and all that, till the time you have finished cleaning up and have everything wrapped up and stowed and have your clothes back on and everything, it's damn near an hour. And at times it's taken longer. Because when you peel that bag off, you try to take a handful of paper, and you know, lead the way in with that, but by the time you get done, you've got stuff spread all over your backsides, and if you're not careful, your clothes, and everything else.

Warshall: Have you ever had an accident where the stuff got out of the bag?

Schweickart: No, because generally speaking it's fairly sticky, so once it's in the bag it doesn't come out, but the problem is making sure it's loose of <u>you</u> when you get the bag off. It just is not a simple procedure, no matter what you do. Well, in any case, that was in Apollo.

In terms of the urine system, that was simple in Apollo. It's just the same as a relief tube in airplanes. It's a tube with a funnel on the end that you urinate into. And at the other end of the tube is lower pressure than at the business end of it. So there's a differential pressure in the outward direction.

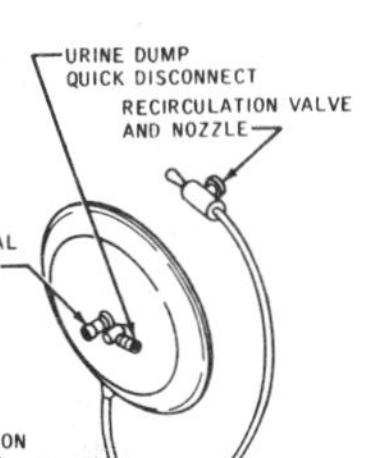

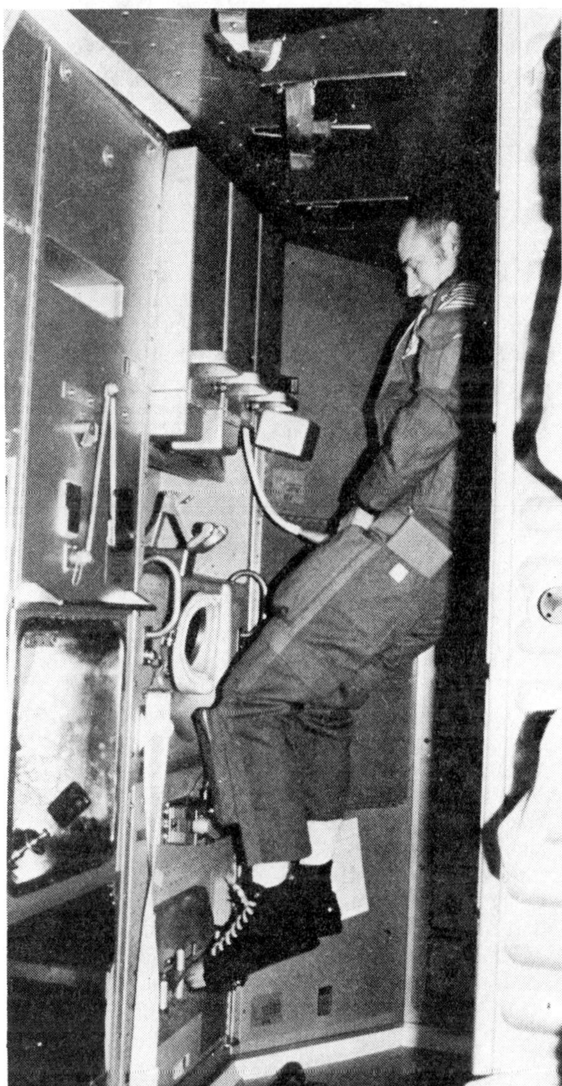
Al Bean urinating in Skylab II. Just in front of his knee is the air-flushed fecal containment system, which you back up to. There's no particular "up" in this picture. The surface nearest Al's head was the floor when the spacecraft was launched. Note soap floating by sink.

Well, we did exactly the same thing, except you know on the other end of the hose you've got a vacuum instead of a couple of psi down or something. So you just basically urinate into a relief tube. There have been various designs so you can use a roll-on cuff to do it or you can just hang it out there in the air and do it. There are a couple of different variations, but basically you urinated directly overboard through a relief tube. And of course, you didn't lose much cabin air, because while the liquid is in the tube, in the hose, no air is going down. It's differential pressure carrying the liquid. So it's only a matter of designing it for the right flow rate.

Warshall: They couldn't do that for feces, some kind of vacuum system?

Schweickart: Well, actually, in Skylab we did something similar to that. But on Apollo the urine then would go outside, and you'd have to heat the nozzle

because, of course, it instantly flashes into ice crystals. And, in fact, I told Stewart this, the most beautiful sight in orbit, or one of the most beautiful sights, is a urine dump at sunset, because as the stuff comes out and as it hits the exit nozzle it instantly flashes into ten million little ice crystals which go out almost in a hemisphere, because, you know, you're exiting into essentially a perfect vacuum, and so the stuff goes in every direction, and all radially out from the space-craft at relatively high velocity. It's surprising, and it's an incredible stream of . . . just a spray of spark-lers almost. It's really a spectacular sight. At any rate that's the urine system on Apollo.

Now, when you come to Skylab, it's a little more sophisticated. We tried to get sophisticated before that, but it never really worked. But on Skylab, the problem was that the medical experiments required that we sample all the urine and return all the feces because we were trying to make a total metabolic analysis. We could no longer dump the urine over-board because we had to measure the volume and take samples. What we did was to substitute a fan for the vacuum, and again use basically the same design, that is use a relief tube type of thing, except that the differential pressure was supplied by a fan which would decrease the pressure inside a bladder. You would pool 24 hours worth of urine. Then, after 24 hours, once a day, every morning, you would measure the volume, shake it up, get it nice and homogeneous and then take a 10 ml. sample, seal it up and throw the rest down the trash airlock.

Warshall: So each man had his own relief tube?

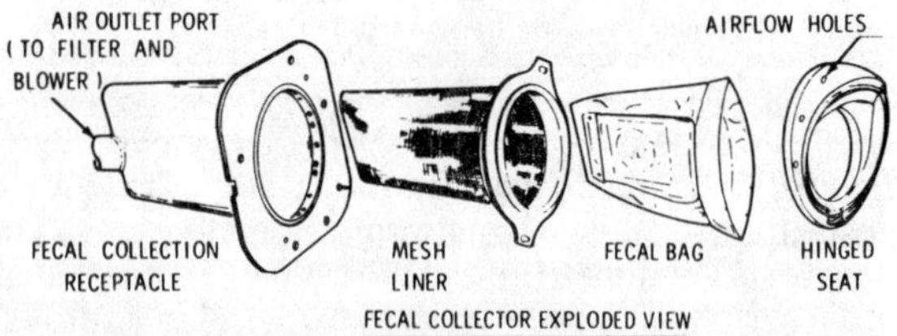

FECAL COLLECTOR EXPLODED VIEW

Schweickart: So each man had his own relief tube and his own collection system, right. On feces, we got very clever. Again we designed a plastic bag, but in this case, one side of the bag — now let me think if I can remember the words there —

Warshall: Was it like a stopper or something?

Schweickart: Well, one side of the bag had a material on it which would pass gas but not liquid. Oh yeah, a hydrophylic filter. And that bag, which again was something like 12 inches long and say 6 or 8 inches in diameter, got pushed down into a receptacle, and the receptacle was made of screen, and you drew a

differential pressure across the bag with a fan so that the air flow was out through one side of the bag and of course sucked cabin air in through the top of the bag. Okay? Now, there was a little seat that folded down over the top of the bag and then you sat on that little seat. Well, sat . . . you floated. In fact, you strapped yourself to the seat to make a fairly good seal there. Then what happened, as you sealed the open part of the seat with your rear end, there were side vents, circumferential vents, just under the seat, which allowed air to flow in from the side, and all of these little orifices — circumferential orifices — were directed at the exit point of your anus. So the air flow now, if you can picture it when the guy was sitting on it, the air flow would be in around the periphery of the seat, but all directed in little streams, like little jets, and then down through the bag and out the side of the bottom of the bag, see. So what happened then, is you ended up with air flow substi-tuting for gravity. And it would cause the bolus, so-called, to separate from your anus. Then the air flow would carry it down into the major volume of the bag.

Warshall: Kind of floating it in air down there.

Schweickart: Yeah. You just use air and air flow to substitute for gravity, and _that_ worked very well.

Warshall: Really?

Schweickart: Surprisingly. And as a result you were able to take care of a defecation with relative ease and a lot shorter time. So the system worked well. But then after you did that, you sort of stuff the bag with wiping paper, seal it, then weigh it to get the mass, the wet mass, and then put it into a vacuum oven to bake, and evacuating all the water. Then you'd stick it in the stack with all the other cow pies and bring them all home for analysis. So all you got rid of was the water. You keep all the solid material.

Warshall: So you just kept it in the same bag and . . .

Schweickart: Yeah, you kept it in the same bag, and in fact, the vacuum system, the vacuum drying sys-tem, used the same port through the side of the bag to get rid of the water vapor.

Warshall: So then you needed only one for the crew, you didn't need one for each person?

Schweickart: Right; only one toilet to collect the solids, but each person had his own relief tube for urine.

Warshall: That's incredible.

Schweickart: It's really wild. And you know, people say, "Why don't you fly women?" Well, Jesus, I'd hate to think about the plumbing. You know, it's really funny, because a lot of the girls at Purdue asked that, "When are you going to take women in the pro-gram?" and I always throw that one out, the part about the plumbing requirement. We haven't had the proper plumbing in the past. Well, the fact of the matter is, we're designing it into shuttle. I don't know what it looks like, I haven't looked at any of the detailed design from Washington here, so I don't know what it looks like, but we are ready for women on this one.

Warshall: A lot of people were wondering what happens when you start dumping things into space. Like, what happens to the . . . do you use toilet paper, for instance?

Schweickart: Well, the only thing that gets dumped into space is the urine, and that no longer is dumped into space, or at least was not dumped during Skylab, but that was during the early, Mercury, Gemini and Apollo. The fecal matter has always been stored on board.

Warshall: Oh, I see. People have visions of fecal matter and urine ruining space.

Schweickart: No, no. The only thing that is left floating around out there is principally water which instantly flashes into ice crystals and then subsequently under the influence of solar radiation, sublimates and ends up in a purely gaseous state and my guess is then, I would suspect, I'm not sure the interaction of sunlight on the gases at that altitude, but they either decay down to the lower atmosphere, or they get blown off. I'm not sure which.

Warshall: So somewhere out in space is just some sublimated water crystals floating around.

Schweickart: Yeah.

Warshall: Well you can see what they were worried about. Some day the sun's rays . . . blocked out by . . .

Schweickart: It's all stored aboard, so they don't have to worry about it.

Warshall: Do you use any kind of special toilet paper?

Schweickart: No, not that I know of. There may be some flame retardant chemicals put into it just so you don't have any unnecessary flammable materials around, but I'm not sure whether that's the case or not.

Warshall: So it's just like any other toilet paper.

Schweickart: It's basically like any other toilet paper.

Warshall: Is it stuck in the bag and then burned, or . . .?

Schweickart: No, it is in the same bag with the fecal material, and in the early missions that was a plastic

bag that you mixed in a disinfectant or actually an anti-gas, oh, what's the word I want, I guess disinfectant would be the best word, which holds down the generation of gas, and you mix that disinfectant liquid all through the fecal material. You mix it in, seal the plastic bag.

Warshall: How do you get it in there?

Schweickart: Well, it's in a small, like a ketchup, a little plastic container like you find ketchup in in restaurants, in a cafeteria or something, it's like that. You tear the slit across the top, being careful not to squeeze it so the stuff comes out, and then you drop that into the fecal container, and then seal the fecal container. Then you squeeze it through the, you know, externally, you know, which forces it out of the container, and then you mix it by massaging the fecal bag. It's really fun when it's still warm.

Warshall: The other thing that I've been asked about is how your diet affects everything; what you're actually eating; how that actually affects, you know, going to the bathroom, and do you do it once a day, since you said it took about an hour when you had . . .

Schweickart: Well, it's not at all clear how or whether the food really has any effect different from what you experience down here. The food does, I'm sure, have different preservatives and that kind of thing in it in order to be able to take it up there in the first place and be able to store it for long periods of time and that sort of thing. But I don't think there was any consistent observation, and there's no way to separate the effect of the weightless environment or other changes, physiological changes that are going on from the effect of the food.

Warshall: I mean, the image people have is you're mostly eating out of kind of toothpaste tubes or the food is being squeezed out, or something like that.

Schweickart: That's a pretty long conversation to try and get into, there are all different types, but the toothpaste-type tube has never been used in space. Well, it was used in early Mercury as a kind of test thing, but we have never in fact had toothpaste types of pastes or anything like it. It's always been freeze-dried food which you add water to to reconstitute, or what are called thermal-stabilized foods, almost exactly the same thing you would get in canned peaches or pears, that type of foods which you don't need to add water to, or in the case of Skylab, we also had about 10-15% of frozen foods, including filet mignon, and lobster and roast pork and vanilla ice cream for that matter.

Warshall: And you don't have to worry about that floating out into the space lab?

Schweickart: Well, if I could show you the movies, you'd see how we handle it. In some cases, like the filet for example, you cut a piece, and as you're taking it off of the fork the main piece of the filet may be floating up out of the plate, but you stab it with your fork and put it back down in, and if you do it carefully, the surface tension of the gravy will keep it in place. ∎

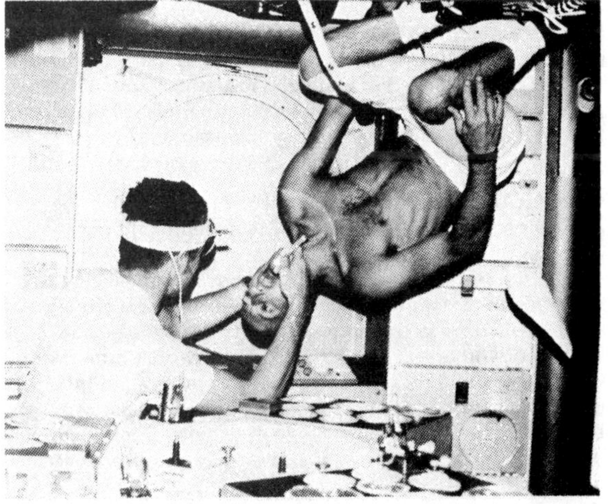

Dr. Joe Kerwin giving Pete Conrad's mouth the once-over.

Controversy Is
Rife on Mars

INTERVIEWING CARL SAGAN AND LYNN MARGULIS

At Carl Sagan's invitation I covered the Viking I landing for CQ. (The last time I stayed up all night was my last peyote meeting years ago.) It was a psychedlic occasion once again as Mars gradually came through all that press equipment and muddle and took over. The first photo of the Martian surface, which came immediately after landing, was electrifying. You felt you could reach out and touch those rocks — how far away?

These two phone interviews were done the weekend of Sept. 3-6, 1976, just after Viking II landed successfully. The first complete set of life-detection experiments by Viking I was over and was busily being interpreted. Lynn Margulis had just returned to Boston from the Viking scene at the Jet Propulsion Laboratories (JPL) in Pasadena, California. She'd been at JPL for several days as a member of the Exobiology Subcommittee of the National Academy of Science Space Science Board, observing the quality and content of the biology going on with Viking.

Lynn has a special interest in Mars because she and her co-author of The Gaia Hypothesis, James Lovelock, have formally predicted that no life would be found on Mars. Their reasoning is that the Martian atmosphere is consistent with strictly chemical processes, in contrast with the Earth's atmosphere which shows chemical anomalies explainable only by the presence of gas-producing organisms. According to the Gaia Hypothesis, the Earth's life effectively modulates the atmosphere, buffering it against major perturbances (Summer '75 CQ).

By now, late 1977, it is the increasing consensus that the Viking biological experiments give no evidence for Martian life — an apparent victory for Margulis, Lovelock, and the Gaia Hypothesis. But in some ways, as Carl Sagan has pointed out, the discovery of the exotic, active, "pre-life" Martian soil behavior is even more instructive than finding familiar bugs.

We're extremely grateful for Carl's invitation to the Viking landing, and we support his campaign for a Martian rover next time.
            —SB

Stewart Brand: Hello Lynn. How are you?

Lynn Margulis: I broke my rib in California.

SB: What?

Margulis: This guy had just got a motorcycle and was riding it, so I thought I could do it too — consummate self-confidence based on nothing. So I pushed the handlebars in the wrong direction, and accelerated instead of stopped, and the whole goddamned bike landed right on my rib.

SB: Well, how's the Gaia Hypothesis, in view of Viking?

Margulis: It's flying. We may be right. It looks like there's no life on Mars.

SB: You'd better get specific. First of all, how does Viking overall look to you?

Margulis: It's an unbelievable success. Everything is working beyond their wildest dreams.

SB: How do the biology experiments look to you at this point?

Margulis: The thing is that if there's any problem, it's a problem in interpretation, rather than in technical problems, because the engineering's worked unbelievably well. They're getting good data, as good as conditions permit. Do you want me to tell you about it from the beginning?

SB: Sure.

Margulis: There are three experiments that are called the biology package, and there's a fourth experiment, more important than any of those for biology, and that is the GCMS, the Gas Chromatograph Mass Spectrometer. It is an instrument that was tuned up to look for organic carbon and organic nitrogen compounds, going up to molecular weights of 400 or so. There's not a piece of soil in the world you can pick up that it wouldn't detect organics in. In fact, you can just run your fingers across a piece of glass, and there's plenty of organic matter in your fingerprint to show up in that machine, because the sensitivity goes from parts per million to parts per billion of organic stuff. This is Klaus Biemann's instrument.

7:30 a.m. on Mars, looking east from Viking I at a dune field. The boulder on the left is 25 feet away and measures about 3 feet by 10 feet.

The chromatograph is very sensitive, and it's hooked up to a mass-spectrometer which allows you to immediately look at the mass numbers, the number of atoms in whatever you see on the gas chromatograph. It's a double control on it — an excellent instrument. The point is that it's detected zero. That's NO organic compounds, no organic nitrogen or carbon compounds. Therefore all the biology experiments have to be looked at in that light, the fact that there's no organics in the samples at all.

SB: How many tests has the gas chromatograph made?

Margulis: As far as I know it's made two. One is a low-temperature run with stuff pyrolyzed — that is it's heated up without oxygen — and that gets almost all of the not-very-tightly-bound organics. That would give off all your amino acids and all kinds of small organic acids, if you ran it on an arbitrary soil sample on the Earth. And then they did a high temperature run, I think at 600° C, which will break up organic goop, tars and things they call kerogen and humic acids. High molecular weight incomprehensible materials get broken down into component parts, and you never know what the high molecular weight stuff is but you know what it yields. Both those runs were done with internal controls on them to be sure the instrument is working, and the instrument is definitely working. For example the instrument is seeing background atmospheric gases the same as the entry probes did, so it's definitely seeing things that are there.

It's essentially seen no carbon except oxidized carbon, CO_2 and CO, so any biology results have to be interpreted in the light of the fact that there's no evidence of organic compounds. Now, such organic compounds may be there, they could be under the surface or they could be hiding somewhere, but they're not in the samples.

SB: If they were in the atmosphere, would they be picked up?

Margulis: Oh yeah. They're not in the air by other criteria. Everything is in a very oxidized state. You know what happens to organic matter, it just burns, so the carbon is in CO and CO_2, and that's very bad for organic material, those conditions.

Most people feel that some organics ought to be on Mars because of meteorites. There's organic matter being brought in by meteorites, and we're not even detecting that. So that means that it's either been destroyed, or it's been preserved deeper down than the samples have gone.

SB: So the finding is even lower than expected?

Margulis: It's incredible. To me, it makes the biology much more easy to interpret, because it's a much more clear-cut negative result than I would have anticipated. I was anticipating a very low organic finding, which would be ambiguous, because you don't know whether that's from meteorites, or whether it's from pre-biotic processes, or what else. But in fact it's a moot point now, because the reading is zero.

[more→]

Viking I's telescoping scoop takes the first sample of Martian "soil" and dumps it in the hopper (right) for analysis.

Terrain at Viking II site

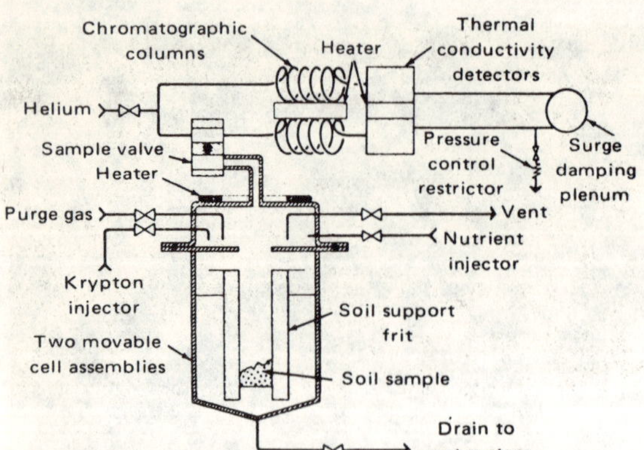

The gas exchange experiment showed an unexpectedly high release of oxygen after the Martian "soil" sample was wetted with nutrient.

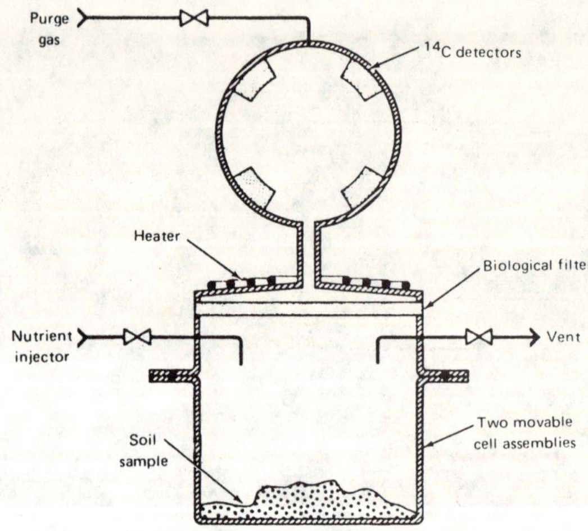

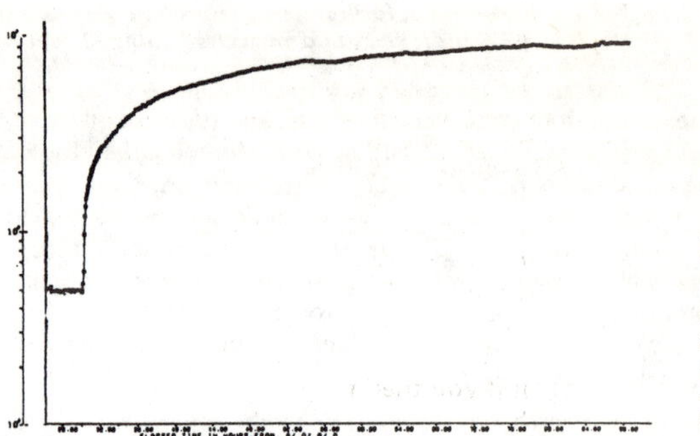

In the labeled release experiment, radioactively tagged carbon-14 in the nutrient showed up almost immediately in gases given off by the wetted sample, climbed steeply, and then leveled off after 9 hours.

SB: Why is everyone interpreting the other three experiments as being still possible for life?

Margulis: I wouldn't say everyone. There's no doubt some fancy chemistry going on on that surface, because you've got the results of the three other experiments — all mutually contradictory, so there's almost no way of getting a single result to satisfy all three of them. There's been a lot of criticism of those particular experiments. They were designed under a much more favorable impression of the Martian surface. They were designed in '68 or something, and that's part of the problem. They were fixed very early.

SB: I remember you said in Florida at the Viking launch that they were probably too hot and too wet.

Margulis: Both, yeah, with the exception of Horowitz's, which is only too warm. You've got three of them. The first is the gas exchange experiment. That's Vance Oyama's experiment. The whole concept of monitoring gases is I think basically an excellent one, but the problem with that particular experiment is that it's only been run by moistening the soil with one of the most complex media I've ever seen. It has I don't know how many components, all kinds of organic compounds and vitamins — dozens of different things in that medium. That's the "chicken soup." They moistened the soil, and then they monitored the gases above it. Now, if you do that on the Earth you get a whole profile of gases being removed for respiration, and being produced, and metabolism — you get a whole lot of gas exchange going on. What they found is that as soon as the moisture hit that Martian regolith (I say regolith instead of soil because to me soil is a very organic thing; that's the word they use for the moon covering) . . . as soon as it hit that, you got this production of oxygen, relatively large, about 15 times what would have been expected, and most people feel that you've got some sort of peroxide, or some kind of oxygen-holding material in the sample that released the oxygen on wetting. In fact Phil Ponnamperuma (he's an incredible guy) may have already been able to simulate this result with very very dry peroxide-containing material. You add water and you heat it up and oxygen is eliminated. The exact Martian simulation isn't known, but most people feel that it's a chemical response to the adding of the water. Also it may or may not have anything to do with the complex medium.

SB: They don't have the ability to do a control with just water?

Margulis: Don't ask questions like that.

SB: Come on.

Margulis: They make me laugh, and laughing is very bad for a broken rib. They do have certain controls, but one of them is not water alone as far as I know. They have a "sterile" soil control, and you get less response after you heat the soil up first. But whether you're sterilizing, or whether you're inactivating a chemically active surface or what the hell you're doing by heating it up is ambiguous. There's all kinds of things that respond to heat that aren't biology, right? Anyway, I think most people feel that the release of oxygen is very interesting, and you're going to learn something about the Martian surface, but it has probably nothing to do with any kind of living response. You see, it stayed constant. It gave off the oxygen and then it stayed constant over quite a long period of time. Which is what happens when you add chemicals and mix.

OK. The second experiment is Gilbert Levin's labeled release experiment in which he has carbon-14 — radioactive carbon-labeled materials — in his so-called medium. Again, hot and moist. Too hot, and too moist. Much hotter and much more moisture than one would expect on the Martian surface. Parenthetically, the gas chromatograph, the GCMS that I spoke about first, measures water, and some water came out at low temperatures, but what was very impressive was that a lot of water came out at relatively high temperatures, implying that the water is tightly bound. It's water of hydration in the minerals themselves.

So it's really dry stuff. Basically we're talking about a slight amount of water vapor, measured in preciptable microns, where on the Earth it's centimeters. Now, the labeled release experiment shows some incorporation. What they're looking for is release of the radioactivity as radioactive $^{14}CO_2$. In other words, they're adding radioactive food and looking for radioactive waste products. And indeed they got radioactivity in the form of $^{14}CO_2$ coming off, and I think they got more released than with the "sterilized" sample. But, again, you don't know whether you're releasing natural radioactive C-14 in the soil, or what kind of chemistry is going on in the components of the medium. This labeled release experiment

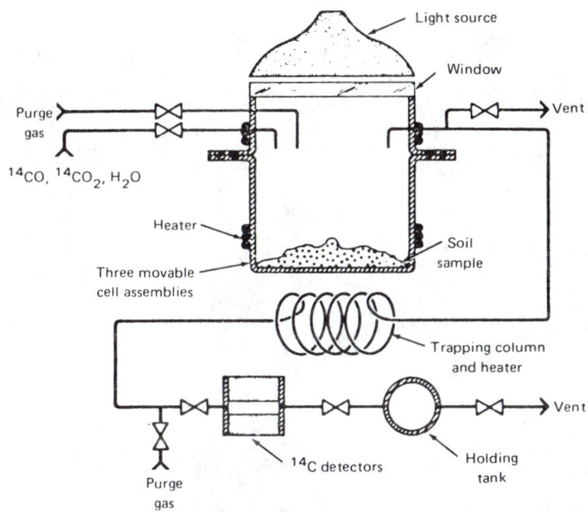

The pyrolytic release experiment put radioactively tagged ^{14}CO and $^{14}CO_2$ gases with a sample in artificial sunlight, let it incubate for several days, and then heated it to see if tagged carbon had been incorporated by the sample. It was, and markedly more than in a "sterilized" control sample.

is another one that can be essentially imitated in the laboratory in a non-living way. You can just add chemicals until you release radioactivity. So both of those experiments are too hot and too wet, and have alternative interpretations, and neither of them are giving you the kind of complex response you get if you tried to monitor even a desert soil on the Earth.

The third one, the pyrolytic release experiment, had a very interesting result, and no one knows how to interpret it. It's got to be interpreted. This is Norman Horowitz's experiment which looks at the incorporation of radioactivity introduced as ^{14}CO and $^{14}CO_2$ — tagged carbon atoms. Now, this is tricky, because if you were on the Earth you'd be convinced by the finding, but the Martian conditions are weird, and you're not sure. What you do is, in the light you add a sample of radioactive ^{14}CO and $^{14}CO_2$. (That carbon monoxide is in much higher percentage, unfortunately, than in the Martian atmosphere.) Now, on the Earth you'd get photosynthesis. You'd get the incorporation of these carbon-14 gases into organic matter in the light. You'd also get non-photosynthetic carbon dioxide and carbon monoxide fixation, which means that these things react and get incorporated into living material even in the dark, although the light reaction is a much stronger reaction on the Earth. Now again, the experiment was run at ambient conditions for the spacecraft, which means higher temperatures than have ever been seen on Mars, at least at the latitude of Viking.

SB: What temperatures are those?

Margulis: I think it's 12° or 15° Centigrade (50° - 60° Fahrenheit). All temperatures around Viking so far have measured below 0°C, below freezing, and this experiment is definitely above freezing. They had to run it at the same conditions as the spacecraft, because the machinery won't run at ambient temperatures. But it's not wetter than ambient. There's no liquid water at all in this experiment — only what vapor there would be anyway. Of all the experiments it's closest to Martian ambient. (One could argue in the other two experiments that you essentially drowned everything, because you've added more water than probably has been seen on that planet in a billion years.) On this one even the temperature's not that unreasonable, because in equatorial regions in Martian summer temperatures do get that high.

So in the pyrolytic release experiment after you've added the tagged ^{14}CO and $^{14}CO_2$ you heat the sample and measure

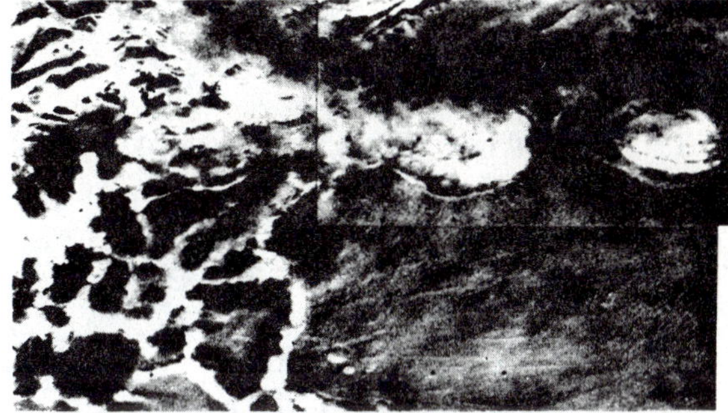

Water vapor on Mars. In the early morning dense fogs form in low areas such as this complex of canyons called The Chandelier.

the radioactivity of what cooks off. The ^{14}CO and $^{14}CO_2$ that was not reacted goes off first. You get rid of that. And then you get only the material that is retained by the column, called the organic vapor trap, which under Earth conditions would mean that it's organic matter. And they did see that there was enhanced radioactivity in this second peak, which implied that the carbon-14 was incorporated into something organic. The results superficially imitate quite handsomely results from pretty barren Antarctic soil. There was something like 96 counts per minute in that second peak, and the "sterilized" control had something like 15 counts, close to the ambient background. So there was an enhanced signal. But we haven't done the same experiment in the dark yet. That's coming.

In some people's minds you've got some sort of positive signal here. In my mind you've got a positive signal that has nothing to do with life at all. That's my particular prejudice, and only time will tell. I know that I'm ready right now to start working on a piece with Jim Lovelock about what this says about the Gaia Hypothesis. We have two controls — Mars and Venus — and Earth is the experiment. Earth has life on it and has all this atmospheric modulation going on which Mars and Venus don't have. I think we're going to get a consensus before the end of the year that there's no life on Mars.

Something else terribly interesting that I realized when I was out at JPL is: the camera's been sitting there surveying the landscape for 40 days, and not one thing has moved in 40 days. Nothing. Not even a dust particle. I just can't imagine sitting anywhere on the Earth for that long and having nothing move except shadows. Even some wind, some weather or something. I'm just asking for some geology, and you're not even seeing that. It's very possible that the surface you're looking at was formed 2-1/2 billion years ago and there's been very little change since then. An occasional meteor impact and an occasional dust storm of very very fine dust.

SB: Do you mean those little trails of sand behind the rocks could have been there for some good while?

Margulis: For thousands if not millions of years. That's something that people want to do is date these things. You do see those river-like channels — everyone agrees that some of those were made by water. There's now no disagreement on that. When you get water-formed channels that have meteorite craters in them it's <u>possible</u> that they're recent craters, but it's much more likely that the channels occurred at very very early stages, when the meteorite rate was higher, and that pushes back the problem to: did life ever arise on Mars? It's certainly possible. Did it arise and die out?

SB: If life were introduced now to Mars, would it find a foothold?

Margulis: If you just set something out, it would die very quickly. It might pop open from the low atmospheric pressure. The cold temperature per se you can survive in, but you can't metabolize in it, and there's only so long you can sit around waiting.

SB: How big a deal would it be to make the planet habitable, do you suppose?

Margulis: A big deal. A very big deal.

SB: If you carted in the stuff for the atmosphere . . .

Margulis: You're not going to increase the gravity, so you'd have trouble keeping it there. You'd have to have closed containers for atmosphere.

SB: In terms of Gaia, you made some predictions a while back that there'd be no life on Mars because of what you knew about the Martian atmosphere. Does the Martian atmosphere still pretty much have what you thought was in it?

Margulis: It has very much what we thought. It's got all those exhaust gases, all those highly oxidized states of carbon. I'm glad they found the nitrogen, 'cause now we have a number on it, but there's no reduced nitrogen in there, which you need. You see, there are no reduced compounds in the presence of the oxidized ones. Which is our clue. Everything's essentially oxidized.

SB: If there were life you'd look for what gases?

Margulis: You'd look for ammonia, methane, hydrogen, or hydrogen sulfide or any of these things that are flagrant contradictions in the presence of oxygen and CO_2. What you have on the Earth is this sort of flagrant contradiction which would, if left alone, go to a Martian type of state. The fact that you have these reduced gases on Earth with its oxygen-rich atmosphere means that they're constantly being produced by organisms. You don't have any of those kind of contradictions on Mars at all. And that's why Jim and I are going to try to capitalize on the interest in this.

SB: What was your impression of the scene at JPL when you were there?

Margulis: People were pretty much ecstatic. There was the possibility of a crash landing, or of everything not working, but everything worked. They're really getting data, a lot of data. Everyone is up to their ears. I know that the biology team has been having 6 a.m. meetings. Everyone's exhausted, but elated. It's a regular high around there.

SB: You sound pretty excited yourself.

Margulis: Well, Mars is interesting to me because it's kind of like a naked Earth. As soon as you get a consensus there's no life on Mars, that's where the ball game begins as far as I'm concerned. Because that's when we'll be able to define the ways in which we know life has modulated the Earth. But as long as someone's holding out that there's life on Mars that's modulating the planet, then we don't know how to subtract it, because it would be a totally different kind of life, and its effects would be unknown, and so on. I can't begin on the exercise. So I'm one of the few people, I guess Jim Lovelock too, who are really excited about a good negative result. What I would call this, if I were writing your article, is "The Three Billion Dollar Negative." I don't know if that's the right price, but it's some huge number like that.

SB: So Earth is alone.

Margulis: As far as the eye can see in the Solar system, we're here by ourselves. When you start talking about life elsewhere, that's highly probable, but it's terribly distant, far outside the solar system.

SB: What does this say about the delicacy of the position of Earth in relation to the Sun? Venus is just a bit closer and Mars is just a bit father, and if they're both dead, that puts us in a very narrow cambium.

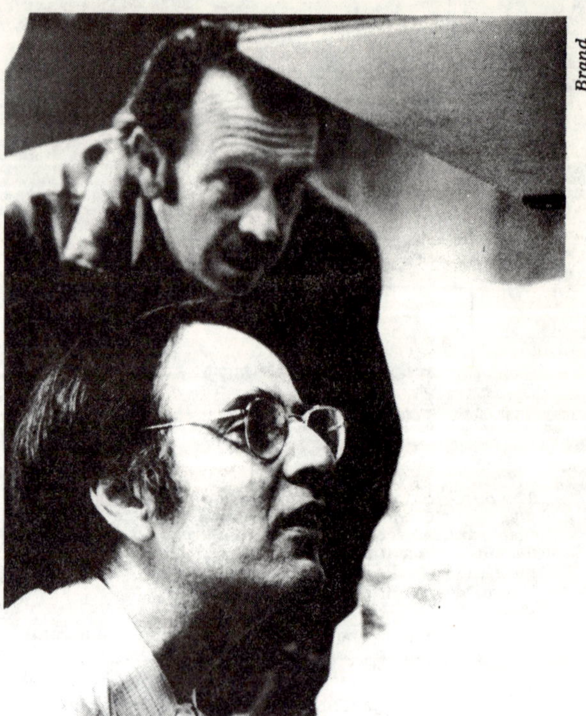

July 20, 1976, Viking I landing. Carl Sagan viewing the first photographs from the surface of Mars on a TV monitor. Behind him, an ABC-TV cameraman. Carl was on various national networks practically all day.

Margulis: That's true, but on the other hand, it's known by anyone who's cared to think about it that for life to persist you've got to have open bodies of liquid water. I don't mean steam, and I don't mean ice, which excludes constant temperatures above boiling and below freezing. Life is essentially a variation on the theme of water.

SB: Do you know anybody else besides yourself and Lovelock who've formally predicted no life on Mars?

Margulis: That's a good question. You find someone else* and I'd be very interested. I never ran across it, but you know i don't really read Martian literature very much except what's sent to me that I have to read. I think that there's a general sentiment amonst astronomers that there probably isn't any Martian life, but I don't know if they've said that in any kind of intellectual framework or in a formal way. ■

*In '67 Hitchcock & Lovelock made this prediction based on what we knew then of the atmosphere of Mars.   —LM

CARL SAGAN

Stewart Brand: How's it look for Viking II right now?

Carl Sagan: Well, we had a hairraising time during entry because we lost communication with the orbiter, so it was sort of wild, but everything seems to be in fine shape there now. As far as I know, there are no malfunctions at all. We landed in a place with the promising name of Utopia. The landing site some people thought would be filled with small sand dunes, but there's nothing but rocks as far as the eye can see. Many of the rocks are quite strangely shaped, they may be aeolian ventifacts, rocks which are sculpted by sandblasting by windblown dust. Some of them certainly seem volcanic in origin.

SB: It looks enough different from the Viking I site to be interesting?

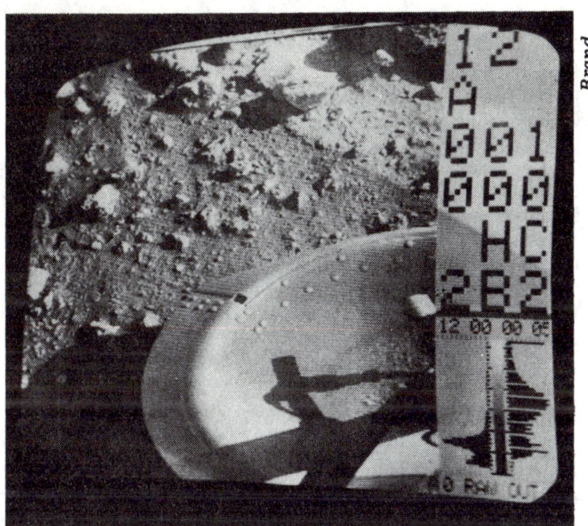

*The first photograph from Mars on a monitor at JPL.
Viking looks down at its own foot. The graph on the
right shows how the contrast has been selectively
computer-enhanced.*

Sagan: It's different, but both Viking I and Viking II landing
sites were chosen fundamentally for their blandness, so we
should not be surprised if we find that close up they're
reasonably bland. We know the planet as a whole is remarkably exuberant and challenging. The reason for the Viking
II landing site is that it's warmer and particularly more
humid, and since two of the microbiology experiments are
oriented towards liquid water it was reasonable to take some
additional risks to in some way increase our chances of finding the sorts of microbiology that the experiments are geared
to detect. We won that gamble. The landing was successful.

SB: What's your current feeling about the life experiments
on Viking I?

Sagan: Well, the results are tantalizing. All three microbiology experiments have given positive results, and what's
more each one different from its control experiment in a
manner that if we had seen it on Earth, we would not have
hesitated to ascribe it to microbiology. But there certainly
is a possibility that there's an extremely active and unfamiliar
sort of surface chemistry going on on Mars, perhaps due to
the fact that ultraviolet light gets to the surface of Mars more
than gets to the Earth, and that there's a lot of water bound
up in the rocks of Mars, as there is not in the rocks of the
Moon. If that's the answer, it's still a remarkably important
conclusion because it means that there is a kind of non-
biological surface chemistry which duplicates in some
substantial detail the biological chemistry of living systems
on the Earth. That's both oxidation and reduction reactions.
If that's the answer, we have learned something quite
important about the origin of life.

SB: Say more about that.

Sagan: Well, if it turns out that (and I stress this is the least
interesting possible result) that the surface chemistry is able
to simulate biological chemistry, is able to reduce carbon
dioxide to organic molecules, well, that's what photosynthesis does; that's what green plants are about. And if the
surface materials are able to oxidize organics to collect
oxygen and make water, well, that's what respiration is
about. If processes which, to some extent, simulate respiration and photosynthesis can occur before there's life on a
planet, that means that all biology has to do is simply use
the pre-existing surface chemistry. Some aspects of biological chemistry would then be not as major an invention
as people usually thought before. So in that sense, these
results make the origin of life easier and clarify an important
question on the origin of life.

SB: The idea being that such might have been the
conditions here before life got going?

Sagan: We would have had ultraviolet light reaching the
Earth's surface before there was oxygen in our atmosphere
and it was still a reducing atmosphere. And we would have
had abundant water on the surface. So in that respect we
duplicated that sort of chemistry.

SB: Is it the case that if abundant liquid water were made
available on Mars, then something might get going?

Sagan: No, I think both for the Earth and for Mars, the
reducing atmosphere necessary for very large scale synthesis
of organic molecules is absent. I think if all life were wiped
off the Earth in some unimaginable catastrophe, life would
not arise again, because of the escape of hydrogen and the
absence of reducing conditions. You might have a sort of
low level and faint rhythm in the bass, so to say — surface
chemistry of both oxidation and reduction reactions which
would to a small extent resemble biology. But the theme
would be gone.

Of course the other possibility is that all of this peculiar
chemistry is indeed due to microbes, and that we have
uncovered a Martian biology. I think this is a more
interesting conclusion, but either way we will have learned
something extremely important.

SB: What do you make of the gas chromatograph experiment
which is not showing much of anything?

Sagan: Well, that certainly is interesting, but you have to
bear in mind that the large amount, almost 1% of bound
water in the Martian soil, poisoned the ability of the gas
chromatograph mass spectrometer to detect simple organics
up to about two or three carbons. So these results were
only for much more complex organics, and the results are
quite negative down typically for a given molecule to
something like 10 parts per billion. That's not very abundant;
that's a very very sensitive test. Some people say that there
are places on the Earth which have an indigenous microbiology but exceptionally low levels of organic matter, and
other people see that as an open question. But suppose it's
true that in the one or two places that we've looked there is
microbiology but there is no organic chemistry. What's the
conclusion? The most likely conclusion is that life on Mars
exists in what Lederberg and I 15 years ago called micro-
environments. That is, in certain specialized zones of the
planet, maybe wet places for example, microbiology thrives;
but it's dispersed, possibly by wind. So you can find the
bugs lots of places, but they reproduce and are concentrated
only in a few. That's at least one possible explanation.

SB: I understand that the GCMS in this experiment was
showing even less organic carbon than would be expected
from meteorites and such things.

Sagan: That depends very much on the rate of infall and
the rate of destruction of the organics, so I don't think that
conclusion can be too clear. There are certainly some models
of the meteorite infall and ultraviolet photo-destruction
which would be consistent with the upper limits of organic
matter detected by the GCMS.

SB: Let me be real clear in my understanding of what that
is not finding. What would have to be there for it to get
a positive?

Sagan: Let's take a molecule, napthalene. You'd have to
have more than ten parts per billion of naphthalene for it to
be detected. On the other hand you could have huge
quantities of propanol or formic acid, say, and you wouldn't
detect it, because those are low carbon number organics,

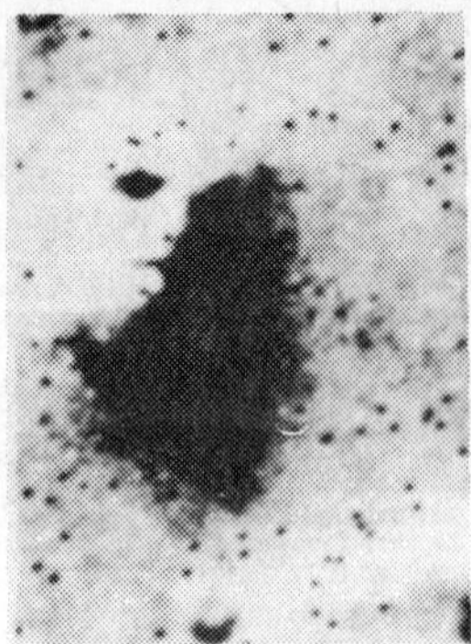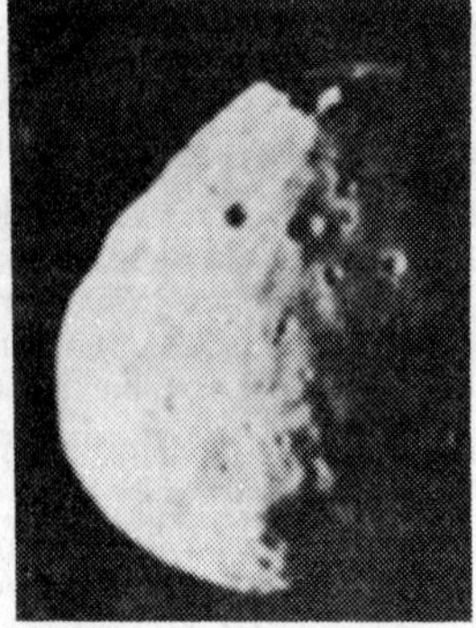

Large Martian? The "face" is a mesa formation one mile across.

Right, Phobos, the inner of the two satellites of Mars, is about 12 miles wide and 17 miles long. In the Martian sky it shines 1/4 to 1/2 as bright as the Moon on Earth. The other satellite, Deimos, is seen as a bright point of light on Mars.

which would be totally obscured by the bound water. That will be avoided by a clever maneuver in the second lander, and so we may have information on the simple organics in the next few weeks.

SB: What other matters look interesting at this point besides the biology?

Sagan: There are a whole bunch of other questions. Why do various landscapes look the way they do? How did they get that way and what does it say about the origins of the planet? How sure are we now that there really was a dense early atmosphere during which water flowed — which looks a lot more likely now than it did a few months ago.

SB: Are they getting a better idea on when that might have been that the waters flowed?

Sagan: Not yet. That'll involve very careful crater counting in the channels — the only available quantitative method of determining the ages of geological features on Mars, and the method certainly has uncertainties attached to it.

SB: What's your activity now?

Sagan: Now that we're done with landing site certification I have a little more time to devote to scientific questions. I'm interested in the physics of wind-blown dust on Mars, the possible influence of ultraviolet light on the microbiology and organic chemistry results, sand ripples on Mars, and a number of other questions. You understand, the Viking mission is a team effort. The results we have were obtained by a large group of very competent scientists and engineers.

One other issue I'm quite interested in is an apparent paradox. There are remarkably well preserved rocks at the two landing sites — sharp angular faces and so on. Yet it seems that the rate of wind erosion in the great sand storms should be phenomenally high. How can we reconcile those two? One possibility is that both landing sites have been only recently exhumed. They might have been covered over with dust for quite long periods of time, and of course there are other places which are in the opposite condition. There may be something wrong with our calculations of abrasion rate and so on, but right now I'm quite excited about this question.

SB: Would it bode well for the biology experiments if those were recently uncovered areas?

Sagan: Interesting question. Maybe. I can only see one model in which that would be the case, which is if there were biology early, and it was covered so the bugs were dormant and protected under a heavy blanket of dust. If the overlying dust has recently been removed perhaps the bugs are now accessible.

SB: Obviously what you'd like to be able to do is drill and make some kind of a core.

Sagan: That's absolutely right, and go to many places — investigate in all three dimensions. I'd like to have a rover with deep-boring core capability. That would be heaven.

SB: Is there hope for such a vehicle?

Sagan: Well, it's interesting. As a precaution against various system failures of the two Viking orbiter-lander combinations, essentially all the pieces of a third orbiter-lander combination were constructed and assembled. So all that's necessary is to put it together, mount it on wheels or tractor treads, make some adjustments in the experiments and launch it off to Mars.

SB: What adjustments?

Sagan: That's still to be debated. At the very least you would want a more elaborate pharmacy.

SB: Namely?

Sagan: I'd like to see more molecules, I'd like to see them added one at a time, and I'd like to have somewhat more control over sequencing, so that if the experiment goes one way you do this, and if it goes another way you do that.

SB: A branching regime.

Sagan: Exactly, a branched contingency tree. Now, that might involve certain additional items of artificial intelligence — which are well within our reach. You know, it's so frustrating, because you see from orbit this fabulous geology, exciting terrain, and then you consciously go and land in the least exciting place. And you're stuck. You can't hop an inch. In the Viking I stereo pictures you see these gently rolling hills with hints of a valley between the hills, and you have this extremely human urge to go take a walk. And you can't do it. But the technology for mounting the whole thing on tractor treads and going off is at hand. You could wander to your own horizon every day quite easily. You'd look around, see the most interesting thing that's far away, go to it, look at it close up, and look around again. And in the likely lifetime of such a space craft you could travel at least many hundreds and perhaps even thousands of kilometers. That's a very big turf.

SB: You're on terrain that treads can handle?

Sagan: Yes. It's a lot easier to wander over rough terrain than it is to land on it. You could have in the vehicle the sorts of feedback loops that you see on toys. When they come to the edge of the table or an obstacle, they know to stop and turn around. It would be able to pick its way among the boulders. If such a device ran into trouble, it could stop and ask for instructions.

SB: Since Viking I landed a month and a half ago have there been any changes in the terrain or the weather?

Sagan: The atmospheric pressure is steadily declining every day. In less than two years it'll go to zero, at this rate. . . that's not likely to happen. What's happening, we think, is that it is winter and one of the polar caps is rapidly growing at the expense of atmospheric CO_2. So the atmospheric pressure is going down. When the season changes the trend in the atmospheric pressure will reverse.

SB: What's the expected lifetime of the landers?

Sagan: It's hard to tell. It might be extremely long, many years. They're not going to run out of power, because they're powered by the decay of radioactive isotopes, so it's just a question of parts failing.

SB: Years being two years, ten years, twenty years?

Sagan: Well, twenty sounds long. Two sounds perfectly possible. The nominal lifetime it's been geared for is 60 to 90 days.

Carl Sagan and Viking Project Scientist Gerald Soffen on "Mars" between ABC-TV news interviews. The color backdrop is by Don Davis, who painted the front cover.

SB: How about Earthside? Is the program up for staying with Viking as long as it's alive?

Sagan: There are plans for an extended mission now.

SB: What's an extended mission?

Sagan: That begins after solar conjunction in late December. There are a number of marvelous things that can be done. One is we can move an orbiter so that it can proceed sufficiently close to Phobos — 50 kilometers — to take pictures of <u>one meter</u> resolution, which should be quite an eye opener. In less than three years operation we could photograph the planet pole to pole at better than 80 meters resolution. But the money for such an extended mission is not yet allocated.

SB: I was looking at an image of Phobos and wondering about its size. Is it large enough that if you stood on it would it hold you to the surface?

Sagan: Yeah, you would be held to the surface, although you could probably do a standing high jump of one kilometer. And you could, if you were at all good, pitch a baseball to escape velocity.

SB: And run yourself into some kind of an orbit? If you could get the footing?

Sagan: Well, let's see, at about 20 miles an hour you could launch yourself into orbit.

SB: Considering the kind of jumps you could take, that might be possible.

Sagan: The whole prospect of inter-planetary olympics arises — different records for different worlds.

SB: One thing I've wondered about looking at the eery familiarity of the Martian domain, how much more habitable is it in terms of humans than say the Moon?

Sagan: An unprotected human being would be in trouble in both places of course, but Mars has so much more in the way of available resources than the Moon does, it's just no comparison. Mars is loaded with water. Not in a liquid state but it is available, and all those puffs of oxygen in the gas exchange experiment show it's loaded with oxygen. Not exactly so that you could stick your head into the ground and take a deep breath, but with a little technology you could collect that oxygen. So I would say if people had a serious reason to do it — and I don't know of any yet — I think it would be possible to make some colonies on Mars.

SB: The atmospheric pressure is the equivalent of what altitude above the Earth?

Sagan: 100,000 feet.

SB: Is the Martian atmosphere pretty much as expected from the early data?

Sagan: Yes, the important findings are a few percent of both nitrogen and argon. The importance we give to the nitrogen is that if you want to imagine a microbiology in any way similar to the terrestrial sort, you have to have nitrogen in the atmosphere. But the main thing is the abundance of isotopes of argon and nitrogen — argon 36 and argon 40, and nitrogen 14 and nitrogen 15, which by two separate arguments give evidence for an early dense atmosphere on Mars.

SB: It is less dense now because why?

Sagan: A variety of reasons. The ones that I lean towards are the freezing out of the atmosphere in the polar caps. I think it very likely that the dense early atmosphere is still there, but it's buried.

SB: What's the sunlight at high noon at the site of Vikings I and II versus sunlight at high noon in Pasadena?

Sagan: Half. It's like what you have here on an overcast day.

SB: How does Mars compare with Venus?

Sagan: Well, despite the good operation of the Soviet spacecraft on Venus it got fried in an hour.

SB: So if you have a choice between fire and ice, take ice.

Sagan: Yes. Although it's not always that icy on Mars. But both planets are exotic by terrestrial standards.

SB: It's assumed at this point that Venus has no life?

Sagan: Well, it would have to be a remarkably hardy form of life to survive that 900° Fahrenheit on the surface. A completely airborne sort of biology may not be totally out of the question.

SB: What's the next planetary number for us now?

Sagan: There are two more approved missions, both approved many years ago. One is called Mariner Jupiter-Saturn, which will be launched next year and go by Jupiter in '79, Saturn in '81. That means 30,000 photographs or those two planets. That's a fly-by mission, preliminary reconnaissance, it's not at all a detailed study. And then there will be launched one year later Pioneer Venus, a small rather modest set of entry probes. And that's it. We have no approved viking follow-on for Mars. Nothing like a lander on Titan, a probe into Jupiter, an orbiter around Venus, a rendezvous with a comet, all sorts of meaty kinds of missions that we're fully capable of and our scientific knowledge cries out for. There are no such missions, and there's a serious question that the remarkable set of talents put together for this sort of exploration is going to just simply dissipate, go into other things. As yet there is no government approval for continuing the systematic exploration of the solar system. ∎

Education for Space work

The following is reprinted from the August '76 National Space Institute Newsletter (membership $9/year for students, $15/year for adults, from 1911 N. Fort Meyer Dr., Suite 408, Arlington, VA 22209).

Many letters come to NSI asking what to study to break into a space-related career, and whether there will be more such work available in the future.

Comprehensive answers to the first question and implications for the second are contained in two volumes which NASA published last January, **Outlook for Space** and **A Forecast of Space Technology 1980-2000**. These present the careful, thorough conclusions of a two-year study conducted by a NASA team which conferred with many outsiders. The authors recognize two needs in a space program: to serve the physical requirements of humanity and immediate problems, and to challenge the mind. A zest for exploring the unknown is linked to national vitality. NSI members will cherish the latter point; letters reveal a straining toward the outer limits of space experience, but then NSI members are "the believers."

Outlook for Space urges that NASA, in its planning, be responsive to national needs. Future programs could elaborate on unique satellite services to the major concerns of food, energy, the understanding and protection of the environment, health care and such. Certainly the value of such programs can be generally appreciated, could bring NASA into more conversation with potential users of space, win support for more Earth-oriented programs . . . and so open more jobs.

According to NASA, the following five subjects will be at the heart of future technological developments, sure careers anywhere. (Incidentally, NASA is not the largest space employer; at the height of activity in the mid-1960's, 33,000 were on NASA payroll among the 500,000 believed involved in total space-related activity.)

1. Electromagnetic properties of solids.
2. Integrated circuit technology.
3. Cryogenics and superconductivity.
4. Microstructures
5. Coherent radiation and integrated optics technology.

Integrated optics is so new it is not even mentioned in **Physics in Perspective**. It means achieving in the optical region of the spectrum an analog of the present coherent microwave technology. According to Dr. John R. Pierce, Caltech, it offers the greatest advance in communications since the invention of the transistor.

The following organizations usually have career information to send:

American Institute of Aeronautics and Astronautics (AIAA)
1290 Avenue of the Americas
New York, NY 10019

Aerospace Industries Association of America, Inc.
1725 DeSales St. NW
Washington, DC 20036
Dr. Wayne R. Matson, Educational Services. Ask for the reprint, "Career Opportunities in the Space Program."

National Aerospace Education Association (NAEA)
806 15th St. NW, Room 338
Washington, DC 20005

NASA Education Programs Division,
Mail Code FE
Washington, DC 20546

The Engineers' Council for Professional Development
345 East 47th St.
New York, NY 10017
Sends lists of institutions for engineering or engineering technology study, those with aerospace departments noted.

NSI can send a list of universities teaching remote sensing, as for interpretation of Landsat images.

For orders to Government Printing Office:
NASA Outlook for Space, 3300-00640-3, $3.60;
Forecast of Space Technology, 3300-00641-1, $4.00
(GPO: Washington, DC 20402).

128

KEY SCIENCES FOR SPACE

Applied math (needed in more than half the following)

Thermodynamics	Organic chemistry
Atomic physics	High temperature chemistry
Nuclear physics	Solution chemistry
Plasma physics	Electrochemistry
Fluid physics	Photochemistry
Coherent radiation physics	Biochemistry
Meteorology/atmospheric physics	Biology
Molecular physics	Exobiology
Solar Physics	Space medicine
Geophysics	Optics EM
Low temperature physics	Particle optics
Biophysics	Optical properties
Elementary particles	Magnetic properties
Macromolecules and polymers	Electricity and magnetism
Colloids	General relativity and gravity
Superfluids	Cosmic rays
Thin films and membranes	X-rays
Superconductivity	Gamma rays
Lattice dynamics	Ultraviolet
Structure of solids	Visible/near infrared
Surfaces and interfaces	Radio
Metallurgy	Celestial mechanics
Thermodyn phase phenomena	Astronomy
Transport phenomena	Planetology
Catalysis	Metrology
Reaction kinetics	Ecology
Radiation effects in solids	Computer science
	Information theory
	Artificial intelligence

The Directory of Aerospace Education

I had no idea there were so many organizations, magazines, books, films, museums, programs, graphics, and services available to serve the kid interested in space. This is a comprehensive, nicely annotated guide, surprisingly exciting reading for a reference book.

— SB

The Directory of
Aerospace Education
1977; 79 pp.
$2.95 postpaid

The Journal of
Aerospace Education
Monthly (Sept.-May)
$10/yr

both from:
American Society for Aerospace Education
806 15th St. NW
Washington, DC 20005

SKYLAB FILMS

In 1973 three crews lived in the weightless environment of the sky-lab Space Station. AAPT, with the support and assistance of NASA, has produced twelve single concept films from film footage taken during these missions. Each film demonstrates a concept in the world of zero-gravity physics. Each film is approximately 3 minutes long. The cost of each film is $14.75 to AAPT members and $21.00 to others.

1. HUMAN MOMENTA — The astronauts float through the orbital workshop. Various initial conditions of motion are shown.

3. ACROBATIC ASTRONAUTS — A complex series of spins, turns, and half-flips looks easy in zero gravity.

4. GAMES ASTRONAUTS PLAY — Skylab recreation included darts, paper airplanes, weight lifting, and a balancing act.

5. REFERENCE FRAMES — Scenes of moving astronauts taken with a still camera and a moving camera are contrasted.

7. COLLISIONS — A sequence of colliding objects, from astronauts to water drops, is shown.

●

ASTRONAUTICS AND AERONAUTICS
1290 Ave. of the Americas
New York, NY 10019
Included with student dues; $28 per year to nonmembers of AIAA, issued 11 times a year; covers technical developments in aerospace.

AVIATION WEEK AND SPACE TECHNOLOGY
1221 Ave. of the Americas
New York, NY 10020
$27 per year for qualified subscribers (involvement directly in aviation and space); $37 per year for nonqualified, issued weekly; hard news in the international aviation and space fields.

Colonies in Space

Of the three Space Colony books around at this writing (Fall 1977 — the other two are O'Neill's and this one), Tom Heppenheimer's is the most focussed and out-of-doubt. He writes knowledgeably, comfortably, with conviction, offering — to my mind — somewhat too clear a view of the future for the future ever to resemble. The book will date quickly as a result, but it'll feed aspirations in the meantime.

— SB

Colonies in Space
T. A. Heppenheimer
1977; 224pp.

$12.95 postpaid
from:
Stackpole Books
Harrisburg, PA 17105
or Whole Earth

Suppose we want to get a ton of aluminum. The "recipe" would be:

"Take ten tons of anorthosite. Melt in a solar furnace at 3200°. Add water and quench the melt to give a glassy solid. Allow to settle in a centrifuge; pipe off the steam from the quenching to a radiator to condense it to water. Remove the glassy material from the centrifuge, grind fine, and mix with sulfuric acid. Pipe to another centrifuge to separate off the aluminum-bearing liquid which has resulted. Mix with sodium sulfate and heat to 400°; pipe to still another centrifuge to allow the resulting sodium aluminum sulfate to settle. Remove it after it has settled, and bake at 1470° to produce a mixture of alumina and sodium sulfate; wash out the latter with water. Mix the alumina with carbon, and react the mixture with chlorine. This gives aluminum chloride. Put the aluminum chloride through electrolysis. Result: one ton of molten aluminum."

•

Water fights will be great fun, carried out at long distances. A double handful or pailful of water will more or less hold together under its surface tension, forming a glistening blob which squirms and wiggles in its flight.

Space Age Review

A brand new and small Space buff publication with appealing zeal. Its regular feature of bite-size reports on wide front of Space news is handy.

— SB

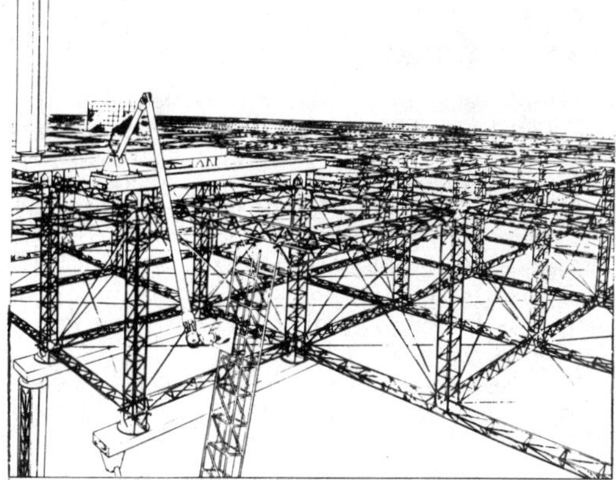

Space Age Review
Steve Durst, Editor

$10.00/yr (monthly)
$1.00/single copy
from:
Space Age Review
378 Cambridge Ave.
Palo Alto, CA 94306

Nichelle Nichols (Lt. Uhura on "Star Trek"):

To black people I say, we'd better get in, sit in, fit in and grow in it as though your very lives depended on it — because they do. Humankind is going into space whether we like it or not. And when we colonize space, we don't want to be there as chauffeurs and tap dancers.

The assemblers will not be robots. They will be linked to small computers, programmed to guide the assemblers through repetitive operations on command. For instance, there will be a computer routine which will direct the assembler to make a weld in a given spot; the operator only has to specify the spot. Rather than looking like robots, the assemblers will be similar to numerically controlled machine tools. Such tools have been used for many years. Some of the largest are at the Boeing Company to rivet and assemble jetliner wings without the need for human riveters.

•

There is one science-fiction cliche which we can immediately dismiss. This is that the homes will use plastic, perhaps even be built entirely of plastic. Plastics are made from hydrocarbons such as oil and these will be quite costly in the colony. The precious stores of carbon will cycle through the plants and food, not be locked up in the walls of a home unless there is a clear need (for example, as insulation for wiring). For the same reason, wood may be a rarity even if it is locally grown. The builders will rely on materials readily available. These include the metals aluminum, titanium, and iron or steel. Also there will be plenty of glass, including fiberglass for insulation or soundproofing. Other building materials will also be available from the lunar materials brought to the colony.

For instance, many lunar soils are chemically similar to clay. It should be possible to mix them with water, make bricks, and fire them in a solar furnace. The resulting bricks will usually be gray or chalky in color rather than being red or light brown. Bricks could also be made of cast basalt — a material used in France for tiles, plumbing pipes, and the like — from lunar basalt melted in the same solar furnace. Quicklime, obtained from lunar plagioclase and mixed with water and dry lunar soil, can be used for cement. It is quite common on the moon, nearly as abundant as aluminum. There would be concrete for sidewalks or walls, concrete blocks for foundations, and thin, watery cement to serve as stucco for houses.

Earth/Space News

Edited by an enthusiast with libertarian bias, this newsletter is primarily interested in space industrialization, space capitalism, and political freedom in space.

— SB

Earth/Space News
Paul Siegler, Ed.

$10.00/yr (bi-monthly)
$2.00/single copy

from:
Earth/Space News
2432 Whitney Dr.
Mountain View, CA 94043

These uses of Skylab are not restricted to the original owner (NASA), for under international law, Skylab is now recognized as a 'Space derelict'. This means the Space station is eligible for salvage claims by *any person or organization* which can board it first. The same holds true for the multitude of other derelict satellites now voiceless but retaining their high capacity for productive work.

129

Space Settlements
Space Manufacturing Facilities

If you want to get up to speed on the detailed progress of Space Colony design and speculation you'll need these books. Both are two years after the conferences they report, so many of the details have been superseded, but also many ideas that later became central doctrine first emerged here and have not been elaborated on since.

Space Settlements reports on the 1975 (the first of three) NASA-Ames Summer Study organized by NASA in California. The style here is group work toward specific design goals. O'Neill and his cohorts participate.

Space Manufacturing Facilities is the proceedings of the first two Princeton conferences organized by Gerard O'Neill, 1974 and 1975 (there have been two more since then). These are individual papers, many of them of considerable quality, on both technical and liberal-arts questions.

— SB

Space Settlements: A Design Study
Richard D. Johnson
& Charles Holbrow, Eds.
1977; 185pp.

$5.00 postpaid

from:
Supt. of Documents
U.S. Govt. Printing Office
Washington, DC 20402

or Whole Earth

Space Manufacturing Facilities (Space Colonies)
Proceedings of the Princeton AIAA/NASA Conference,
May 7-9, 1975
Jerry Grey, Ed. — 1977; 266pp.

$19.50 postpaid

from:
American Institute of Aeronautics & Astronautics
1290 Ave. of the Americas
New York, NY 10019

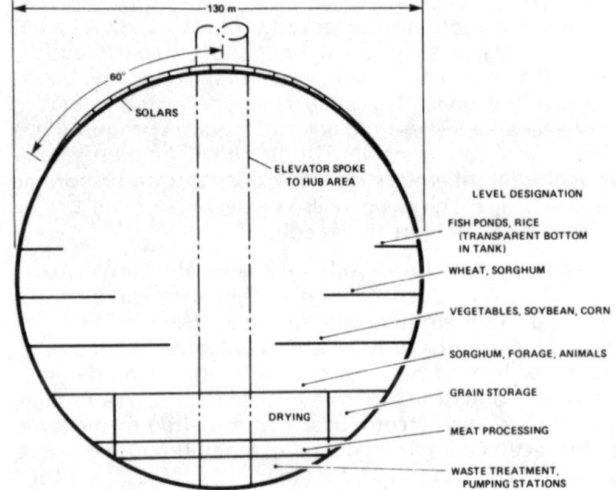

Cross-section of agricultural region.

— Space Settlements

Christians have a seven-day week. On the other hand, the Balinese have several types of weeks running concurrently: 2-day week, 3-day week, 5-day week, 7-day week, etc. These different cycles "heterodyne" at regular intervals (though not at different frequencies, but at common multiple intervals), and you have a 105-day anniversary, 210-day anniversary, 420-day anniversary, etc.

Most Americans eat three times a day. There are cultures in which the number of meals per day is 1, 2, or 5. In France and Italy, the largest meal of the day is the noon meal, and the "lunch break" lasts 2 to 3 hours.

Heterogeneity between settlements not only increases the probability of the survival of the human species, but also increases the speed of the cultural evolution, as well as enrichment of human life. Furthermore, since we do not know which social systems work well and which do not until we have actually tried them out, we need to experiment with several different social systems.

Magoroh Maruyama
Space Manufacturing Facilities

Spaceflight
JBIS

These two publications from the British Interplanetary Society (BIS) supply the very two things lacking in American Space journals — articulate English and disciplined grand speculation. For the Brits Space is an adventure, and their attitude is catching. And their perspective on American and Russian activities (bemused, delighted, appalled) can help prevent deadly ethnocentrism in our planetary endeavors. Loveable mags.

— SB

Spaceflight
Dr. A. R. Martin, Ed.
$22.00/yr (monthly)

JBIS (Journal of the British Interplanetary Society)
Kenneth W. Gatland, Ed.
$16.00/yr (monthly)

from:
The British Interplanetary Society
12 Bessborough Gardens
London SW1 V 2JJ
England

The Spaceflight Revolution: A sociological study
By William Sims Bainbridge, John Wiley & Sons, 1977, pp. 294, $21.

This is a thoroughly professional book, yet fascinatingly readable and easy to understand. As its title indicates, it deals mainly with the people involved in the Spaceflight Movement, their backgrounds, hopes, aspirations and deeds, and the impact of the Movement on past, current and future history. However, the hardware receives attention adequate to the unfolding of the history of what is undoubtedly the greatest step forward since the Industrial Revolution, besides having far greater ultimate significance than anything since Man's discovery of the Genie of the Bottle — Fire.

The book is in every respect a penetrating and sensitive survey, and there is ample relevant meat therein to provide scores of thinking and talking points. Overall, the book tells how not the public will, but personal fanaticism drove men to the Moon. There are a number of interesting Tables and two Graphs, but no pictures. The inside histories of the great pioneers — Tsiolkovsky, Goddard, Oberth, Sänger, von Braun and others make absorbing reading, with new angles explored.

— Space Flight

JBIS
CONTENTS (MARCH 1977)

SPACEFLIGHT, COLONIZATION AND INDEPENDENCE: A SYNTHESIS
COSMIC RAY SHIELDING FOR MANNED INTERSTELLAR ARKS AND MOBILE HABITATS
DETECTION OF STARSHIPS
FACTORS LIMITING THE INTERACTION BETWEEN TWENTIETH CENTURY MAN AND INTERSTELLAR CULTURES
MATERIALS IN INTERSTELLAR FLIGHT
SIGNALING OVER INTERSTELLAR DISTANCES WITH X-RAYS

. . . space flight is, quite literally, the expansion of intelligence; that it is one of the keys to the long-term survival and growth of the human species; and that it will change us irrevocably from what we are today.

— JBIS

Concise Atlas of the Universe

It's more important to have a recent atlas of the universe than a recent atlas of the Earth — the Earth is pretty well known and politically stable compared to the constant change of information coming in from astronomy and Space exploration. When you're very ignorant, as we are about the universe, a little information is a big change.

So, this is the most recent, comprehensive, and spectacular tome on what used to be called the heavens. I use it to cure confusion (where is the asteroid belt?), boredom ("Star Wars" is only a movie), and egotism (there are only 30 galaxies in the local galaxy group — a small group).

LOCAL?!

— SB

The Rand McNally Concise Atlas of the Universe
Patrick Moore
1970, 1974, 1978; 190pp.
$30.00 postpaid

from:
Rand McNally & Co.
PO Box 7600
Chicago, IL 60680
or Whole Earth

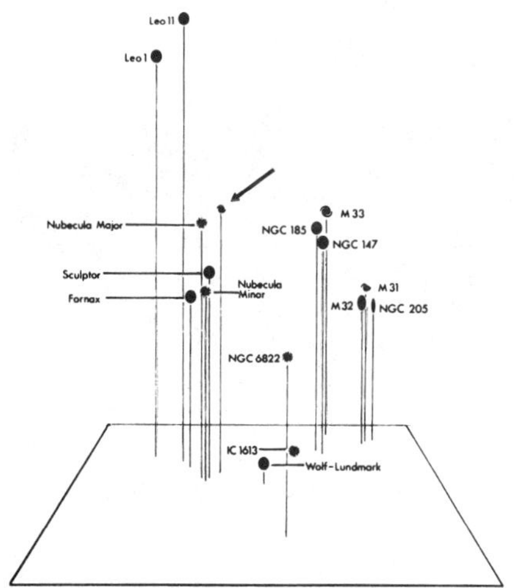

Local Group of Galaxies

Our Local Group is a small cluster, having less than 30 known members: the spirals M31, M.33 and our Galaxy; the Clouds of Magellan, and smaller elliptical and irregular systems. These galaxies are so close to us that their own peculiar motions mask the effect of the red shift/distance relation.

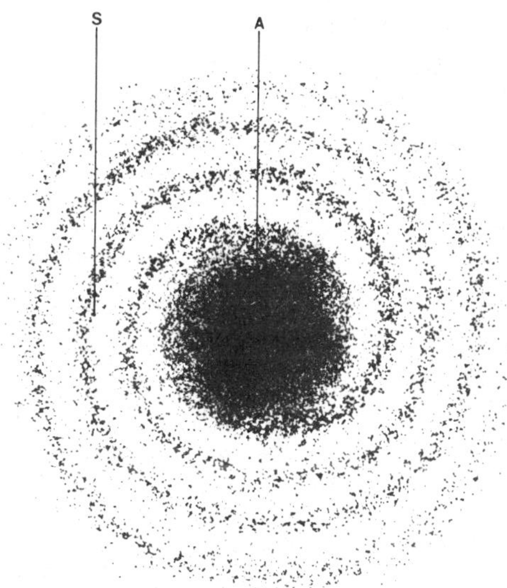

Our Galaxy contains 100,000 million stars. It has an overall diameter of 100,000 light-years; the maximum breadth is 20,000 light-years, and the Sun, with its planets, lies close to the main plane, 32,000 light-years from the galactic centre.

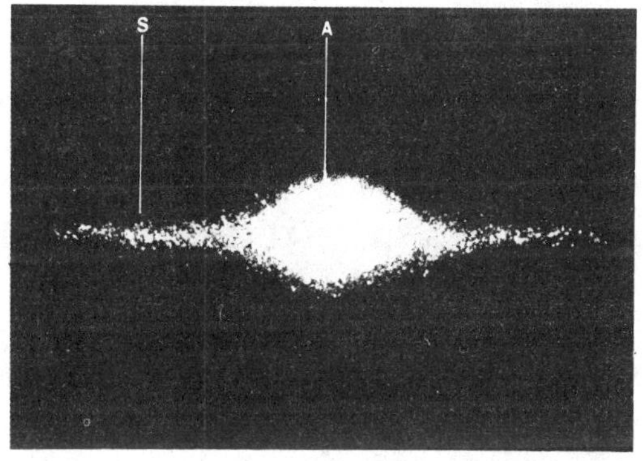

Sky and Telescope

All good space cadets have first hand observational experience of the sun, moon, planets, and stars, and they use this magazine to keep up with new equipment, discoveries, and gimcrackery such as meteorite rings ("the only rings on Earth with stones from outer space"), planet photos, etc.

— SB

Sky and Telescope
Joseph Ashbrook, Ed.
$10/yr (monthly)
$1/single copy

from:
Sky Publishing Corp.
49-50-51 Bay State Rd.
Cambridge, MA 02138

Lost for 41 years, the famous minor planet 1936 CA Adonis has been found again. This tiny, faint object is observable only when it makes a fairly close approach to Earth, as in 1936, when the minimum separation was about four times the moon's distance.

Astronomical Calendar

I'm afraid that this is going to sound like one of those bar bets, but if one was asked to name the best annual astronomy guide it would have to be Guy Ottewell's Astronomical Calendar.

It has only been around for three years so it hasn't yet received the attention that it should, but already it's replacing the Royal Astronomical Society of Canada's Observer's Handbook.

The Calendar *is published annually and contains all one wants to know and more about the coming events for the year. Guy himself does the cover illustration (15'' x 11''), and produces, publishes and distributes the book himself. Somewhat like a Shakespearean play it seems to work on all levels: the casual amateur or the serious astronomer can read through it and find something to appreciate on every page.*

Guy has spent most of his life sleeping out of doors, and as a result the book lacks the cold detachment that turns one away from other similar books.

—Laurence Kent Sweeney

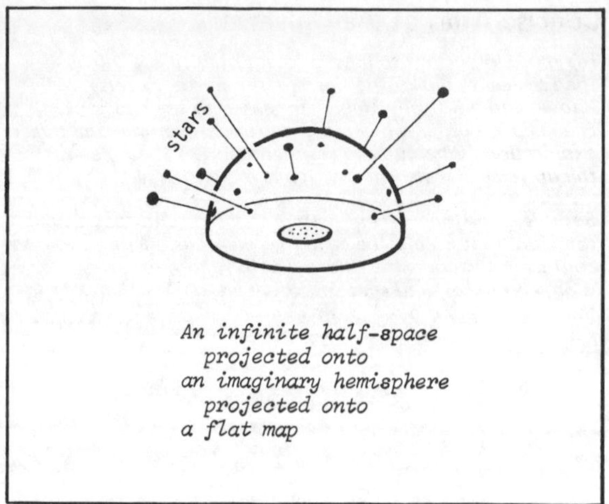

Astronomical Calendar
Guy Ottewell
55 pp.

$4.95 postpaid

from:
Astronomical Calendar
Dept. of Physics
Furman University
Greenville, SC 29613

Astronomical Calendar *is ready each October. A fine Christmas present.*

— SB

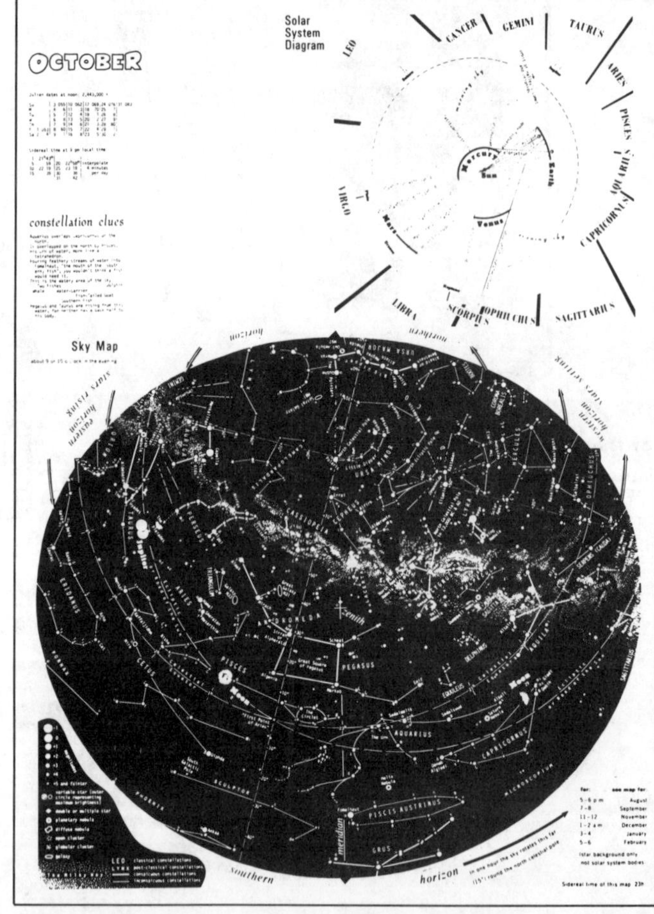

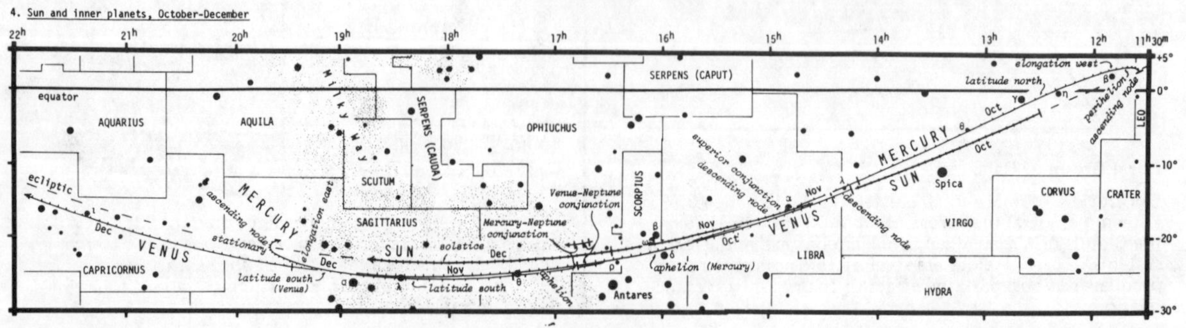

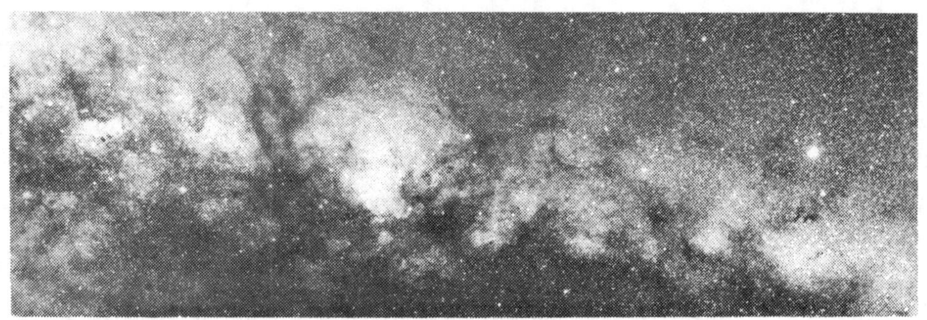

Starfield in Serpens-Aquilla region. Photo by Mount Wilson and Palomar Observatories.

OLBERS' PARADOX

And I heard the learned astronomer
    whose name was Heinrich Olbers
speaking to us across the centuries
  about how he observed with naked eye
   how in the sky there were
       some few stars close up
and the further away he looked
      the more of them there were
  with infinite numbers of clusters of stars
   in myriad Milky Ways & myriad nebulae
So that from this we can deduce
  that in the infinite distances
    there must be a place
     there _must_ be a place
       where all is light
  and that the light from that high place
       where all is light
   simply hasn't got here yet
  which is why we still have night
But when at last that light arrives
 when at last it does get here
   the part of day we now call Night
   will have a white sky
    with little black dots in it
    little black holes
      where once were stars
And then in that symbolic
    so poetic place
     which will be ours
we'll be our own true shadows
  and our own illumination
       on a sunset earth
      — *Lawrence Ferlinghetti*

133

Illustrations by Eric Drexler and Don Ryan

Solar sails are a way of moving things around in space, from one orbit to another. After a year's work, they are beginning to look like the best means of space transportation for a wide range of uses: they may be both cheap and fast. Before discussing clipper ships vs. canoes, however, we should first discuss boats and the ocean.

Space is big. It would take as many Earths to fill the solar system (500,000,000,000,000,000) as elephants to fill the sea (an unpleasant prospect). The Earth's orbit around the Sun is 23,000 times the Earth's circumference. Driving to the Moon (1/400 of the distance to the Sun) would take six months, at 55 mph of course. Driving to the nearest star would take 50,000,000 years, and so on. Space is Big. To get anywhere you have to go fast.

But, you say, since there is no air resistance in space, perhaps a patient traveler (or load of freight) could start out slowly and simply take whatever time was needed, drifting along. But, alas, gravity is in control. Objects in space don't really go anywhere, if left to themselves; they simply go around in orbits. Unless you kick something so hard that it stops completely (in which case it falls into whatever it was orbiting) or kick it so hard in the other direction that it can fly away despite gravity, never to return, the object will simply grunt at the kick, and shift its orbit somewhat. To get from one orbit to another generally takes at least two pushes: the first to put the object onto an orbit that crosses the orbit you're trying to reach, and another at the crossing point, to make the object start following the orbit you want, instead of the transfer orbit that the first push put it on. Another way to do the same thing is to push gently for a long time, and slowly twist and stretch the first orbit until it matches the second. Either way, you can add up the change in velocity that all the pushing would produce, in the absence of gravity, play around with different directions and times of push, and find that the total velocity change needed has a minimum that can't be beaten for a given trip. This requirement is usually measured in kilometers per second (1 km/sec is about 2,200 mph). One of the lowest requirements of any interest is 2.4 km/sec: the velocity needed to get off the moon.

Rockets have limits, because they must carry mass to throw away. A rocket can reach the same velocity as its exhaust fairly easily; not much fuel is needed to reach a few kilometers per second. The problem is that fuel has mass, just like the payload. Let's say you have a rocket with enough fuel to reach 1 km/sec, and to take a ton of payload with it. How much fuel would you need to reach 2 km/sec? Enough fuel to take the ton of payload to 1 km/sec, <u>and</u> enough fuel to take the fuel needed for the <u>second</u> km/sec to 1 km/sec. The total fuel mass needed turns out to increase exponentially with the velocity reached, just as population has been increasing exponentially with time. Both increases can gobble up more resources than you can afford to provide. Using the Saturn V moon rocket as a first stage, and piling up rockets from there, we could have reached 30 km/sec with enough payload to drop one haunch off an elephant into the Sun (an unpleasant prospect).

Rockets burning chemical fuels run out of ability fast when measured against the solar system, although they were decent for getting us as far as the Moon. The exponential curve that gets rockets into trouble can be made less steep, however, if more energy can

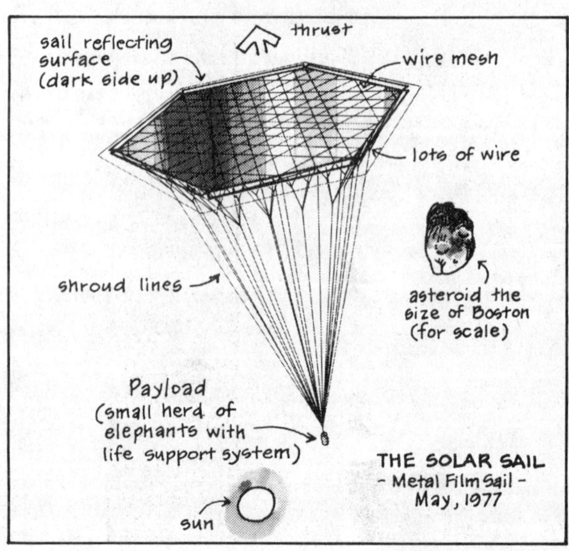

be put in the exhaust. This is the principle of the electric rocket; by soaking up solar energy in space and using it to throw small amounts of mass away <u>fast</u> (a mass driver is particularly efficient and versatile at this job), payloads may be pushed around the solar system in a reasonable way. The main problem is the cost and mass of the solar power plant. To use it efficiently accelerations must be low and trips long. Costs are also low: freight rates from Earth orbit to Mars orbit might be as little as $ .20 per pound.

Solar sails don't work on the rocket principle, but on light pressure. Like stage magic, this trick is all done with mirrors. Because $E = mc^2$, energy, including light, has mass. For light in particular, that little bit of mass moves very fast through space; when it is bounced off a mirror it exerts a force, just like fast ping-pong balls bouncing off a wall. If you wanted that wall to move quickly, even without friction, you'd want it to have little mass and be hit by <u>many</u> ping-pong balls. Similarly, the mirror that makes up a solar sail should be very thin and lightweight, and have a <u>large</u> area — a square mile of reflected sunlight exerts enough force to support the weight, not of a building, not of a car, a person, or a large dog, but of

a medium-sized cat. The name of the game, then, is to maximize acceleration by minimizing the mass per unit area of the mirror.

People have looked at this problem, off and on, for about 20 years. They set themselves the problem of stuffing about a square mile of folded reflecting surface into the nose of a rocket, of launching it, and of making it unfold and stretch into a reasonably flat surface in space. A design for a kite-like sail, with thin, aluminized plastic film for the reflecting surface, has finally reached an advanced planning stage at the Jet Propulsion Laboratory in Pasadena, California. (See illustration on inside back cover.) Their design can accelerate at about 1/7,000 of a gravity, which is actually fairly good: the sail can reach 1 km/sec in about eight days. This lets you get around, and because it needs no fuel, and no fuel to help carry fuel, and so on, it doesn't peter out at high velocities like a rocket does. They want to use it to reach Halley's comet (an object which is going around the sun the wrong way compared to the Earth; a huge velocity is needed): the flight would take four years. They may not get to do it, because solar electric rockets, mentioned above, still look good by

Asteroid mining facility with moored sail. Top, right: solar sail (10 km diameter). Top, left: Bernal Sphere colony (½ km diameter). Bottom, left: asteroid (1 km diameter). Bottom, center: industrial complex. Behind asteroid: mooring tower with shroud lines extending to sail in the distance. The pit on the right side of the asteroid has supplied enough material to build this colony, the industrial complex, 50 power satellites, and many, many sails like the one shown. The solar system contains thousands of similar asteroids. The sail shown is one of a fleet used for asteroid mining; when loaded (2,000 tons) it will depart for a 2 year trip to Earth.

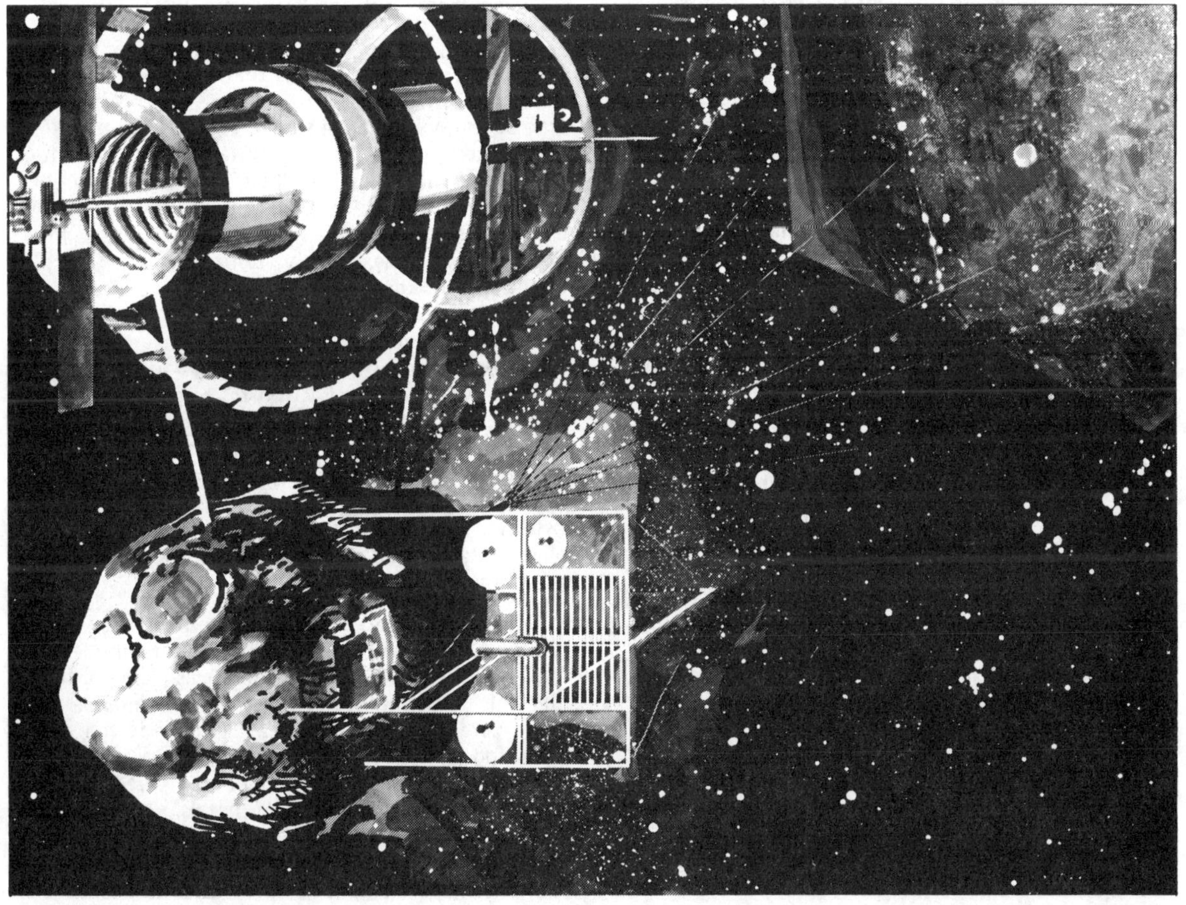

comparison (1/7000 of Earth's gravity isn't spectacular) and because these rockets have been sitting in
everybody's "come on, let's do it" file for many years.
They have seniority.

Can solar sails be made better? The answer seems to
be yes, if you forget about folding them up and launching them from the ground. I came to suspect this in
the summer of 1976, and now, a year later, it looks
as if it may be true: solar sails can be made in space,
not as aluminized plastic sheet, but as aluminized
nothing, which weighs far less. Designs now worked
out on paper use aluminum foil as the reflecting
surface, but foil 1/1000 the thickness of the kitchen
kind. These sails are over 40 times as light, and
therefore over 40 times as fast, as previous designs.
This is spectacular.

If I had to draw a sail today, it would be a hexagon
about six miles across, and weighing 20 tons. This is
somewhere between the size of Manhattan and San
Francisco, but the metal of the sail could be wadded
up to the size of a Volkswagen bug. They could be
made both much larger and much smaller. The sail
itself would be a spinning (to keep it taut) metal mesh
with long, parallel strips of very thin metal foil glued
to it. At regular intervals across the front, wires
would come up, and be bundled to form groups,
with each group having a wire coming from it, with
these wires, in turn, bundled to form groups still
farther in front of the sail. After this bundling and
re-bundling has concentrated the load of light pressure
on the sail enough (that's what the wires are for),
shroud lines take the concentrated force to the
payload (see drawing).

The sail would be made on a large, lightweight framework, like a loom. Wires would be laid down, and
fastened to each other where they crossed. As the
wires go down, a device would travel back and forth,
producing thin metal foil by vapor deposition on wax,
vaporizing the wax for recycling, and laying the foil
on the wire mesh. The whole process would take
about six months; building the "loom" would require
several flights of the Space Shuttle. Additional sails
require about one flight apiece to provide needed
raw materials. Eventually, sails would be built from
extraterrestrial materials.

What can you do with a solar sail? First, how can
you "tack"? Boats can go in any direction by using
both wind and water; solar sailing vessels can go in
any direction by using both light pressure and the
Sun's gravity. Light pressure on a mirror is always at
right angles to the mirror's surface, even when the
mirror tilts and bounces the light at an angle. As the
mirror tilts towards being edge-on to the light, the
force becomes smaller and approaches zero. This
means that the mirror can collect some force in any
direction that would take it farther from the light
source, in this case the Sun. So how can a solar sail
reach, say, Venus, which is closer to the Sun than
Earth? By using light pressure to slow down in its
orbit, then letting the Sun's gravity pull it in. Solar
sails can go anywhere in the solar system, and, in
the inner solar system (where we are), they can get
there faster than almost anything proposed.

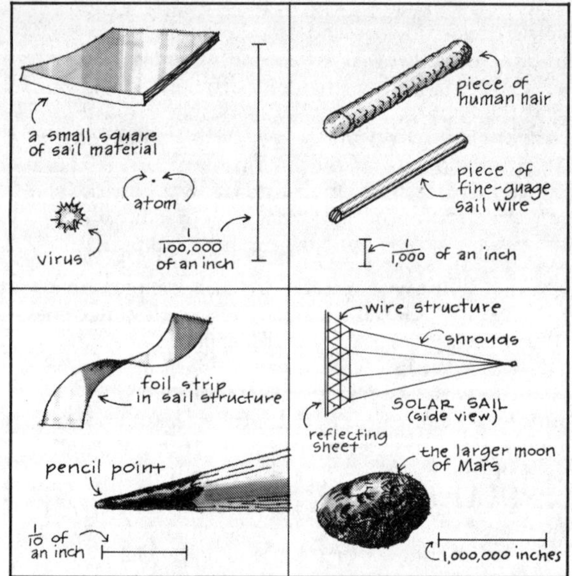

The 20 ton solar sail mentioned above could take 180
tons of payload to any place in the solar system, stop
(not orbit, but stop) and hang there on light pressure.
With 800 tons of payload to slow it down, it would
finally have the same acceleration as a plastic film
sail with no load at all. With 6 tons of payload, it
could fly to Pluto in one and a half years. Pioneer 11,
launched over four years ago, won't reach Saturn until
two years from now, and Saturn is only 1/3 the
distance of Pluto.

Rough cost estimates suggest that solar sails will cost
between $ .03 and 1/3 cent a square foot. Kitchen
wrap costs about $ .01 a square foot. If nobody
throws them out of the solar system, toasts them too
close to the sun, or crashes them into something, they
should last for thirty to three hundred years. Maintenance costs should be about nil (you don't fix the
sail at all, and there are only about two dozen reels
for the shroud lines to keep track of). Each sail,
without fuel expenses, can cruise around the solar
system almost indefinitely. While a rocket must be
built differently for almost every mission, the same
sail that flies from low Earth orbit to geosynchronous
orbit and back can do a perfectly good job of flying
twenty times the freight to the asteroid belt, out
beyond Mars. Not only that, but the sail costs above,
with 10% real rate of interest on capital, can give
costs like $ .10/lb for transportation around the
solar system. And sailboats have always had little
environmental impact. . . .

How do (apparently) good ideas like this arise? Well,
they never seem to be new. Solar sails are an old idea.
The literature even contains references to metal film
solar sails, although not of the high performance discussed here. It even contains a reference to the idea
of making the material in space. When I first had the
idea, my reasons were not to seek high performance,
but to try to make a sail out of metals, which are
readily available in space. My background was
oriented towards manufacture in space, towards
materials properties, and towards vapor deposition.

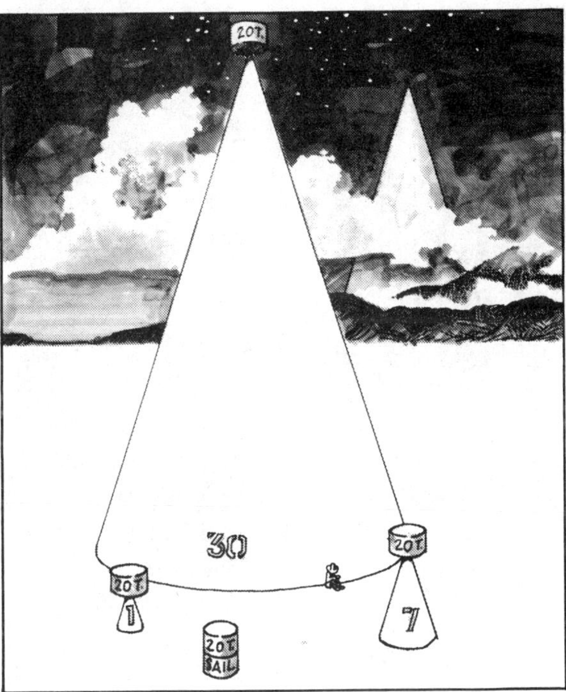

Solar sails and chemical rockets. *Rocket mass increases exponentially with the velocity to be reached. In the foreground is a 20 ton payload sitting on a representation of a 20 ton solar sail. To its left is a similar payload resting on a cone (labeled "1") representing the mass of the chemical rocket needed to reach the velocity that the sail could reach after a day's acceleration. On the right, the cone labeled "7" represents the mass of a rocket to equal the velocity given by the sail in 7 days. Similarly with the cone labeled "30" (note man napping at base). In the background, behind Mt. Everest, is a 50-mile tall cone representing the mass of the rocket needed to equal the velocity with which the sail can throw 20 tons out of the solar system. An electric rocket would be considerably smaller.*

Previous workers concentrated on hauling sails up from the ground, but metal film sails are too delicate for that, so they were never studied. The few who considered making the films in space considered inappropriate manufacturing techniques, which either didn't work or produced films about 500 times as thick as optimum.

Many questions come up about the new design:

● Can films be made that thin (thinner than a soap bubble)? Yes.
● Do they have decent strength? Yes, I've made some, and taken them from coast to coast in my luggage.
● Are the films still reflective when so thin? Yes.
● Will fast moving atoms from the sun knock enough atoms from the film to destroy them quickly? No.
● Will this stream of atoms exert destructive forces on the sail during solar storms? No.
● Will electric charging or magnetic fields cause problems? No.
● Will the films be able to stand the temperatures they reach in the inner solar system? Yes.
● Will meteoroids destroy the films quickly enough to matter? No.

● Will they destroy the wire mesh that is the structure of the sail? No.
● Can the films be mass-produced in space at low cost? Yes
● How much does the device for making the films weigh? It's manageable.
● What trajectories can the sails fly on? Name it.
● Could the sails possibly find use for near Earth transportation, as well as deep space? Maybe.
● Will the sails ripple under light pressure and gradually tear themselves apart? No.
● How well can the sails be turned to point where you want them to? Well enough.
● Will stresses from changing temperatures cause problems? No.

And so on. Work of this kind never really stops until something has been built and running for a while and people take it for granted. Solar sails of this design have two things going for them: they have passed many tests, and no one has examined the idea and rejected it in the past. At the time of this writing, formal publication and widespread examination and criticism are about to take place. Time will tell, but the chances seem good that we're on to something interesting.

How interesting? As interesting as cheap space transportation, free of fuels and complex maintenance. As interesting as moving Earth's industry into deep space and scattering her life to the Sun's light. Arthur Clarke said: "If man survives for as long as the least successful of the dinosaurs — those creatures whom we often deride as nature's failures — then we may be certain of this: For all but a vanishing instant near the dawn of history, the word 'ship' will mean — 'spaceship.' " And, those ships may yet have sails. ■

Heliogyro solar sail

The Jet Propulsion Laboratory in Pasadena, California, has proposed a spinning solar sail where centrifugal force keeps the twelve 4-mile-long sail blades flat. The blades would be 1 mil thick aluminized plastic 25 feet wide for a total area of about 600,000 square yards and a payload capability of 1100 pounds. The blades would be variable pitch so that sunlight would spin the vehicle and allow you to control attitude and acceleration and deceleration. Such a vessel could rendezvous with Halley's Comet in 1986.
— SB
[Sent by Mike Liebhold]

Operating on the Edge

RUSSELL SCHWEICKART

TALKING

Schweickart 1977

Jane McClure: Did the idea of becoming an astronaut cross your mind when you were a young boy?

Rusty Schweickart: Well, I was born in the country, in New Jersey. We lived on a farm, and at that time it was really the outback. The road wasn't even paved. My mother and father and I and my sister would walk along, and chat back and forth, eat blackberries and look at the stars and the moon, listening to the night sounds.

My real two dreams were to become a pilot, especially a fighter pilot. I used to watch the airplanes over the house dogfighting, in mock combat. I would lie on my back and watch them fly around the clouds and chase each other. Then for some reason the other thing I wanted to be was a cowboy.

McClure: How old were you when you went away to school? Did you stay around New Jersey?

Schweickart: I graduated from high school when I was 16, and went from there out to MIT.

McClure: Did you have a special interest in high school?

Schweickart: Yeah. I had a ChemCraft chemistry set when I was a kid. It was given to me for Christmas. I can remember clearing out an old farm house and making one dusty corner my chemistry

laboratory. I would sit up there mixing my different color liquids and having 'em change and bubble off things, so I wanted to be a chemical engineer when I first went to MIT. Then I had one year of chemistry.

McClure: That changed your mind?

Schweickart: That changed my mind.

McClure: So what direction did you go after chemistry?

Schweickart: Well, I knew I liked engineering. Then my love for airplanes kinda took over and I went into aeronautical engineering.

McClure: Did you go into graduate work afterwards?

Schweickart: Well, after I finished aeronautical engineering at MIT I went into the Air Force. I was in ROTC, so I was commissioned on graduation, and was one of the first ones in the class to go on active duty in July of '56. I went through training in Georgia and Texas and all across the South where people learn to be Air Force pilots.

Stewart Brand: You went straight into jets?

Schweickart: In those days you went to propeller airplanes first no matter what your final destination was, and so I went through primary training in propeller airplanes in Georgia.

Brand: Was flying a natural or an acquired taste?

Schweickart: I took very naturally to it. I really loved it, from the beginning.

I remember when I was a little kid I would come home from school and turn on the radio, and listen to a whole series of 15-minute soaps — Tennessee Jordan, Superman, and Captain Midnight, Hop Harrigan and Tank Tinker.

Brand: Jack Armstrong, all-American boy?

Schweickart: Jack Armstrong, the all-American boy! Sure. And Tom Mix. Take a tip from Tom, go and tell your Mom, shredded Ralston can't be beat. . . . They were great. I think it was the image of being that free in three dimensions — the kind of freedom that you have in a fighter plane where you're not just flying straight and level with cargo in the back but where you're free to move around. If you take that principle to its limit, it's weightlessness, where there's absolutely no restriction. There's no up, no down, the total spatial domain is open, and there's no preferred reference or orientation. So that's a very interesting sequence, that what I saw and expressed as a kid led to space as the epitome.

Brand: When we were talking the other day you said that you got tossed around a lot as a child and that's what makes you comfortable with the idea of being in space.

McClure: Yeah, I like anything that takes me off the ground. Whether it's being thrown back and forth, or falling, jumping, carnival rides, anything. I won't get off them, just go round and round.

Schweickart: What's your earliest memory of the flying experience?

138

McClure: My father and his friend throwing me back and forth across the kitchen. I loved it! Those moments of weightlessness. Rusty, don't you get that for short periods in an airplane when it's flying a parabolic arc?

Schweickart: Yeah, in a fighter, if you put a pencil or something up on the dashboard and then push forward on the stick, you can just lift it off and keep it hanging there just by controlling the airplane. And that's exactly what you do in the big airplanes that we fly weightlessness trajectories in. Get going fast, pull up into about a 60° climb, and then the pilot pushes forward on a stick until the airplane's in weightlessness.

Brand: Sounds like it would flush your head.

Schweickart: It'll flush more than your head! They don't call it the ''vomit comet'' for nothing!

Brand: Do you get some washouts at that point in the training? Do people find they're just too unhappy in zero G?

Schweickart: No, because it doesn't necessarily correlate to what happens in space flight. We've had some people who get very sick in the C-135 and who haven't had any problem at all in space and vice-versa.

Brand: Was that you?

Schweickart: No, I tended to get a little sick in each.

Schweickart 1968

McClure: When you were talking about controlling the pencil in flight I suddenly related that to. . . All your life you're fighting for control, whether it's of yourself, your surroundings, everything around you. They've found that it starts in newborn infants. They're always struggling for survival, for control. I was wondering when you're in the spacecraft, whether there's a sense of control of your destiny.

Schweickart: I think it's somewhat the other way.

McClure: You mean out of control?

Schweickart: Being close to being out of control, anyway. Operating on the edge, where with a little bit of a slip on one side you can be out of control. It's interesting. What's attractive about that? Some people express it as a death wish, but to me that's far too simplistic an answer.

Brand: Is the use of space equipment as edgy as fighter planes or is that a whole different kind of discipline?

Schweickart: No. The normal day to day kinds of things are very safe. You go out of your way to minimize all risk, and you don't leave anything out near the edge. The suits and all of that are very very conservatively designed. You don't even ask questions about money when something regarding safety is involved.

McClure: Where were you, and how old were you when you heard that NASA was hiring astronauts?

Schweickart: I was at MIT, in graduate school, finishing up my master's degree. I was doing research in atmospheric physics and flying in the Air National Guard. I guess I was around 27. Actually I was in the Air Force twice, because after I got out the first time I went in the Air National Guard when I was getting my advanced degree at MIT, and then the Berlin Wall went up and Kennedy activated the reserves and so we went back on active duty.

That was about '61, when John Glenn flew. I can still remember the morning when I heard on the radio that he'd made it into orbit. I was looking at it in the paper in a coffee shop in eastern France and thinking, that's where I'd like to go. It seemed to match with some rather deep vibrations inside.

Brand: What do you think inspires interest in space exploration today?

Schweickart: Well, O'Neill, and all the things that he's doing. Of course the roots go back far before Gerry, but he's the input point for all the cosmic energy that's getting into the system right now.

McClure: He makes space settlement seem very feasible and exciting.

Schweickart: Is that how you got interested?

McClure: Yeah, when I got through reading all his articles in **CQ** I thought, boy, this is really possible.

Schweickart: It'll all work out, that's the point. There are challenges but not impossibilities.

Jane McClure

McClure: O'Neill oversimplifies it to make it sound exciting enough for people to really get interested in it, which is serving a real purpose, because I wanted to jump right in after reading his articles. But then I read something in your article which just turned my head around. Stewart asked you, 'What are the mechanical problems in space?' and you said theoretically nothing. You explained that it's frictionless and they're having a little trouble figuring out how to oil things. But then you said that at the end of one of the flights an inertia wheel fell off.

Schweickart: It started to, actually one fell off and the second one was on its way out.

McClure: And you said they didn't understand how it happened.

Schweickart: Well, the covered wagons were okay, but there were Indians too. There were inherent problems of the environment.

Brand: What about safety, Jane. If it's hazardous in Space would you rather wait 'til it's not?

McClure: I don't know, for myself. There's many young people my age who would just do it.

Schweickart: It's very interesting when you ask this question. The yardstick that's held up by the people without it being asked directly is the safety aspect, whether they perceive it as being risky, or perceive it as being fun and the risk worth it. It's a very interesting response, and what it tells me is that if risk were not an issue then almost everybody would like to go out in space. And the risks aren't all that great in reality. If you compare the time you spend on an airplane, space travel is actually safer.

[more ➔]

And as an astronaut, I would say the risk of my being killed in a car accident is greater than my risk of being killed as an astronaut. So it's the perceived risk that gets to people, it's not what the actual numbers are. When you really get into being an astronaut, you realize that just a tremendous effort, to the point where it drives you absolutely buggy, is put into safety.

Brand: Rusty was telling me the other day, Jane, about when things went wrong with Skylab and they had to jump 10 years ahead in the program to fix it. They were doing things that they wouldn't have considered for years in order to make the thing fly.

Schweickart: Well, that's always the secret with anything in life. If you want to move forward into new territory, what you do is put yourself near that ragged edge, because that's where things are moving the fastest. If you're learning how to ski, the optimum learning rate is where you're not standing up all the way down the hill nor are you totally wiping out every time you stand up, but where you're right on that edge of being out of control but you're just barely in control. That's where you learn the most rapidly. If what you're interested in doing is learning, then what you do is place yourself in that kind of a situation.

Brand: Astronauts so far, are they people who love learning or love risk?

Schweickart: If I look at the objective behavior of astronauts, what they do is spend a tremendous amount of time minimizing risk. So it's difficult to say that they love risk.

Brand: But they're in a situation where there's considerable risk to be minimized.

Schweickart: What they're doing is they're operating near the edge but in the safest possible way. You're operating responsibly near the ragged edge. Okay? I mean there are people who are way back from the edge in the society, who are living totally decadent lives. I mean decadent in the sense that they have died even though they are ultra-secure. It is their demand for security which has killed them, and they operate so far back into the middle that life becomes a uniform experience.

McClure: Another form of death.

Schweickart: Sure. Now, you talk about death wish. Well, where is the death wish expressed? Up near the edge or back in the middle? It's a life wish to me to go out in space. In whatever it is you're doing, there's a life wish. I mean you are placing yourself as a life form at the place where that life form is evolving the most rapidly and you are manifesting that desire, that tendency in life, willfully.

Brand: By the time NASA gets into hiring third and fourth generation astronauts, as they are now, what are the real selection principles? What do they look for, what do they discard?

Schweickart: It depends on what the perception of the requirements for the

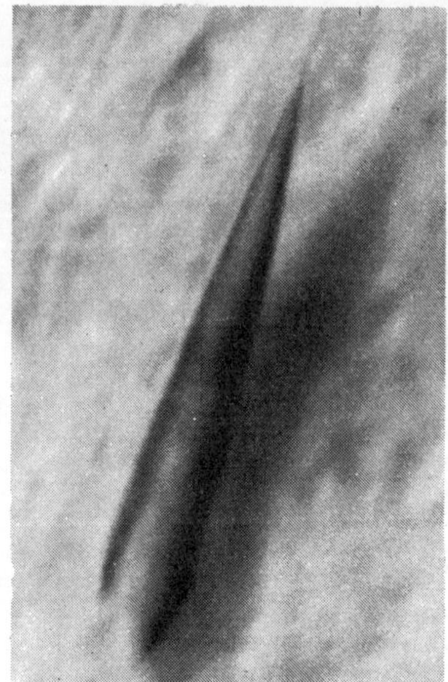
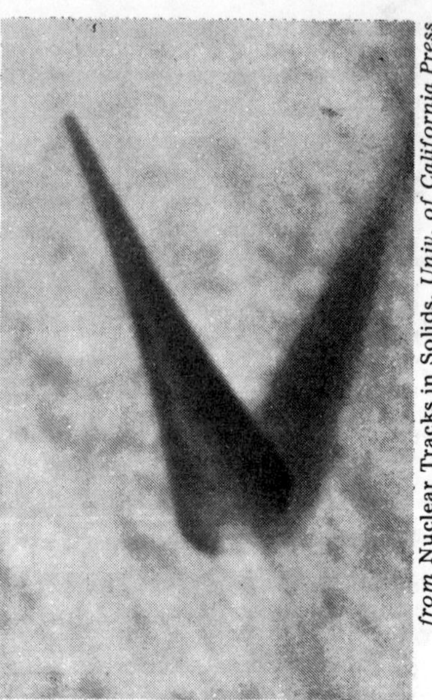

Cosmic ray tracks in an Apollo space helmet, amplified 10-millionfold by chemical etching.

upcoming missions is. In the past you really could not afford to have anybody on board either Gemini or Apollo, or Skylab for that matter, who could not fly. There was always the chance that things could go wrong and anyone around would have to be able to bring the ship back. Now with the space shuttle coming up, it's quite clear that not everyone will have to learn how to fly, and in fact one of the three astronauts on boards the space shuttle, the mission specialist, will very likely not know how to fly.

Brand: Recently NASA got 6000 astronaut applications. Out of that, what, 30 will be selected?

McClure: I read a clipping in the paper which said that most of the women who were interviewed were on the average aged 30. Here I am thinking about it at 21. Why is it that the average age came out to be about 30? Is it because more schooling is required?

Schweickart: What you want is someone who has a broad background. The mission specialist on a space shuttle flight will be overseeing the what you might call on-orbit operations. That is, the mission specialist is basically an assistant or passenger almost during the launch and during re-entry. But once you get on orbit and the operation begins, the mission specialist is pretty much in charge. If there are specialists who are researchers and just up on that specific flight to do whatever the research task is, the mission specialist will act as a kind of coordinator and communicator, to interface with the ground people and the vehicle and the support systems of the orbiter to get the job done. And because there may be a whole variety of things going on — biology research of medicine or astronomy or atmospheric physics or Earth resources research or communications research or all kinds of things — the mission specialist needs to be

someone with a very broad background.

McClure: How many people would be able to be transported in a shuttle? How many people would the mission coordinator be responsible for?

Schweickart: Well, the vehicle can support up to 7 people. Three of them are the commander, the pilot, and the mission specialist. They would normally be the astronaut crew, and then you can have up to four payload specialists who are not astronauts.

McClure: So these 30 prospective people being chosen are going to be the mission specialists, is that the idea?

Schweickart: There are two groups. The mission specialists in one group and the pilot astronauts are the other group. And there'd be about 15 each in this present selection. Far and away the majority of applications came in for the mission specialist opportunity — that was over 6000. I don't know how many came in for the pilot astronaut but I would guess on the order of 500, something like that.

McClure: Not everyone has had pilot experience.

Schweickart: Sure, that's why in the past we haven't had any women or blacks, or Jews for that matter, in the program. The test pilot population is not a population where minorities have aspired. That's been pretty much a WASP activity.

Brand: With no sex or age discrimination now does that mean that you're permitting older and older astronauts with each group?

Schweickart: Yeah. Deke Slayton was 51 I think when he flew. I think we're obviously moving to the point where in the reasonable future anyone will be able to buy a ticket the same way you do on an airline. Now, I don't know exactly when that'll happen, but. . .

McClure: Honeymoon on the moon!

Schweickart: It'll happen, it'll happen. You can picture a space station with a small part of it probably leased to a hotel chain for very exotic tourists. There's not a substantial market there yet, but it'll happen.

Brand: Not much is said about the payload specialists. Where are they going to come from? Do they go through any kind of training? At what point do they come into the program? For someone who is interested in flying before the next hiring of astronauts, can they look at the payload specialists as a way in?

Schweickart: Well, we just talked about the fact that the mission specialists should be characterized by a broad experience background and especially experience in field operations as opposed to theoretical kinds of things. With the payload specialists, though, I think it's quite the opposite. The payload specialist is someone who might very well be quite vertical in interest and experience and in fact could even be quite theoretical or academically oriented. If we have for example ultraviolet telescopes on board, as we did on Skylab with the solar physics experiments, a really skilled theoretically oriented observer could extract a maximum from any one period in orbit.

Brand: You want someone with theory there who'd know what to look for.

Schweickart: And really understand what is unusual and what is not unusual, and capitalize on the what-is-not-expected aspects of the observation. That's the difference between the skilled scientific observer and the generalist.

Brand: I remember listening to the tapes of some of the earlier astronauts, gooning over everything they saw. I can imagine there was some frustration by geologists on the ground saying, "C'mon, forget about the scenery, all we're trying to find out about is those very hard-to-see fault zones."

Schweickart: Well, scientists often get lockjaw on something too. You know they'll develop this great fixation on what they think is very unusual and somebody who's not that familiar with styles in science says, "Hooo, isn't that interesting?" and it damn well *is* interesting. That's one of the differences between the manned and unmanned operations in exploration. When you deal with an unmanned thing you have to design it, of necessity, for what it is you want to see and what you want to seek and permit to interest you. The unmanned, the remote controlled unit cannot present things to you which it was not designed to present. Whereas the human observer can totally redesign its input filters in real time, depending on the circumstances, and that's what's so exciting.

McClure: You're comparing an unmanned craft to a narrow-minded scientist?

Schweickart: Well, to anything which is designed and then operates remotely.

Although there are remote things which are now reprogrammable, the limits of the reprogramming are always inherently built into it, and are considerably less than the limits to programming of the live biological form of life.

Brand: Who was it on one of the early moon flights that noticed the flashes on their retina? And who figured out that it might be cosmic rays? Maybe you should explain to Jane what the experience was.

Schweickart: It must have been on either Apollo 10 or 11. In any case what was noticed was flashes of light when the eyes were closed and either getting ready to sleep or just resting. It may very well have been Buzz Aldrin, as a matter of fact, on Apollo 11. Anyway, it was noticed that there were light flashes and that they weren't just self-created. They began to watch these things and identified several different types, you know, streaks, starbursts.

McClure: There was an observer watching this objectively on someone else, or...?

Schweickart: No. Closing the eyes and watching it yourself, but you could tell whether it was in your right eye or your left eye and so you could sense the direction and type and things like that.

Brand: Was there any self-inquiry as to, "Am I seeing visions? Going crazy?"

Schweickart: Yeah. But after you watch it carefully for a while and decide it is worth it, you tap your friend on the shoulder and say, "Uh, how 'bout closing your eyes and..."

McClure: "...telling me what you see..."

Schweickart: "...if you don't see some flashes of light." And so it was done and confirmed...

Brand: How often did they occur?

Schweickart: Several a minute. The theory which has been tested on the ground and appears to be the legitimate one is: heavy cosmic ray particles come through the wall of the spacecraft and leave a track in the retina of the eye as they pass through, and this track releases radiation or stimulates direct triggering of the retinal nerves which then cause these perceived flashes of light. It had never been discovered before because the Van Allen belts protected you in lower Earth orbit from this phenomenon...

McClure: What's the Van Allen belt?

Schweickart: The Van Allen belts are the belts of radiation around the earth that are trapped there by the magnetic field of the Earth, and they form a kind of screen if you stay within them. A lot of particulate radiation, ionized radiation, is deflected away and doesn't penetrate the lower atmosphere. So we thought that they did not appear in lower Earth orbit, but then on Skylab we tested it and found out that, yeah, they really do. And so we started trying to correlate them with the South Atlantic anomaly which is a place near

the polar regions where the Van Allen belt dips down closest to the Earth, and it turns out that in fact when you get near the polar regions you do have higher frequency of retinal stimulation.

Brand: If these things are going through your head enough to make a flash, are they going through your head enough to make damage?

Schweickart: Sure. But you and I and Jane are sitting here with things whistling through us and creating damage on a continual basis. It's just that they come in certain sizes and shapes when you sit down below the atmosphere and different sizes and shapes when you're up above. I think the rate of damage without any question is certainly higher above the atmosphere that it is below it. That's one of the functions of the atmosphere...

Brand: The Skylab crew was out for 84 days and they still seem to be getting around some way or another, I take it they weren't disabled by this.

Schweickart: Oh, no, it's a long way from being disabling.

Brand: So far, but what about with continuous habitation and exposure to solar flares?

Schweickart: Well, it's like being in Alabama or Mississippi or Missouri or somewhere. You can live there for a long time before a tornado gets you, but by god if a tornado comes over your house you'd better watch your tail. And if you have a strong solar flare, you know you better get to someplace where you can get some shielding. So you need to have a warning system and a shelter and head for it whenever the radiation exceeds a certain level.

Brand: Lead BVDs and a helmet or something?

Schweickart: You need some form of protection. So you can protect the whole environment or you can build storm cellars. Storm cellars are obviously a lot lighter because you don't have to survive there very long.

Brand: Is this being figured into the shuttle program? The guys out there, if they get a flare, they got what, 10 minutes, till they start getting bombarded?

Schweickart: The shuttle is I think fairly well shielded anyway. The primary potential damage from a solar flare is filtered out by the Van Allen belts, and the space shuttle will be operating below them. That hazard really becomes much more real when you get outside the protection of the belts.

Brand: What about these big antennas and solar power satellites? Aren't they going to be in geosynchronous orbit, with is 22,000 miles out?

Schweickart: They're outside the Van Allen belts. However the construction work may be done in a lower Earth orbit, you see, and then after you get the construction done, then you boost it up to high orbit. The shuttle can do both. It's a very simple device. It's like a truck or a bus which is waiting for hire in 1980. It runs on a charter basis and you can buy a share in the flight. You may have your favorite satellite to do research, or provide some communication service, or whatever you want to do. You can buy a ride on the space shuttle and you're charged a pro rata share depending upon how much of the size, weight, and resources that you use. Because the payload bay is 15 feet in diameter and 60 feet long, you carry a tremendous amount of equipment up there for a relatively low cost.

McClure: In the past, once a satellite has been orbited, it takes years for it to come down or sometimes not at all, depending on how it is launched?

Schweickart: And most importantly, the last time anybody can do anything with it is on the pad on the ground before launch. The most severe part of the lifetime of most satellites is the launch phase, the lift-off and the vibration during the trip into orbit. Once it's up there it's in a very benign, relatively constant environment. The space shuttle now allows you to run your checks and tests, and stroke it on the solar panels or whatever it needs, before you let it loose to do its job.

Brand: If the shuttle had been in existence at the time of the Skylab, would it have been a lot easier to fix the problems you had?

Schweickart: Yeah, because you could have gone up and rendezvoused with it and taken up special equipment and people.

Brand: How many shuttle flights will there be a year? Do we know?

Schweickart: No, we don't know, but there are projections of traffic which are people's best guesses depending upon assumptions on how fast certain markets will build and that sort of thing. What are called mission models are generated which take into account the phasing of capability in terms of things like additional launch pads required by the fall of '83 or whatever. It depends on capability in terms of ground support, flight equipment, crews, and to some extent even the technological development schedules. To start with, I would say something like one flight every two or three weeks in 1981. About 1984 or '5 there would be a build-up to about one flight a week, or 50 to 60 missions per year.

McClure: That includes how many shuttles?

Schweickart: Eventually 5 orbiters. And they're shared by NASA and the Air Force.

Brand: They're built for 100 flights or so?

Schweickart: Right.

Brand: Okay, just looking at that, there's 5 orbiters each designed for 100 flights. That's 500 flights, with on board anything from 3 to 7 people — times 500 in one sense or another (plus

> ## Put yourself near the ragged edge, because that is where things are moving the fastest. If you're learning how to ski, the optimum learning rate is where you're not standing up all the way down the hill but neither are you wiping out every time — where you're right on the edge of being out of control.

certainly a lot of repeaters) that takes you to something on the order of 1000 people in space within the next decade.

Schweickart: Decade or a little more. Yeah.

Brand: People are going to start knowing people who know people in space.

Schweickart: That's right. To me one of the real turning points in the program is that soon you don't have to make flying in space your whole life. You can get into space in order to do whatever your thing is — astronomy, exobiology, advanced communications work, processing of new materials, research on crystal growth, or all sorts of things. People from many walks of life will now be flying in space and to me that's one of the most exciting parts of the program.

Brand: Jane may wind up in space without even trying.

Schweickart: That's why I tell people who are interested in specific sciences or specific activities and who at the same time are curious about space, that what they're doing is the best way to go. Don't try to be a mission specialist, that's silly. What you really want to do is take your thing — crystallography or whatever — and find who in the crystal research activities in space is doing the most advanced work, and go to work for that person in his or her research laboratory.

McClure: The thing with the space program, though, is that it doesn't seem at this point that there is much room for a biologically oriented scientist.

Schweickart: Well, I'm not sure. If you start thinking about the kind of knowledge base that's got to exist before anybody can reasonably commit to a space colony, you've got to have an extremely firm biology base. There's no way to get there without beginning to really seriously address the questions of biology in space, especially things like plant growth, germination processes, and the fundamental relationships between . . .

McClure: I was thinking specifically of the shuttle.

Schweickart: But you've got to start that research on the space shuttle. That's the vehicle that you have to be using to enable you to do that research.

Brand: I understand that at the third NASA-Ames study group on space

colonies going on now south of here, the largest population of anything in particular that they have is biologists.

Schweickart: To me that's one of the biggest challenges in space right now. The whole life process is going to be experiencing a change in one of the fundamental things which it's always had to live with, and evolve within, and that's a gravity field. Here there is for the first time an option to sustain life under less than 1 gravity. What will be the effect of that fundamental change on the evolution of life forms, nobody knows.

McClure: You could even get psychologists up there to watch the astronauts.

Schweickart: Well, that's nothing new.

McClure: When you were in the spacecraft, what were relations like with the people you were with? Did you not allow tensions up there because of the situation? How did you all get along?

Schweickart: I would suspect not very differently from the way people get along in any kind of operational situation. It's very clear that your survival depends upon your performance and the performance of the very limited number of people around you. There are tensions because your personality still exists and the other people's personalities still exist, and you *have* to account for them and you know them very well, but you know you're in a situation where you can't let the clashes between personalities — the differences and sensitivities — affect the behavior. You just avoid behavior which will stimulate defensive reactions on the part of the other person.

McClure: How do you think things would change if there was a woman on board? Do you think it would cause a great change in the way men would act with each other?

Schweickart: It depends so much on the people. In terms of people within the astronaut core I think you'll find very little change in behavior there, simply because you do so much simulation, so much work together, that whoever you are, you get past the point of the transitional behavior. I mean, I know you squeeze the toothpaste tube in the middle and I've adapted to that. But in the case of the payload specialists, I think that's a different situation. There the challenge is not going to be the fact that the payload specialist may be a woman, it's going to be the fact that the payload

specialist is someone who has not been well integrated into the behavior of the rest of the crew.

Brand: How long do payload specialists train?

Schweickart: There's a kind of idealized image where we're going to minimize costs, and we'll sort of swoop down and pick whoever the local expert is off the street or out of the laboratory, slap their tail in the space shuttle, and off they go to do extraordinary research. In fact, I think we'll start out being quite conservative, and the first payload specialists will have to have a considerable amount of training.

Brand: Presumably as time goes on and you get more and more people in space who do or don't work out, your selection principles will get more refined.

Schweickart: You hope so.

Brand: You've not said much about that and I'm not sure that there's much to say yet.

Schweickart: Well, there isn't. Clearly the pilot-astronauts are people who are going to be faced with decisions where quick reactions are needed and where the actions that are taken reflect directly on everybody's safety. That's what happens during launch and reentry. On orbit you don't usually face those kinds of situations. I mean the only two things on orbit that you really worry about in that sense are fire and decompression. Other than that you don't usually have to react quickly.

Brand: Who's the mission commander? The mission specialist?

Schweickart: No, the commander.

Brand: The guy in the left front seat.

Schweickart: Yeah. It's like an aircraft carrier. The captain of a ship is the one in charge even though the Admiral may be on board. In terms of who runs the ship and does it or doesn't it turn right or whatever in an emergency situation, it's the commander who decides.

Brand: What I hear is that all these things we're talking about, including space colonies, are basically vehicles in the sense that a ship is a vehicle. So the Star Trek image of a bridge, the commander, and the lieutenants, is really the mode you're talking about, even up to large space habitations. Someone has to be in charge .

Schweickart: Yeah, but that's true here on the ground. When you have a small town there is a mayor and the mayor is the captain of the community.

Brand: I don't think those are comparable. The vehicle I suspect is quite different in the severity, maybe the clarity, of the focus that there's someone in charge and there's a hierarchy of relations. The mayor of a town can be a joke and often is, but the town can flourish splendidly with a joke mayor, but in a vehicle where decisions have to be made, whether it's an aircraft carrier or a space colony, you can't have a joke mayor, I don't think.

Schweickart: I suspect that's a problem of perceptions. I think, relatively speaking, you do have joke captains of ships. I haven't been in the Navy, but I'm certain that just because you're dealing with human nature that there are destroyer captains who are jokes, for example. And it's the executive officer and the other people around that person who permit that particular joke to survive.

Brand: Point taken. I guess what I'm looking at is, I'm trying to see some of the differences of a manufactured environment versus an evolved environment, where you can relax, pull over to the side of the road, and things'll balance out. In one sense or another in these manufactured environments, at least until they get extremely large, you can't just relax and let things take care of themselves because they won't. You have schedules that have to be met, keeping the algae growing . . . .

Schweickart: I think the principle difference is the response time of the system. When you're dealing with something like a community on the Earth, or the whole Earth as a system, the ultimate consequence of any particular action may be many many generations in the realization. Whereas when you're hanging ten on a surfboard or when you're right on the edge of crashing down the side of a mountain coming down a most difficult slope skiing, the characteristic response time of the situation is totally different, and your actions are immediately fed face to you in terms of results. That's a difference, but the principle is the same.

Brand: If you're saying that on the Earth it may take a catastrophe ten years to develop, where on a medium-sized space colony it might take a month to develop, still those differences are *small* differences. . . .

Schweickart: No no no, they're not small. That's one of the most interesting things, that we will be evolving, living, interrelating, (I hesitate to say natural, but approaching natural) systems in space, where that feedback time is much shorter inherently and therefore the consequences of actions within living ecosystems becomes a much more immediate thing, and the nature of the responsibility of the human as part of an ecosystem comes home very clearly. That then translates back into the total planetary environment in terms of recognition of responsibility.

Brand: I think there's a flip going on here. On the planet's surface the ecosystems manage the people to a large degree, and here you're talking about a flip where the people are gonna have to manage ecosystems to a larger degree, and that may be paid back in better handling of the Earth, and better respect for the Earth, or whatever. But it's a really fundamentally different situation. It would seem to me that we're now talking about a much higher role of consciousness in biology. You can turn off the human consciousness of northern California, and things will naturally come to a better balance rather quickly. With these situations you're talking about, if you turn off the human consciousness you may wind up with just brown paste within a couple of months — the kind of brown paste that you find in terrariums after a while.

McClure: Scum. Organic scum.

Schweickart: I think we're both really saying the same thing on different ends of the perceptual spectrum. I mean, there in the space environment it's so obvious that you're in control that you forget that you're really not in control. On the Earth you are so clearly out of control that you forget that you are in control. And in reality you're both. In all situations.

Brand: Amen, brother. So you speculate that experience with biology in space will make us better Earth dwellers?

Schweickart: Oh yeah.

McClure: I think it would. Any time you expand yourself and look at something from the outside it makes you appreciate what you've left behind.

[more ➔]

143

Schweickart: Sure. It's part of Alan Watts' observation: "What is it I call me? I call me that which is within this membrane called my skin. However what's within this membrane that I call my skin is also that which is not within what is outside my skin." In other words I could play the reference either way. And I will in fact learn more about what's inside my skin if I understand what it is that's not outside my skin. We'll learn more about ourselves as human beings in studying the ground. We tend to focus on the figure.

Brand: So studying the ground of space we got the figure of the Earth?

Schweickart: Yes.

Brand: What then is the ground of which space is the figure?

Schweickart: Life. I mean, it is life which perceives space.

Brand: Go on.

Schweickart: What is it that perceives or conceives of the universe? It's life, consciousness, awareness, or intelligence, or God. You can express it in many ways, but it is that which is the ground on which space or the universe is the figure.

Brand: In orbit, are you nearer my God to thee?

Schweickart: I think so, in the sense that your perception is certainly enlarged. And I would say the larger the perceptual context the nearer my God to thee.

Brand: No atheists in foxholes or in orbit?

Schweickart: You can fight it. If for some reason you want to fight it, it can be fought.

Brand: There's a wonderful passage in C. S. Lewis somewhere about the light in space being the pure light of God's countenance, by contrast with the scuzzy stuff you find on the surface of planets.

Schweickart: Well, that's an easy image to understand. On the Earth or on the surface of any planet, a good portion of the light is reflected, either from the atmosphere or from the solid surface. Whereas in space, especially when you're outside the spacecraft, the source of the light is very very clear. It's not quite a point because our local star is close enough that it's an extended object, but it's very clear that that's where the light is coming from, and the background is totally black, and yet it's bright, I mean. . .

Brand: Beg pardon?

Schweickart: Well, the fact is that the sun is shining through that blackness, and that any time you want to look at something it's very bright because what you want to look at is reflecting the sunlight.

Brand: Is the shadow side extremely dark also then?

Schweickart: Well, the shadow side is very dark if there is no local surface reflection on it, like the Earth's surface. If you're in the shadow you are cold, and when you are in sunlight, you are very warm. So the source of energy, of the fundamental driving energy which runs everything, is very clear.

Brand: Like being near the campfire on a cold night.

Schweickart: That's right.

Brand: So you continually baste yourself.

Schweickart: Round and round.

Brand: Well that's news. I haven't heard that from anyone. Has everyone who's been in orbit had some perception of that experience?

Schweickart: I don't really know, but I think the experience is there.

Brand: Most journalistic pursuit has been, "What did you learn about the Earth?" "Well, it's this gorgeous precious jewel, especially when seen from a great distance. It's home, it's whole, it's holy, it's all these good things." Very few people have talked about the Sun.

There's something else no one will talk about. War in Space. How about that, amigo? Will we have Star wars in near-Earth orbit?

Schweickart: I don't know about near-Earth orbit, but there's no question it'll happen. That's only a matter of time. We carry the human spirit wherever we go. I think that's the kind of gut reason that some people react against the space program — the feeling that here is a pristine environment and we can't take the present fallible human character up there. We've got to perfect ourselves before we enter this new domain.

That's not the way life works. I hesitate to say it, but it's a kind of idealistic messianic vision that we will perfect ourselves, and then move out into this new and sacred environment. It's moving out that, to me, is an integral part of the process of moving more toward responsibility. Not the other way around.

Brand: They're saying responsibility first, travel later; you're saying travel first, responsibility later.

Schweickart: I'm saying that the two are interlocked. The life experience is part of the process of moving toward understanding which is the foundation of responsibility.

McClure: Rusty, are you going to fly again? In Space?

Schweickart: I'd love to. I don't think that it's in the cards in the immediate future because I'll be working here in California. But I certainly don't perceive that my chances of flying again have been shut off at all.

McClure: I was just wondering personally if you would like to?

Schweickart: Very much.

Brand: Why?

Schweickart: It's still the place in the physical environment where life is headed. It's the evolutionary path.

And at the moment I'm interested in affecting the perception that Space is an element of the cutting edge. But I would like to cycle back to where, instead of operating on that perception within the public, what I'm doing is operating on myself.

Brand: How could someone interested in going into Space, find out what we know so far about Space. Where do they go? Are there three really good books about being an astronaut? Or the history of the Space program? What are the magazines you read to stay current?

Schweickart: I think the best book that's been written on the surface level of the astronaut experience, is clearly Mike Collins' book, **Carrying the Fire.**

Brand: Also the Mike Collins Museum.

Schweickart: The Smithsonian National Air and Space Museum in Washington that he directs, sure. In terms of what's actually happening with the nuts and bolts, the two mass magazines are **Aviation Week**, which is sort of a trade journal, and **Astronautics and Aeronautics**, which is a publication of the American Institute of Aeronautics and Astronautics *(see p.128)*. Those are very technical. Now, in terms of the conceptual, there's another book which is very interesting, though I don't think it's "a good book", it's called **The Fourth Kingdom** by Carl Sauber.* It deals with the whole concept of life emerging from the womb in a sense — the fourth kingdom being the synthesis of the human and the machine, and this permitting an evolutionary move into a new physical environment.

Brand: Most of the environmentalist types, and maybe myself included, would be sort of appalled by that, but you're not appalled by that thought, by the thought of real symbiosis with the machine.

Schweickart: No. I've lived it. I've had relationships with machines which are just pure things, they're just beautiful. And others which are pure horror. But there are relationships where you and a machine are integrated through good design in a relationship which approaches those, you know, precious moments of fulfillment. You and that machine can perform actions or can enter into a space which is very high.

Brand: Can you afford an example?

Schweickart: Certainly there are examples in music, where the instrument played by the musician becomes a higher form.

Brand: You're being abstract. I wanted an example of yours.

Schweickart: Well, it happens that the computer on board the Apollo spacecraft in both the command module and the lunar module were two versions of the same computer. They were almost like twins of a family. We had to evolve the programs within the computer which permit you to solve the equations

---

*$6.95 from Aquari Corp., Box 1966, Midland, MI 48640

Soothing the human/machine interface. Jane was trying out a biofeedback device which converts galvanic skin response into an audible tone — high if you're tense, low if you're easy. Jane kept getting a high nervous squeal from the thing. "Here," said Rusty, cupping her temples, and the tone eased right on down.

of rendezvous, entry, abort during launch, or navigation programs. And the interaction between you, the human observer-controller, and the machine — the telescope, the sextant, the attitude control jets, the main engine for thrusting, the orientation of the whole vehicle in Space or with respect to the Earth, the pointing of the antennas — these things are all part of the programs of the computer, and you control then that total organism. In a sense you *are* the command module, through interface and conversation with that computer.

And it speaks to you in terms of nouns and verbs. If for example you wanted to maneuver to an altitude which would allow the telescope to point at Canopus, to align the guidance system, you would call the computer awake by verb 50 noun 25 which would say, "Please perform the following program," and then it would say, "Which program would you like?" And I would say, "The alignment program." And it would say, "Okay, which option within the alignment program?" And I would say, "The one that establishes a new Earth reference." Then it does some operations and it comes back and it says, "Which reference would you like? The local vertical one or the one perpendicular to the orbit plane?" And I'd say, "What are the angles?" And it says, "The angles are the following, a, b, and c." And I say, "Okay, go ahead and maneuver." And it says, "Do you like the present rate at which I apply the maneuver?". And I'd say, "No, I'd like to save a little fuel so slow it down to a slightly slower rate." And it says, "Okay. Is this rate all right?". And I say, "Fine, proceed." And it then

fires the jets just the right amount so that the vehicle begins to maneuver over to the new attitude. When it gets there it says, "Okay, we've arrived, do you want to proceed to pointing the sextant at the star?" And I say, "Yes", and it slews the sextant over to Canopus or where it thinks Canopus is and it says, "Please sight on the star.". So I look through the sextant and. . .

Brand: Finally *you* do something.

Schweickart: Sure. Because it asks you to please mark on the star. So it has placed the sextant toward the star as well as it can. And it's the error now with which it does that, that I want to remove from its memory. So I look through the sextant and Canopus is there but it's slightly off the center, so I ask for manual control and then I move the hand controller so that the sextant points exactly at Canopus and I press a "mark" button and it says "Thank you, now go to the second star." And I say, "What stars do I have to choose from?" And it says, "Here are three within the pointing capability with this attitude." And I say, "Okay, point toward Sirius," and it says, "Okay," and it slews the sextant over to Sirius. Now it says, "Okay, please mark on Sirius." And I look through the sextant and I see it's also slightly off and I say, "Let me have manual control again," and it gives me manual control and it says, "Please mark when you've got it in the middle." I get it in the middle and it says "Thank you, here's the correction that's brought about by those marks. Does it sound reason-

able?" And I look at it and if it said 45° I'd say NO! But if it says 2 arc minutes error in its basic reference system I'd say, "Yeah, that sounds like that was a reasonable error that we just corrected. Go ahead and torque the gyros over to the new attitude." It says, "Okay," and it does that. It says, "Would you like to check some other star to make sure it's right now?" And I say, "Sure, I'd like to check." So now I say, "Point toward Rigel." And it says, "Okay," and it points toward Rigel and I look and sure enough there's Rigel right in the middle of the sextant with no error because it's now been corrected. It says, "Is it okay?" And I say, "Yeah, it's okay." And it says, "Good, I'll record it permanently," and it goes Vffft! and goes back to sleep. And now I've got a new reference system.

That's the kind of conversational flow back and forth with a computer. At first you're talking with a computer, but as you go through hours and hours of training with it and operating it under all kinds of different circumstances, you develop a relationship where you know the kinds of questions it's going to be asking you and the kinds of information that looks reasonable and doesn't look reasonable, and it becomes a process of which you are a part and the computer is a part and when you're really in operating mode, really functioning, you're not conscious of which is you and which is it. You are together doing something.

Brand: So you have visited the fourth kingdom and you like it.

McClure: I've had some experience like that just on a minor level. I put off physics and chemistry an infinite amount of time. I mean when you're a biology major that's the first thing you take, right? That's the last thing I've taken at the last possible minute. At first I thought there was something wrong with me because I couldn't do it, I couldn't understand it, the concepts were too abstract. Now I have an excellent physics professor *and* I have a calculator. I'm not problem solving now, I don't have to worry about whether I added right or subtracted right. I have this machine.

Schweickart: Right. There it is.

McClure: So now I'm able, instead of spending all my time figuring out the math, to understand concepts that I did not understand before. My physics teacher said, "I don't mind teaching math to young kids anymore, they have calculators and they can depend on them. I love teaching children new mathematical concepts I would have never attempted before." And here I am, a college level student, and there's a whole new realm opening up for me because of this little love affair that I'm having with my calculator. I'm in control, I'm setting up the equation. It's just solving it. So I don't feel that I'm not learning or something's being taken away from me. I feel like it's kind of an interaction.

Schweickart: It stops being "that thing" and it becomes part of you. ∎

145

JERRY BROWN— from Limits on Earth to Possibilities in Space

To celebrate California's leadership in space (51% of NASA's procurements in 1977 went to California — the next state was Alabama with 8%) and the occasion of the first free flight test of the Space Shuttle, Governor Jerry Brown hosted a "Space Day" on August 11, 1977, at the Museum of Science and Industry in Los Angeles.

The event was organized by Russell Schweickart, on loan to the Governor from NASA, and co-sponsored by the state and the aero-space industry. It got widespread news coverage because 1) Brown had always been seen as strictly Mr. Era-of-Limits, 2) the Carter administration was giving signs of reducing the NASA budget, 3) it was the summer of the phenomenal success of the film "Star Wars", and 4) Brown was the first major political figure to offer a national vision of space adventure since President Kennedy.

Speaking at "Space Day" were all of the major NASA leadership (including the new NASA Adminstrator Robert Frosch) and also Gerard O'Neill, Carl Sagan, Jacques Cousteau, and Robert Anderson — head of Rockwell International, which built the space shuttle.

To end the program former "beat poet" Michael McClure read a new work, "Antechamber", against the silent showing of a film made of the most spectacular NASA footage. Michael, who is a proponent of space exploration but not of space colonies, jotted some poems during the course of the day's talks and gave us permission to print them here.

Next morning, August 12, most of the "Space Day" participants were at Edwards Air Force Base, along with 68,000 other kibitzers, to see the smoothly successful first atmospheric flight of the space shuttle Enterprise.

— SB

JERRY BROWN, 1977:

"Ecology and technology find a unity in Space."

•

"When the day of manufacturing in Space occurs and extraterrestrial material is added into the economic equation, then the old economic rules no longer apply. Going into Space is an investment. It's not a waste of money, it's not a depleting asset, it's an expanding asset, and through the creation of new wealth we make possible the redistribution of more wealth to those who don't have it."

•

"Awareness of limits leads to awareness of possibilities."

•

"As long as there is a safety valve of unexplored frontiers, then the creative, the aggressive, the exploitive urges of human beings can be channeled into long term possibilities and benefits. But if those frontiers close down and people begin to turn in upon themselves, that jeopardizes the democratic fabric."

•

"As for Space Colonies, it's not a question of whether — only when and how."

Brand

Michael McClure

PHILOSOPHICAL POEM #1

"SCALE IS THE PROBLEM."
All things
have
no
size.
We arise
from the heaviness
of nothing
and
burst

and
collapse

as
STARS:

We can't be stopped by love!
We can't be slowed by wars!
We're the structure of our tries!
We're the molecules of sighs!

—McClure

Bruce Murray, head of the Jet Propulsion Laboratory in Pasadena, avid reader of Shumacher and Illich, puts appropriate technology in space perspective. Behind him, from the left, Russell Schweickart (chairman), panelists from the Brown administration — Richard Silberman (Business and Transporation Agency), Don Vial (Labor Relations), Huey Johnson (Resources Agency), and Stewart Brand (Consultant to the Governor, panel moderator).

Maurice Manson

SPACE DAY

EMOTIONAL VITALITY MEANS
that we — we Californians —
(cousins of the sabertooth)
look at these plans
for
rockets
like old Florentines examining
some paintings
or the casting
of a cannon.
—*McClure*

Captain Jacques Cousteau and Governor Brown

Governor Brown makes his closing remarks to the 1100 invited guests at "Space Day." Directly behind him is Gerard O'Neill's working model of a mass driver.

Brown is briefed by the pilot of the 747 that carried the space shuttle up to 30,000 feet for its first free flight.

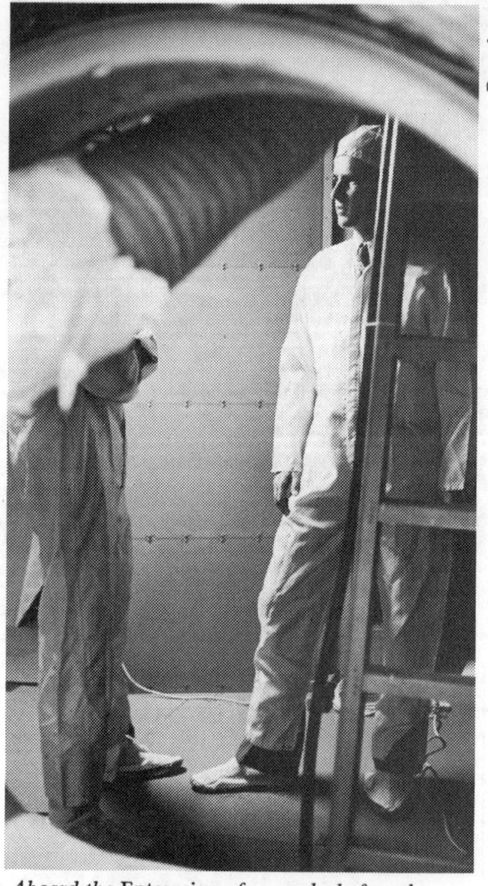

Aboard the Enterprise *a few weeks before the free flight test, Brown talks with astronaut Deke Slayton. The "bunny suit" is to prevent anything on the visitor getting loose in the space craft.*

PHILOSOPHICAL POEM #2

LIVE BY YOUR WITS!
The universe sings
in
fits.
We dance
in the glance
of a dead man's
eyes.
Yet we're merry
as cherries
blown by the winds
in the boughs.

—McClure

Flanked by Russell Schweickart, Brown watches the 200 mph final approach of the Enterprise *— the first wheeled landing of a space craft on Earth.*

PHILOSOPHICAL POEM #3
Beginning with words by J. Cousteau

"WE HAVE TO GO ALL
THE WAY
IN ALL DIRECTIONS."
Our
centers
are mere
reflections
of the edges
that we laugh
and dance

up-
on.

Now I'm here.
Now we're gone!

—McClure

EVOLUTION-COMPLETION IDEA

Each of the following sequences is an evolution. Complete each sequence according to the rules of change by which each example becomes the next.

1. <u>transportation</u>

    A. walk   B. ride   C. fly   D. ________

2. <u>economics</u>

    A. hunt   B. pasturage   C. industry   D. ________
       gather   farm   commerce
                                   services

3. <u>abstraction</u>

    A. words   B. calendars   C. periodic table   D. ________
       numbers   maps   unified field theory

4. <u>information media</u>

    A. stone tablet   B. ink/paper   C. global TV   D. ________
       carved bone   paint/canvas   telephone
                                    network

5. <u>self-orientation in universe</u>

    A. egocentric   B. lococentric   C. geocentric   D. heliocentric   E. ________

6. <u>social organization</u>

    A. individual   B. family   C. city   D. empire   E. ________
                      tribe   nation   united nations

7. <u>self-reproduction</u>

    A. asexual   B. whole from   C. single   D. muscle   E. ________
       replication   part   organ   nerve
       of whole   regeneration   regeneration   regeneration
       (single cell)   (annalids)   (crustacea)   (mammals, man)

8. <u>matter complexity in universe</u>

    A. astronomical   B. geological   C. chemical   D. biological   E. ________

Conceptual artist Donald Burgy completed this in December '73, it appeared in Artforum in '74 (minus paragraphs 6 and 7), and Robert Horvitz sent it to us. I agree with Horvitz: the mere stating of the question in these terms puts a kind of pressure on us from the universe or on the universe from us.

— SB

Instead of Frictionless Elephants

TALKING WITH GERARD O'NEILL

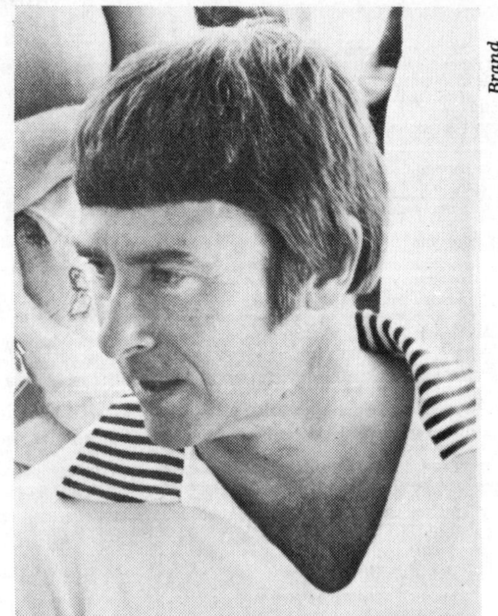

Gerard O'Neill

Jane McClure: What gave you the idea to give that space problem to your students?

Gerard O'Neill: It was the year of the Apollo landing, and I was teaching freshman physics. The course had never had a theme as such so I chose the Apollo project rather than the classical physics problems of pushing frictionless elephants up inclined planes and so on as they always had. To a special seminar of ten or twelve people I suggested the overall question, "Is it possible for a technological civilization to do its natural expansion in space rather than on the surface of the planet?" Or, another way around, "Is the planetary surface the right place to be in the long run." As to why I asked that question, I have a feeling that it probably came out of an automatic approach to problem solving based on 15 years or so of having worked in experimental high-energy physics up to that time.

Stewart Brand: You seem to have some habits now of how to approach the right question. What is the structure of that?

O'Neill: Yes, there are a few prejudices that I've built up over time, based on the experience of seeing experiments that seemed to work easily or that had problems. One of them is, I found that if you've got a complicated problem, it always seems to help if you break it in pieces and then solve each piece separately. The tendency of a fairly green designer is to try to take a problem and come up with one grand solution that will do everything at the same time. That's usually a mistake. There's another habit of thought, and that is if you're looking at a very tough problem, it's often useful to try to stand back from it and see if you're locked into some thought pattern.

Brand: Like assuming you're stuck on a planet.

O'Neill: Right. You know the six-match problem? This is the problem where you take six equal matches and without breaking them or bending them or anything, you make four identical triangles out of them. I don't think I'm a particularly good problem solver, but it happens that when I was shown that problem I could solve it very quickly.

McClure: Solve it! *[He did:*  *]*

Brand: Jane, I wonder if it'd be interesting to probe a little bit the nature of the younger generation of designers and maniacs that Gerry's attracted. Like the ones who built the mass-driver.

O'Neill: They're nice kids, aren't they? Well, two of the students at least at pres-ent are hoping to come to Princeton and graduate school within another year or so.

McClure: Where are they now?

O'Neill: One is at the University of Michigan, that's David Kaplan. He's just finishing his fourth year of a five-year program in which you get a combined degree in aerospace and computer science. And Jonathan Newman, who has just finished his junior year at Amherst in physics, with a fairly strong law background. At the moment he's hoping to come to Princeton either in aerospace engineering or physics.

McClure: How did they get connected with you in the first place?

O'Neill: Well, David because I gave a lecture at the University of Michigan a little over a year ago and I mentioned that we were having a summer study and asked anybody who was interested to see me. I was terribly impressed by David because he proceeded to skip a meal and by the end of the day handed me a completely typed up, one page, highly condensed resume of his background, experience, interest, and all that. So I hired him for the '76 summer study, at which he did a beautiful job. We stayed in touch after that and then he came back to the '77 study as well.

Kevin Fine came to MIT from a physics background, going into graduate work in aerospace engineering. Professor Miller, who was the head of the department and my host, offered to have some graduate student work with me during my year there and it seemed to us that Kevin was the best candidate.

McClure: Out of curiosity, how many women do you get interested?

You should know that all the interviews with O'Neill in this book took place after midnight — the man is a model of grace-despite-fatigue. This time it was after a full day of Governor Brown's Space Day, at which O'Neill spoke, and a night-piloting job to a desert airport (details p. 155). Jane McClure had recently attended the press conference following the third (1977) NASA-Ames Summer Study on space colonies and had some questions.

— SB

O'Neill: Lots.

McClure: Lots? I didn't see any working with you or on the mass driver.

O'Neill: Well, it's the statistics of small numbers at this point. The work I've been doing this past year is very specialized in the sense that it's largely on this question of the development of these electromagnetic mass drivers, so that tends to mean someone with a fairly strong physics background. There are not many women in that field.

Now, at the NASA-Ames summer study we just finished we had three women professionals who were quite senior and all of them were extremely good. One was in life sciences area — closed cycle agriculture, another was in astronomy — she's an astronomer at Palomar, and the third is a lawyer, who was involved in the international law aspects of the project.

McClure: How would someone in my position get involved in some sort of research or design problem in your line, within about two years. I have most of my schooling behind me and all I have to do is specialize. At this point I could go in any direction.

O'Neill: I'd mostly try to find what you feel you yourself are best at. What are the things that you most enjoy doing?

McClure: Biology. And that doesn't really fit in . . .

O'Neill: It does. Probably in the long run some of the most interesting and least understood problems associated with all of this are biology problems. For the physics and electrical engineering and rocketry there's no problem finding people who can do those jobs. There's lots and lots of such people.

Brand: What would be your list of biology-type applications?

O'Neill: The biologically-oriented study group was the largest single task group —about 15 people — that we had in this year's summer study. There are obviously going to be space medical questions. The most important being, "What is the range of atmospheres in which it's possible for people to live comfortably without developing respiratory problems?" For example, if most people could exist indefinitely without harm living in a pure oxygen atmosphere with an oxygen partial pressure equivalent to say 5000 foot elevation, that's terribly cheap to provide because the Moon is mostly oxygen, and it's going to be literally a waste product of space manufacturing. On the other hand if it turns out that we have to carry along almost Earth normal air complete with all that nitrogen, it's very expensive because it's not found on the Moon. People have already lived for several days at a time in pure oxygen atmospheres. It was done in the Apollo project. Once the guys on the lunar module finished with their first EVA, they dumped whatever nitrogen there was on board and lived in pure oxygen the rest of the time. In some cases that was a week. So we know that there aren't any drastic effects associated with pure oxygen atmospheres. But we don't know whether there's something that builds up with the passage of a long period of time, whether there's subtle effects. That's pretty much physiology; you almost have to be a doctor to handle that.

Brand: How about microbes and plants? Won't they need atmospheric nitrogen?

O'Neill: That's still a big open area, a tremendous amount of research to be done. In the very beginning I think that we're not going to be using closed cycle agriculture. We'll just carry up the same sort of dehydrated food that people are used to in space programs.

McClure: At the press conference after the NASA-Ames summer study, there was someone on the panel who, in response to a question from the press on the psychological implications of space travel and living, said there are three questions which they found the most often asked by people — "Who's in charge?" "Who's paying for this?" and "Aren't people going to go crazy?" I thought those were three pretty definitive questions.

O'Neill: Well, we have of course, a tremendous amount of data on the issue of isolated groups and their power structure. If you look at the situation of sailing ships in the days when they were out for months or years at a time, it was a very clear-cut situation. It's been found that a dictatorship is what works. The very important difference being that the dictator, the ship's captain or the captain of an airplane, is not there for life. He's there for a voyage and he comes back. Whatever the guy does, he knows that he's going to come back, his actions are going to be reviewed, and if he has done a bad job he won't be reappointed.

In the very beginning where you're dealing with let's say 100 people who are up there for a tour of duty of 6 months or a year or something like that, they would probably end up with something fairly equivalent to say an Antarctic colony, where there's always a well-defined chain of command.

McClure: Also, I have the strange feeling that probably you would quickly revert to a kind of a natural pecking order. In a situation like that, pecking orders are a kind of survival mechanism.

O'Neill: It's a survival mechanism essentially because it reduces conflict. There's nothing that produces conflict more than an ill-defined situation of authority.

Brand: You know I wonder about the uses of all these Earth-based examples and metaphors, such as we heard in the Space Day speeches today. A reporter asked me whether I thought space exploration was like Stonehenge, or Queen Isabella and Columbus, or the Nile and the dream of the seven cows. I told him I think we're probably up against the limits of the usefulness of metaphor and those things are going to lead us astray rather than inform us.

O'Neill: That's a good point. We may just have to face up to the fact that here we are doing some things that have new features. If you come back in 100 years there're probably going to be some new words.

Brand: There'll be new things to name, different degrees of solar storm, different kinds of cabin fever.

McClure: What do you think of Cousteau's view that space should not be inhabited, just as he thinks the underwater is not to be inhabited, even though he was experimenting at one time with the possibilities of living underwater, and you're experimenting now with the possibilities of getting in space . . .

O'Neill: Well, I think it's an emotional response and you know he's got a complete right to his opinion on that. In the case of the ocean it's a bit different, Jane, because there are real reasons why the ocean is a relatively harsh and unfriendly environment to inhabit. The reasons are, first of all, that the pressure problem under the oceans means that the amount of space that's available is much more sharply limited. Once you go down far enough that you're away from surface storms, you're down so far you're in a region with pressure really way up. Pressure vessels are normally unstable under external pressure. You take a sphere, deform it a little bit under external pressure and its tendency is to collapse all the way. Whereas in space you're dealing with internal pressure; you take a sphere and deform it a little from spherical shape and it'll return to the spherical shape because that's the stable case, as long as the pressure is on the inside rather than on the outside. And you're dealing with only 5 or 6 pounds per square inch instead of the hundreds of pounds per square inch which you get to very quickly under the oceans. Also, under the oceans you're cut off from your energy source — the sun — even more than you are on land, whereas in space you're opening up your energy source for full-time use at high intensity. It's just a fundamentally different case.

McClure: When I first read your articles I was very excited at the possibilities of space habitation.

O'Neill: I hope you still are.

McClure: I vacillate back and forth — I can't help it. There's an emotional response and then a practical response.

O'Neill: You should, because we are dealing with some potentially explosive issues. Anytime you get into something powerful it always has the capability of good or evil, which is a point that I make several times in my book and I'll return to it. That's one reason why I'm so interested in trying to guide it to the best of my ability because I can imagine the whole thing wandering off.

Brand: What are some of your dark visions of how it might go bad?

O'Neill: Maybe I'm too positively oriented to dwell on them a great deal,

151

Stewart, but one possibility would be that the space habitat business might get done under complete military domination. It might be that one country goes out and starts building habitats in space and arms on the Earth at the same time in a threatening way which makes it impossible for any other nation to go out into space. That would certainly be a very dark vision if that were to occur, because it would put a clamp on our human race which is an even harsher clamp than what we've got right now. I really hope that doesn't happen.

I think it's completely unrealistic to expect that space colonies are going to develop in some way which I lay out in a formula. Obviously. If it were going to do that then there's something wrong with it. It's not worth doing. I have a fundamental faith in the good sense of rather ordinary people, and the less the intellectual pretensions of the people, the more faith I have in their good sense. That's just based on my own experience ever since I was 17 years old in the U.S. Navy. I'm thinking of people who were in some cases semi-illiterate but whose basic common sense was far superior to lots of extremely distinguished types that I've known since who have 14 degrees after their names.

Brand: You trust the noncommissioned officers more than the commissioned officers?

O'Neill: Well, I'm biased — I was a noncommissioned officer myself. But I trust the ordinary seaman perhaps even more.

Brand: You know, some of the papers from your conferences are coming out in books now, but they're always a year or two behind. What are you working on currently?

O'Neill: We're beginning to work out a real program scenario in which you look at specific sets of lights and how much goes up on which and all the rest of it. There was just the beginnings of that in '76. In '77 I had a sabbatical year, which was a great help, and with the kindness of MIT I spent it there. This has been a very productive year mainly because of work on two areas — one was the mass driver, the other was to follow-up the '76 study work on program development, trying to really look at the minimum investment ways of getting to a big productive payback. That's in an article which will be coming out in a month or two.

Brand: Where?

O'Neill: Astro and Aero.* And the Summer Study felt it made good sense to develop a program for how you eventually realize space manufacturing. That's what John Shettler from GM did. He ended up with a program in which you make the first shuttle flight in 1985, and you're getting the first power satellite on line in 1991 with a very rapid build up after that, with Island One I suppose coming in the mid-to-late 1990s.

Astronautics & Aeronautics (see p. 128).

Brand: So far you've hung all of your investment payback arguments on the solar satellite energy.

O'Neill: I think if you're honest you really have to conclude that at the present state of our knowledge that's the only big economic driver.

Brand: How about materials collected from asteroids for use on the Earth?

O'Neill: You're up against a much tougher economics there. Apollo had launch costs on the order of $1000 a kilogram even up to low orbit. The shuttle is $20 million a flight. Let's see — $20 million for 30,000 kilograms of metal, so that's like several hundred dollars a kilogram right there. The price on the surface of the Earth is more likely to be $1 or $2 a kilogram. The price of raw materials is always in that range. Now, if you huff and puff and make an enormous launch vehicle which can reduce the launch costs far below what the shuttle can do, maybe you can get down into the range of let's say $50 — $70 a kilogram to geosynchronous orbit. What that means is that if you can build something whose end use is in geosynchronous orbit, no matter how wasteful and inefficient you are building it, you will have an inherent added value of $70 per kilogram because that's what you would have had to pay to bring that material, that equipment, up from the surface of the Earth. Now, if you return material to the Earth's surface, you no longer have that advantage. You don't have that automatic added value. Now you're competing with Anaconda Copper and U.S. Steel, who are selling their things for $1 or $2 a kilogram.

The fact is that although we are working on something which I think has a tremendous amount of potential, from an economic viewpoint it is very dicey in that it's a one-crop economy. Energy. If someone finds a solution to the energy problem which is better, quicker and cheaper than satellite solar power done this way, then we're wiped out.

McClure: At some point there is bound to be a real energy crisis, and I get the feeling that you will get little support until that time. Then all of a sudden you will be launched into space to mine the Moon and asteroids and build solar satellites.

O'Neill: I doubt if it'll be that automatic. For one thing the whole energy problem is one which has enormous vested interests right now. And I'm thinking not of the utilities, investers, and so on. They don't have the technological hang-ups in that respect. I was very impressed by the utilities people and the investment people when we had a first meeting of our USRA (University Space Research Association) advisory panel last month. Those people said, "Look, we read this article, we see that with some fairly reasonable set of assumptions that you may be talking about energy for $500 a kilowatt or something like that. We don't care whether you're getting it from the antennas or little green men from Mars. If you can get us energy at $500 a kilowatt we'll buy it. No problem. Just show us that it can be

done." This is one issue that's very clear. You're not going to get private capital until the technological risk has been reduced almost to zero.

McClure: What kind of implications are private investments going to add to the whole space program?

O'Neill: In fact we discussed that point. One of the men who was on the panel is from an insurance company. He happens to have been very heavily involved over a number of years in the financing of the Quebec Hydro, which is the largest hydroelectric project ever done in the western world. Quebec Hydro has a whole set of horrendous political problems that I won't go into. Basically, Churchill Falls is a source of energy for which the installed cost is around $1600 a kilowatt, but once you have it it's a free source of energy from there on because it's hydroelectric. The financing involved is like 10 billion, 15 billion dollars, not peanuts by comparison with what we're dealing with — it's very similar. But he made the point very strongly that the technical issue of whether if could be done was never raised. There were 20 volumes of study reports written on all of the details, but the fundamental point was that everybody knows that water falls downhill. We are a long way from space manufacturing and satellite solar power being so proven and so taken for granted that they're equivalent to saying that water runs downhill. And it's not going to be until we have that degree of assurance that we're going to get private capital. It's just going to have to be governmental up to that point. Unless by some incredible magic one can get an enormous broadly based consortium of people all over the world to do it, which is probably completely ridiculous but it every now and then crosses my mind in a wistful kind of way.

I don't feel that in the long run the details of how this all gets started is going to be as important as the fact that it's getting started. I think I could even adjust to the possibility that it would be done and the United States might have no part in it. Even if it meant the beginning of a decline or acceleration of a decline of the United States, because I'm sure that whatever nation or group of nations first makes really vigorous use of nonterrestrial resources almost certainly has got a lease on the next 100 years.

Brand: Japan has I'm told something like $2.8 billion a year now in space, compared to our $4 billion.

O'Neill: I don't know what they're doing. I was told that there had been a Japanese delegation which made an approach to NASA to take part in exploration of satellite solar power, but that they had been rebuffed. That's all I know.

Brand: Any indication of Russian interest?

O'Neill: They soak up everything that they can find out on the subject. Articles of mine have been translated without my knowledge into Russian newspapers. Then some academician of the USSR gives an answer with the classical and

The fabled mass driver built from spare parts by MIT students. In 0.11 second a cooled bucket accelerates at 33 gravities to a speed of 81 miles per hour in a distance of 8 feet. Left to right, Henry Kolm of MIT, Gerard O'Neill, and student Kevin Fine, who wrote his Master's thesis on the instrument.

expected statements — first of all that, yes, these things will happen but not at all in detail in the way that this guy from the West says, and besides it was all done first by the Russians.

I'd certainly point out that Tsiolkovsky did in fact have most of these ideas many many years ago.

Brand: *The High Frontier* has been translated now into what? . . .

O'Neill: Japanese, German, Dutch, French. I think Italian's the last one I've heard about.

Brand: What's the increase in public awareness? Linear growth?

O'Neill: I think much more than that. I think we're seeing an acceleration of the acceleration. It's getting completely out of hand as something which can be coped with without some sort of organization.

Brand: Pure nuisance for you, I imagine.

O'Neill: I wouldn't say that about something that I basically love, but yeah, it can absorb 100 hours of every day, and 100 becomes 200 hours if you wait two months. It's one reason that we started the Institute for Space Studies at Princeton.

Brand: Say more about that.

O'Neill: At the moment it's not a great deal more than a gleam in the eye, but it is a nonprofit, tax-deductible corporation in the state of New Jersey and it's now getting federal incorporation also. It is also probably going to become a so-called Nongovernmental Organization of the United Nations, which is important because we want to be able to make an input to the UN deliberations on such things as treaties about the Moon. We don't want things

to be bargained away which we may very much want to be able to use later on. We have very good connections within the United Nations and our friends there assure us that it will be a real value to have that kind of status, and that it would be quite easy to get. The corporation has as officers people like myself, Tasha, Brian O'Leary, Steve Cheston . . .

Brand: Family.

O'Neill: Family. All the officers work with no compensation. The organization has no overhead because it has no staff and one fundamental article of faith of my own is that, whatever happens to the Institute, we will live within whatever budget we have at the moment. If we have no money, we will spend no money, so we are not going out and hiring permanent paid staff at this stage of the game at all. If we can afford to bring in a Kelly Girl occasionally to help with the typing that's about as far as we go.

McClure: What are the issues that are being discussed in the UN? Wasn't something mentioned about control of geosynchronous orbits between nations?

O'Neill: Well, that's an example of one of these dark prospects that you were asking about, Stewart — if there were to be a binding set of treaties that were to go into effect that would for example make it impossible for satellite solar power to exist at all. It could happen.

Brand: Because of synchronous orbit being already full of satellite at some time?

O'Neill: Well, legally full. From the technical viewpoint I don't think there is a serious issue, because I can see all kinds of ways around the problems of

filling up geosynchronous orbits. But it could be a very real issue from a legal viewpoint.

McClure: But you're afraid that if we start signing treaties now before we know the possibilities of space and space technology that we're going to end up limiting ourselves while trying to keep that peace.

Brand: Like the Law of the Sea meetings. It's just endless argument. Nobody's mining the sea because of the arguments.

O'Neill: That could happen. Or, it could be that we in the United States would give away in a Moon treaty any possibility of our taking part in a consortium which would make use of the lunar resources in the future. That would be enough to delay the whole thing by 20 or 30 years. Ultimately I'm sure that there would be some nation or group of nations who would go out there and do it whatever the treaties said, but it might not be us.

Brand: Gerry, of all the designing that's gone on so far it seems to me that the most demonstrably exciting thing is the mass driver. It's new, it works, and it's ingenious as hell. What was the genesis of that?

O'Neill: The first notions that I had about taking material off the surface of the Moon involved these centrifugal launching machines, which as far as we can tell could probably be made to work, it's just that we think we have a better way. Toward the end of 1973 there was an article which appeared in *Scientific American* by Henry Kolm and Artie Thornton on the subject of magnetic flight in connection with urban transit systems which they had been working on. Henry had built a model of an urban transit device which operated over a 300 foot track. It was a joint project of MIT and Raytheon. It involved a superconducting set of coils and demonstrated honest magnetic flight.

When I saw that, it filled in a third place where there were two places already full and I was looking for a way to fill the third. If you want to have an efficient launching machine it's got to have several characteristics. First of all you don't want to throw away anything expensive. In all the old-fashioned magnetic guns you threw away the moving coil in the course of a shot and that obviously wouldn't do. The second thing is that the acceleration by magnetic fields is pretty easy. Germans even as far back as World War I tried to make magnetic guns. The idea that you could accelerate something by having a coil with a current in it is a very old-fashioned idea. However, the stumbling block was that as soon

153

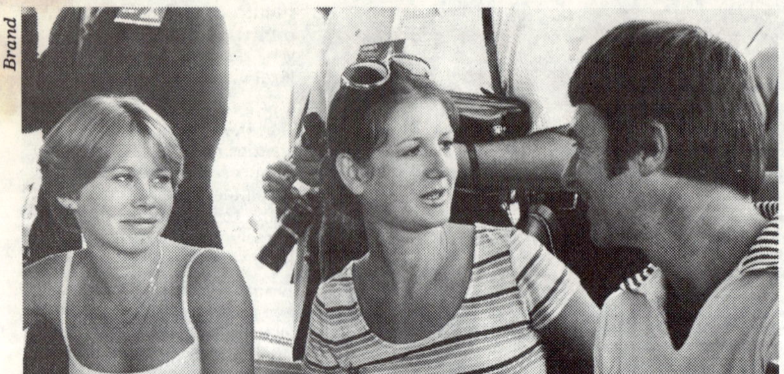

Jane McClure, Tasha O'Neill, and Gerard O'Neill at Dryden Flight Research Center

as you bring something up to a high speed, how the heck do you hold on to it? You can do the calculations, and you find that wheels fly apart at a tiny fraction of the speeds which you want on a mass driver. You can't tolerate friction with speeds like a mile per second.

Henry Kolm's and Artie Thornton's article filled in that big gap by pointing out to me something I had simply never known before, which was the existence of the phenomenon called magnetic flight. It turned out it had been around for 60 years and was demonstrated to Winston Churchill in 1914 at the Paris Exposition.

Brand: As a possible what?

O'Neill: As a system for moving packages rapidly around the city of Paris. So, I did the initial calculations for a mass driver, probably 20 or 30 pages of calculations, just to convince myself that it did make sense and that the numbers worked out in a reasonable way. I described it at the 1974 Princeton conference.

Brand: Does this qualify as an invention?

O'Neill: Probably, from a legal point of view, I did inquire of the research corporation which thinks about patents for Princeton University as to whether it would be patentable. They said yes it would be patentable, it qualified as an invention. However they were not interested in patenting it because it was obvious that if it had application it would be either implemented by a government or a consortium of utilities.

Brand: Say a little about what's new about the device.

O'Neill: In a synchronous motor like this the information is fed back from the moving object. The critical thing about the mass driver is that it's self-timing. What happens is that as the bucket goes careening down the track it crosses microswitches, and every time it crosses a microswitch for one particular coil, it triggers that coil, and that gives the push at the right time to keep it going. In the case of the real mass driver, that would be the interrupting of a light beam rather than a microswitch. You're not dealing with an unstable situation of riding a sort of magnetic wave which is

remorselessly going to move ahead whether the bucket is there or not. The bucket does its own timing as it moves.

Brand: Do you know yet whether there is an optimum mass for the mass driver?

O'Neill: We know that there isn't.

Brand: Infinity has just entered the discussion here.

O'Neill: Not really, no, it's just that particularly as a result of this very intensive work during the last year we have now all the sets of optimization formulas so we can work out masses. There are now computer programs in which you plug in the conditions that you want, the velocity that you want, how many kilograms per second that you want to pump out, what's the diameter of the coils, and one or two other things. You can punch all of those parameters into an HB 67 program and it churns away and comes out and tells you all the parameters of the mass driver, including the masses of all the components involved, how much tonnage of coils, how much tonnage of capacitors, all the rest. You can go down and practically order it off the shelf. One of the things that we investigated was the question of how sensitive this optimization is to the diameter of the machine because if we change the diameter by a factor of ten we change the mass of the payload by a factor of 1000. So we went over a factor of 1000 range of payload mass and asked how much did the mass of the machine change. And it turned out it only changed by 50% over that very large factor.

Brand: How about the energy requirement?

O'Neill: Very little change because it's extremely efficient. On the lunar mass driver, our latest calculations show it should be 97% efficient. The lunar machine has, as I recall, about 100 megawatts of power input. Of that, 97 megawatts appears as the kinetic energy of the payloads that are going off the surface of the Moon. Only 3 megawatts go into radiator panels that are having to radiate away the heat from the coils.

Brand: That's awesome.

O'Neill: The machine that we hope to use as the shuttle upper stage to get

from low Earth orbit to lunar orbit, the one that would make use of the shuttle tankage's reaction mass, that is about 75% efficient. It's less efficient because it's being stressed a lot more. We're running it at 1000 gravities acceleration instead of 100 and it's going up to a much higher speed. It has to go up to 10,000 meters per second.

Brand: How about this little one you've been demonstrating?

O'Neill: Surprisingly enough it's about 50% efficient. I like to go back and recall that in my old **Physics Today** article where I expressed the concept of the mass driver, I was estimating 29 gravities acceleration. This model that you saw operate today runs —if you dunk it in liquid nitrogen to make the core really cold —it runs at 35 g's. And that's built from scrap.

Brand: Is the space environment conducive to a very effective mass driver?

O'Neill: It's far easier to make one in space than it is anywhere else. For many reasons.

Brand: You mentioned the need for chilling it. You get that naturally in space?

O'Neill: No. The average temperature in space is about what it is on the Earth. It's the same equilibrium temperature that you get if you toast one side and let it cool and toast the other side. An artificial satellite is a small planet; therefore it heats and cools like a planet. In the old days before light meters people with cameras used to take pictures of the Moon and wanted to know how to set their cameras, and everybody said, "It's very simple, it's an object in bright sunlight."

McClure: Gerry, in my thoughts about space settlements I get excited and then I get disturbed at the thought that 10,000 people are up there and what if there's a slip in some sort of mechanics . . . I see everyone falling out into space.

O'Neill: Oh gosh!

McClure: It's just a visual image I get —how vulnerable you are in space.

O'Neill: It's probably the same way that Columbus's sailors felt.

McClure: Well, you gotta start doing it. It's the only way to find out. You're not gonna go build this huge satellite first and then put the people in it to see how it works.

O'Neill: You build up step by step.

McClure: I wanted to tell you one thing that a reporter said right after your conference at NASA-Ames. I saw a reporter in a phone booth. He was writing away frantically and speaking into the phone giving directions for a typist on the other end. Obviously it was going to be a big news story. He got ahold of his editor and said, "Either this guy is just totally out of his mind or he's _really_ on to something!"

O'Neill: That's about as charitable a view as I'm asking for at this point. ∎

The Long View

The shocks of this Age are the shocks of pace. Change accelerates around us so rapidly that we are strangers to our own pasts and even more to our futures. Gregory Bateson comments, "I think we could have handled the industrial revolution, given 500 years."

In 100 years we have assuredly not handled it. We manufactured an "Age of Discontinuity" (Peter Drucker) whose time horizon forward is terrifyingly close — 4 years in politics, 10 years in major corporations. I feel serene when I can comfortably encompass two weeks ahead.

That's a pathological condition.

But I think it will pass, partly from its pure unworkability, partly from the move of some of humanity into space. The project of space exploration, industrialization, colonization, and migration is so big and so slow and so engrossing that I think it will bring the rest of human activity into its pace. If you want to inhabit a moon of Jupiter — that's a reasonable want now — one of the skills you must cultivate is patience. It's not like a TV set or a better job — apparently cajolable from a quick politician. Your access to Jupiter has to be won — at its pace — from a difficult solar system.

With the first color photograph of the Earth from its Moon, the whole Earth became a political idea and ecology a political movement which has continued to strengthen in its 11 years so far. Though I have thought at times that the health of space perspective on Earthly activities has gone as far as it can, recently I'm not so sure.

The reach of human intelligence to the stars is an enormous undertaking. When I grasp the reality of that, not just the words, but the actual project, a religious scale of presence that spans centuries comforts me. Feeling comfortable and curious that far forward, and therefore that far backward, I begin to feel at home again. Interested in events longer than the ego's prison of "my lifetime", I'm free to care for other large continuities such as the life of the Earth and the drama of human culture. Previously overwhelming urgencies, like the deadline on this book for me, fall into microcosmic place — worth doing, connected, but not urgent.

Religious-scale projects — and their comforts — have often scourged humankind. I'm thinking of Egyptian pyramids, Moslem jihads, Mongol hordes, Christian crusades, the Third Reich, world Communism, maybe science itself. Part of their hazard is that they become their own universe — an infinite regress of self-reference grounded nowhere.

Space exploration is grounded firmly on the abyss. Space is so impossible an environment for us soft, moist creatures that even with our vaulting abstractions we will have to move carefully, ponderously into that dazzling vacuum.

The stars can't be rushed. Whew, that's a relief.

— SB

Thanks

By way of thanks to Gerry O'Neill and Rusty Schweickart, who are essentially co-authors of this book, let me leave you with the two images of them that abide with me. Both are flying.

Gerry is in the left front seat, his wife Tasha in the right, of a Piper Cherokee. It's night; we're at about 6,000 feet. The two of them are navigating like mad — twitching dials, jotting figures, muttering into the transceiver which is shrieking back, glancing down at the town-lit coastal fog that has slightly complicated this flight of ninety miles from Los Angeles to Mojave. All of us (there's Jane McClure as well, wide-eyed on her first small plane flight) are grey with fatigue from Governor Brown's "Space Day", facing still the small-hours interview you just read, a wink of sleep, and a dawn drive to Edwards Air Force Base for the space shuttle free flight test.

Here at 6,000 feet we're in dreamtime. Gerry opines where we are. Tasha checks a map, gently corrects him. He thanks her, queries the radio, jots the response on his kneepad, confirms with Tasha, takes a new heading, she switches maps, and they discuss whether the Mojave airport might have its lights on. They've done this for several hundred thousand miles, clear to Venezuela one time. From the back seat it's ballet to watch them, pas de deux.

In Rusty's case he and his wife Clare and I and a lady were camping on a desert island off Nova Scotia, devouring in the late twilight sweet mackerel that Rusty had just caught. Rusty and Clare were drifting easy, free for a couple of days from their five kids who were camping elsewhere. Rusty glanced at the sky — "There goes the first one."

Sure enough, a satellite glowed north in polar orbit among stars that Rusty knows by name from years of celestial navigation and keeps an eye on. While Clare asked how many satellites you're likely to see of an evening any more, and Rusty opined a calculation, I began to watch him funny, a personal sense of his Apollo experience at last forming in me. Clare knocked a gold spike in it a moment later.

"You know, Rusty, I never did get out at night to see you when you were a star."

— SB

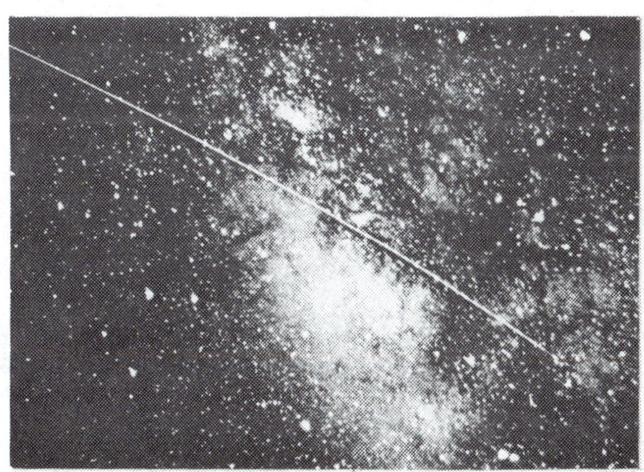

An American satellite crossing the star fields of the Milky Way.

Index

A

Aerospace Corporation, 94-95
Aerospace Education, The Directory of, 128
Aerospace Education, The Journal of, 128
Agriculture, 12, 14, 19, 21, 60, 70, 90, 94-95, 107
  cross section of agricultural region, 130
  needed research, 29, 43, 71, 92-93, 151
  problems of space, 42, 43, 49, 52, 94
Air currents, 63
Algal reactors, 60
Allen, Joe, 25
Aluminum
  manufacturing, 65, 66, 129
  moon, 14, 20
Aluminum-foil
  mirrors, 13, 40, 62-64
  solar sails, 136-137
Anderson, Dave, 25
Anderson, Jim, 96
Ant Farm, 47
Antennas, space, 86-89, 94-95, 98-100, 141-142
Architecture, criticism of, 44-45
Asimov's Laws of Robotics, 88
Assemblers, 129
Asteroids, 9-10, 13, 14, 20, 31, 115, 135
Astronaut selection, 77, 140-141
Astronauts, books on, 91, 115, 144
Astronomical Calendar, 132
Atlas, of universe, 131
Atmosphere, 12, 14, 15, 21, 29, 70, 151
  problems in closed system, 49
Automation, 10, 18, 129

B

Baer, Steve, 40
Bainbridge, William Sims, 130
Ballast tanks, 67
Ballester, Antonio, 92-93
Barghoorn, E. S., 92-93
Beers, Jonathan, 60
Bearings, 20, 78
Benefits of space colonies, 7, 11, 28, 36-37, 39, 43, 80, 97, 146
  better living standards, 10, 11, 97
  to developing nations, 11, 18, 97
  ecological, 43, 97, 143
  energy, 11, 13, 16, 18, 39, 50, 97, 107
  in lift costs, 11, 16, 17, 64
  manufacturing, 13, 28, 88
  metallurgy, 53
  population reduction, 11, 18, 27
Bernal sphere colony, 105, 135; *see also* Island One
Berry, Wendell, 36-37, 82-84, 85
Beutler, John, 53
Beyers, Robert, 49
Beyond the Planet Earth, 30
Biology, space, 142
  needed research, 29, 43, 49, 71, 92-93, 151
  see also Responses, critical, biology
Biology, space, symposium on, 49
Books, on space colonies, 30, 31, 69, 129, 130
Botkin, Daniel B., 92-93
Boulding, Elise, 85
Brand, Julia, 60
Brand, Stewart
  book reviews, 31, 91, 129, 130, 131
  on closed ecosystems, 94
  interviews: Jacques Cousteau, 98-103; Lynn Margulis, 120-124; Gerard O'Neill, 150-154; Carl Sagan, 124-127; Russell Schweickart, 74-81, 138-145
  introduction, 4-7
  on Mars experiments (interviews), 120-127
  response to Wendell Berry, 84-85
  on responses, 33, 85
  on space colonies, 5-7, 72-73, 84, 155
  on space planning, 86-87
  on surveillance and freedom, 87-88
Brautigan, Richard, 51
Brayton cycle turbogenerator, 14
Bricks, 129
British Interplanetary Society, 130
Bronstein, Jan, 35
Brower, David, 45
Brown, Jerry, 146-148
Building materials
  for homes, 129
  for space colonies, 9-10, 13-14, 20, 28, 64
Burgy, Donald, 149
Butler, Pierce, 55

C

Calcium, 14
Callahan, Michael, 6
Carbon, 14, 20, 52, 62, 129
Career information and possibilities, 80, 128, 142, 151
Caron, Jonas, 42
Carrying the Fire, 144
Caulkins, Dave, 50
Cement, 129
Centrifugal force. *See* Rotation rate
Clark, Wilson, 38
Co-evolution, 4, 37, 85
Collins, Mike, 144
Colonies in Space, 129
Communications network, 86-88, 89, 94-95
Conrad, Pete, 91
Cooper, Henry S. F., Jr., 115
Cooper, Thomas S., 90
Cornwall, Timothy C., 42
Cosmic rays, 10, 70, 90, 140-141
Costs and benefits, 15-18, 19, 21, 32, 88, 152
  criticisms of, 54, 62, 64-65, 67, 68
  of travel to Earth, 15, 30
Cousteau, Jacques, 87, 98-103
Cox, Jeff, 89
Criswell, David, 28
Cryogenic magnets, 65-66; *see also* Mass-driver

D

Davis, Hugh, 28
Debate. *See* Responses
Defecation, 116-119
Delay, problems of, 18, 21
Design, 5-6, 12, 14, 17, 20-21, 23, 66, 104-105
  problems of, 62-63, 66-67
  specifications, 20-21
Developing nations, benefits to, 11, 18, 97
Diversity, 28, 42, 69, 108, 130
  criticism of, 54
Drexler, Eric, 25
  on asteroids, 90, 115
  on ecosystems, 90, 107
  on solar sails, 134-137
  on space colonies, 104-108
  on vapor deposition, 32
Durkee, Steve, 39
Dyson, Freeman, 25

E

EARTH/SPACE, 69
Earth/Space News, 129
Eckholm, Erik, 53
Ecological considerations, 19, 43, 83, 90, 92-93, 143
  closed ecosystems, 29, 43, 48-49, 52, 60, 90
  construction materials, 11, 13, 14, 20
  criticisms of, 36-37, 82
  environment, 10, 12, 14, 30, 43
  heat, 39, 53, 62, 64-65
  location of industry, 13, 15, 21
  recycling, 14, 21, 60, 64, 67, 70, 129
  see also Solar energy
Environment, 12, 13, 14, 30, 40, 43, 67, 70
  artificiality of, 40, 63, 64, 67
  problems, 63
  see also Ecosystems, closed, problems
Ecosystems, closed, 14-15, 48-49
  needed research, 21, 29, 93
  problems of, 43, 48-49
  response to, 107
Education for space, 80, 128, 142, 151
Ehrlich, Anne, 43
Ehrlich, Paul, 5, 42, 43
Energy. *See* Solar energy
Energy transmission, 11, 21, 39, 63, 67, 68; *see also* Solar energy
Enterprise, 146, 148; *see also* **Space shuttle**
Escapism, 34, 39, 45
Eschatological concern, 56-60
European Space Agency, 76-77
Evolution, 35, 37, 46, 51, 55, 56-60, 88, 90, 108, 130, 144
Evolution-completion, 149
Evolutionary paths to space, 86-87
Exercise, in space, 78
Extraterrestrial communication, effects of, 31

F

Fallaci, Oriana, 91
Feces, 116-119
Feiman, Jonathan, 61
Feinberg, Gary, 25
Ferlinghetti, Lawrence, 133
Fine, Kevin, 150, 153
Fire danger, 29-30
Fleming, Dean, 36-37
Food supply
  agriculture, 12, 13, 21, 70, 90, 107
    needed research, 29, 43, 71, 92-93, 151
    problems, 42, 43, 49, 52
  dehydrated, 15, 49, 60, 70, 151
Forecast of Space Technology, 128
Fourth Kingdom, The, 144
Frontier
  benefits, 8-11, 19, 51
  negative effects, 36-37, 40, 61
Frost, Robert, 44
Fuller, R. Buckminster, 6, 55

G

Gaia Hypothesis, 120, 123
Garcia, Jose, 40
Gas Chromatograph Mass Spectrometer, 120-121, 122, 125
Gas exchange experiment, 122
Genetic pool, 58-59
Geometries. *See* Design
Girouard, Roger C., 90
Glazer, J. Henry, 6
Goal logic, 71

Gold, B., 90
Graham, John, 44
Gravitational well, 9, 13; *see also* Lift
    costs
Gravity, 12, 13, 14, 15, 45
    moon, 28
    problems, 14, 78
    zero, 71, 78-79, 91, 115; and
        elimination, 116-119; for industry,
        12, 15, 53
Grey, Jerry, 130
Gropius, Walter, 45

H

Hannah, Eric, 25
Hardin, Garrett, 54-55
Hazelrigg, George, 25
Heat
    for industry, 13
    needed research, 29
    waste, 13, 53, 65; problems of, 39,
        53, 62, 64-65
Heliogyro solar sail, 137
Henderson, Hazel, 41-42
Henson, Carolyn Meinel, 95
Heppenheimer, T. A., 32, 63-68, 94-95
High Frontier, The, 31
Himoe, Albert, 69
History of concepts, 22-27, 30, 104-105,
    150
Hoffman, M. A. W. D., 52
Holbrow, Charles, 130
Holladay, Martin, 42
Holmboe, Graham, 97
Holt, John, 62-68, 96
Holton, Gerald, 42
Horowitz, Norman, 122, 123
Horvitz, Robert, 149
House in Space, A, 115
Hudson Institute, 32
Hughes, Barry, 38
Hydrogen, 14, 20, 28, 52

I

If the Sun Dies, 91
Industry. *See* Manufacturing
International Comm. for a New Planet,
    46-47
Iron, 14
Island One, 5, 9-11, 20, 28, 105
    agriculture, 12, 14, 19, 21, 60, 70, 90,
        107, 130
    atmosphere, 12, 14, 15, 21, 49, 70
    construction, 10, 13-14, 20, 66-67
    construction time, 10, 17
    costs and payoffs, 15-18, 19, 21, 53,
        88; criticisms of, 54, 62, 64-65,
        67, 68
    design, 5, 14, 17, 20-21, 23, 66,
        104-105;
        problems of, 62   63
    environment, 12, 13, 14, 30, 40, 43,
        63, 64, 67, 70
    limitations, 93
    manufacturing, 10-11, 12-13, 15,
        20-21, 28, 53, 90
    productivity, 10-11, 17, 65, 66-67, 70
    rotation rate, 14, 15, 21, 23, 30, 62-64
    size, 14, 21, 22, 66-67
    social structure, 10, 11, 19, 45, 54,
        68, 70, 93, 107, 143, 151
    see also Benefits; Responses; Solar
        energy

J

James, Peter N., 69
JBIS, 130
Jennings, Lois, 85
Johnson, Richard D., 130

K

Kaplan, David, 150
Kelly, Walt, 38
Kelp farming, 102
Kesey, Ken, 34
Kolm, Henry, 153, 154

L

L5, 10, 15, 20, 63, 68
L-5 News, 32
L-5 Society, 32
Labeled release experiment, 122-123
Lagrange points, 10, 15, 20, 32, 63, 68
    defined, 32
Landsat, 100
Le Brun, Marc, 53
le Corbusier, 45
Leary, Timothy, 96
Letters. *See* Responses
Levin, Gilbert, 122-123
Libration points, 10, 15, 20, 32, 63, 68
    defined, 32
Lift costs, 11, 16, 17, 64; *see also* Space
    Industrialization
Lift vehicles, 13-14, 19, 20, 21
Limitations in colonies, 50, 97
Limits to growth, arguments against, 8,
    9, 10-11, 13, 50, 97; *see also* Solar
    energy
Linear electric motor. *See* Mass-driver
Logic chains, 71
Lord, Chip, 47
Lovelock, James, 92-93, 120, 123
Low, George, 99, 100-103
Lubrication, 78
Lugo, Ariel, 49
Lunar materials. *See* Moon

M

McClure, Jane, interviews: O'Neill,
    150-154; Schweickart, 138-145
McClure, Michael, 146-148
Magazines on space, 27, 32, 129, 130
Magnesium, 14
Manufacturing, 10, 11, 13, 20, 28, 90,
    129
    gravity benefits, 12, 13, 15, 53
    of high orbit products, 28, 86-87, 89
    productivity, 10-11, 17, 65, 66 67, 70
    of satellite power stations, 11, 16, 17,
        18
    shipbuilding, 28, 32
    of space colonies, 10, 13, 17
    work areas, 12, 15, 21
    work force, 17-18, 19, 67
    see also Metallurgy
Margalef, Ramon, 92-93
Margulis, Lynn, 35, 92-93
    on Mars, 120-124
Mariculture, 98-100, 102
Marlin, Alice Tepper. *See* Henderson,
    Hazel
Mars, 120-127
Maruyama, Magoroh, 130
Mass-driver, 10, 13, 14, 20, 28, 107,
    153-154
    criticisms of, 62, 65-66; response to,
        106
    illustrations of, 16, 147, 153
Mead, Margaret, 85
Meadows, Dennis, 40
Metallurgy, 14, 20, 29, 53, 64-67
    aluminum, 65, 66, 129
    asteroid facility, 135
    problems, 62, 64-65
Microorganisms, 43, 60
Microwave power transmission, 11, 21,
    39, 53, 63, 67, 68
Military uses, 96, 144
Mills, Stephanie, 38

Mining. *See* Asteroids; Metallurgy;
    Moon
Mink, Doug and Missy, 52
Minorities, 54, 129, 140-141
Mirrors, 13, 40, 62-64
Mission specialist, 140-141
Model One. *See* Island one
Model III space colony, 34
Momentum wheels, 78
Moon
    construction materials, 10, 13-14, 20,
        28, 29, 64
    soil, 14, 20, 60, 70, 129
    station, 13-14, 66
Moore, Patrick, 131
Moral objectons, 36-37, 44-45, 68,
    82-84
Morrison, Philip, 91
Motion sickness, 79, 102
Muggs, J. F., 53
Mumford, Lewis, 34
Murphy's Law, 45
Murray, Bruce, 147
Mutants, 48, 58-59
Myers, Gerald E., 89
Myers, Jack, 49

N

NASA/Ames-Stanford Summer Study
    on Space Colonization, 19, 26, 151
National Space Institute, 32
National Space Institute Newsletter, 32
Neoclassical economics, 41
New Alchemy Institute, 48-49
New Planet, International Comm. for a,
    46-47
Newman, Jonathan, 150
Newsletter (O'Neill's), 27
Nichols, Nichelle, 129
Nitrogen, 14, 15, 20, 49
Nitrogen-fixation process, 49, 52
Nitrogenase, 52
Nommisto, Douglas, 69
Norcia, Anne, 53
Norman, Gurney, 69

O

Ocean farming, 98-100, 102
Odum, Howard, 49
O'Hara, Dee, 91
O'Neill, Gerard, 8, 13, 21, 70, 95, 155
    High Frontier, 31
    interviews with, 22-30, 150-154
    on needed research, 14, 18, 29-30,
        70-71, 150-151
    newsletter, 27
    on space colonies, 9-11, 12-21, 28-29,
        70-71, 151-152; birth of concept of,
        22-27, 104-105
    testimony to Congress, 12-21
O'Neill, Tasha, 28, 154, 155
Ore processing. *See* Metallurgy
Oregon State U. symposium, 49
Oro, Juan, 92-93
Ottewell, Guy, 132
Outlook for Space, 128
Oxygen
    atmosphere, 29, 151
    closed ecosystems, 48-49, 52, 62
    Martian, 122, 127
    Moon, 14, 62, 64, 70, 71
    for rocket fuel, 14, 21, 64, 71
Oyama, Vance, 122
Ozone layer, 101-102

P

Payload specialists, 140-141, 142-143
Phillips, M., 69
Phillips, Michael, 46-47
    interview: O'Neill, 24-30
Phosphorus, 52, 90

Photosynthesis, 52, 123
Photovoltaic cells, 17
Plastics, 129
Ponnamperuma, Phil, 122
Population, effect on, 11, 18, 27, 43, 53
Princeton conferences, 19, 24-26, 28, 130
Privacy. *See* Solitude; Surveillance
Productivity, 10-11, 17, 65, 66-67, 70
Public response, 19, 71; *see also* Responses
Publicity, early, 25-26
Pyrolytic release experiment, 123

Q

Quatrone, Phillip, 95
Questionnaire responses, 33; *see also* Responses
Quicklime, 129

R

Radiation, 10, 70, 90, 140-141
Rain, 63, 67
Rand McNally Concise Atlas of the Universe, 131
Raven, Peter, 4
Raymond, Richard, 35
Read, Paula, 43
Research needed, 14, 18, 29-30, 39, 49, 70-71, 92-93, 151
Responses, 34-73, 83-85, 88-97
  critical, 38, 39, 40, 42, 43, 45, 51, 52, 53, 61, 69, 89, 90
    of biology, 42, 43, 49, 52
    as diversion, 35, 41, 43, 68
    of economics, 35, 38, 40, 41, 54, 62, 64-65, 67, 68
    of energy, 37, 38, 39, 42, 53, 62, 68
    as escapism, 34, 39, 45
    moral objections, 36-37, 44-45, 68, 82-84
    political, 50, 54-55, 82, 85
    of technology, 62-68, 83, 97
    of totalitarian features, 36, 37, 54, 68
    vulnerability, 50, 54
  positive, 34, 36-37, 43, 46, 50, 53, 69, 70-73, 85, 90, 96
    abundant energy, 39, 50, 97
    diversity, 42, 73
    ecological, 43, 97
    evolutionary step, 35, 55, 56-60, 88-89, 90
    frontier, 51
    international cooperation, 39
    see also Benefits
  public, 19, 71
  to questionnaire, 33
Robotics, Laws of, 88
Rocket fuel, 14, 21, 64, 71, 134
  solar, 135-136
Rotation rate, 14, 15, 19, 21, 23, 30
  criticisms of, 62-64
  problems, 78
  solar sail, 137

S

Sagan, Carl, 42, 114-115
  on Mars experiments, 124-127
Sanitation in space, 116-119
Satellite antennas, 86-88, 94-95, 141-142
  ocean monitoring, 98-100
  surveillance, 86-88, 89, 94-95
Satellite solar power stations, 11, 16-18, 21
  criticisms, 39
  see also Solar energy
Satellites, derelict, 129
Sauber, Carl, 144
Sauer, Carl, 85

Schirra, Wally, 91
Schumacher, E. F., 37, 38
Schweickart, Clara, 155
Schweickart, Russell, 34, 81, 92-93, 102, 155
  on astronauts, 138-141, 142
  on ocean monitoring, 98, 99-100
  on orbiting Earth, 110-113, 143-144
  on Skylab, 78-79
  on space shuttle, 75-78, 140-143
  on Skylab, 78-79
  on space shuttle, 75-78, 140-143
  on synthesis with machine, 144-145
  on urinating and defecating in space, 116-119
  on what to study, 80; *see also* Education for space
"Scientific Optimism and Societal Concern," 42
Scrivener, B. Alan, 51
Seasat, 98, 101
Sender, Ramon, 51
Sherrill, Peter, 46
Shettler, John, 152
Shetzline, David, 61
Shipbuilding, 28, 32
Siegler, Paul L., 69
Silicon, 14, 20
Sky and Telescope, 131
Skylab, 14, 78-79, 115, 118
  as derelict, 129
  films, 128
Sleep, 79
Smith, David, 92-93
Smith, Eric Alden, 39
Snyder, Gary, 69, 85
Social structure, 10, 11, 19, 45, 54, 68, 70, 93, 107, 143, 151
  and authority, 143, 151
  and control, 50, 54, 143, 151
  and diversity, 28, 42, 130
Soil
  Martian, 120-127
  Moon, 14, 20, 60, 70
Solar energy, 9, 11, 13, 16, 18, 19, 39, 50, 97, 107
  and heat, 39, 53
  satellite, 11, 16-18, 19, 21, 97, 103, 107, 152
  transmission of, 11, 21, 63, 67, 68
    problems of, 39, 63
  turbogenerator, 13, 14
Solar flares, 102
Solar sails, 53, 134-137
Solberg, Gordon, 30
Soleri, Paolo, 45, 56-60
Solitude, need for, 50
Soviet Conquest from Space, 69
Soviet Union, 69, 73, 76, 152
Space, useful attributes, 106; *see also* Manufacturing; Metallurgy; Solar energy
Space Age Review, 129
Space biology, 142
  needed research, 29, 43, 49, 71, 92-93, 151
Space biology, symposium on, 49
Space Day, 146-148
Space derelicts, 129
Space Industrialization, 32, 88-89, 90, 152; *see also* Manufacturing
Space Institute, National, 32
Space law and treaties, 26, 153
Space Manufacturing Facilities, 130
Space Settlements: A Design Study, 130
Space shuttle, 14, 15, 77, 102, 109, 142, 146, 148
  cost, 15
  crew, 76, 140-141, 142-143
  no. of flights, 103, 142
  uses, 75-76, 77, 80, 96, 140
Space sickness, 79, 102

Space wars, 144
Spaceflight, 130
Spaceflight Revolution: A Sociological Study, 130
Spacelab, 77
Specialists. *See* Payload specialists; *also* Education for space
Spurgeon, Bud, 89
SSPS. *See* Satellite solar power stations
Steel, 65, 115
Steindl-Rast, Brother David, 50
Substitutability, 97
Sulfur, 65
Summers, Claude, 39
Sunflower, 17, 21; *see also* Island One
Surveillance, 86-88, 89, 94-95
Swain, T., 92
Sweeney, Laurence Kent, 132

T

Taylor, Edward, 45
Temperatue, 29, 154; *see also* Heat
Third World. *See* Developing nations
Thompson, William Irwin, 44
Thornton, Artie, 153, 154
Three-D flashes, 30
Time tables for space, 71, 86
  for space colonies, 10, 17-18, 71
Tishler, Del, 28
Todd, John, 48-49, 92-93
Todd, Larry, 40
Todd, Nancy, 92-93
Totalitarian features, 36, 37, 50, 54, 68, 84, 89
Transportation costs, 15, 30
Tsiolkowsky, Konstantin, 30

U

Underdeveloped nations, and energy benefits, 11, 18, 97
Universe, atlas of, 131
Urban effect, 59-60
Urination, 116-119

V

Vacuum, benefits of
  for mass driver, 28
  for metallurgy, 28, 53
  see also Gravity, zero
Vajk, J. Peter, 96, 97
Van Allen belts, 141
Vapor deposition, 32
Venus, 101
Viking missions, 120-127
Von Puttkamer, Jesco, 79-80, 86-89, 106

W

Wald, George, 44-45
Walker, Charles D., 43
Warshall, Peter, 52
  interview about elimination in space, 116-119
Water, 60, 67, 70
  as coolant, 62
  fights, 129
  Martian, 122-123, 125-127
  producing, 14, 70
  purifying, 60
  as rain, 63, 67
  recycling, 64, 70, 93
Weightlessness, 14, 78-79, 91, 115, 128, 138
  and elimination, 116-119
Wilcox, Howard A., 102
Wilderness, 30, 43, 44; *see also* Environment
Women
  astronauts, 118, 140-141, 142-143
  and space colonies, 150-151
Woodwell, George, 42, 92-93 ■

CoEvolution Books

This is the first in a series of books from the editors of The CoEvolution Quarterly. *The next one will be* Soft Tech, *edited by J. Baldwin and myself, available early in 1978 for $5, followed by* Watersheds, *edited by Peter Warshall, available in mid-1978.*

The arrangement with Penguin Books is — we publish, they distribute. They are buying 20,000 copies of Space Colonies *from us at $1.25 apiece ($25,000 total) for distribution in the U.S., Canada, and England. (We are printing an additional 2,000 copies for mail order sale to our subscribers at a discount — subscribers get the $5 book for $4.)*

So where the $5 you pay at the bookstore goes is:

```
$2.40 to the bookstore
 1.35 to Viking-Penguin
 1.25 to us, of which 0.24 is profit
$5.00
```

That 24¢ profit is critical for us since without it and the hoped-for profits from the other CoEvolution Books we would have to go out of business. The reason is that The CoEvolution Quarterly *has a persistent habit — it may be incurable — of losing about $10,000 per issue. It did that when we were selling 20,000 copies per issue a year or so ago and it does it still now that we're printing 55,000 copies.*

The CoEvolution Quarterly is what gives us the material to print a book at this low a cost with this high a density, so it all balances out. Meanwhile we put out some other items such as color postcards ($2/10 cards) and posters ($3) of the back cover of this book and items such as our "World Biogeographical Provinces Map" ($3) to help keep ends meeting.

In the event that an additional 20,000 copies of Space Colonies *is bought by Penguin, then we will make about $7,000 additional ($25,000 from Penguin minus the $18,000 for printing). That would be 35¢/book. Then we go through the Rolodex for all the old contributors in this book that were in* The CQ *originally and arrange some additional payment for them. The editing comes "free" since I'm paid by* The CQ *$15,000/year to do whatever happens in the year.*

As for the 2,000 copies we're offering to subscribers at $4, we make a bit less than $3/copy on those when you include our costs for shipping and handling.

The CoEvolution Quarterly, the Whole Earth Catalog *and* Epilog, *and these CoEvolution Books are all part of a non-profit foundation called POINT. POINT once gave away over a million dollars that came in from sales of the* Whole Earth Catalog *— $600 of that covering the costs of O'Neill's first (1974) space colony conference at Princeton. In the unlikely event that POINT gets a lot of money again, we'll give some away again.*

Meantime, solvency is our goal. Thanks for your help. You're welcome for the book.

— SB

Cover art for the next CoEvolution Book, Soft Tech, *ready Spring 1978 for $5.*

SPACE COLONIES
production and manufacture costs

New contributors	$ 600
Staff	3,694
Supplies	230
Index	500
Printing (22,000 copies)	17,261
total	$ 22,285
Unit cost (22,000 copies)	$ 1.01

'Or Whole Earth'

under the access of an item means you can mail order it from:

> Whole Earth Truck Store
> 558 Santa Cruz Ave.
> Menlo Park, CA 94025

Anything else — letters, articles, orders for **The CoEvolution Quarterly** →should be sent to us here at:

> **The CoEvolution Quarterly**
> **Box 428**
> **Sausalito, CA 94965**

DETAILS ARE VULGAR

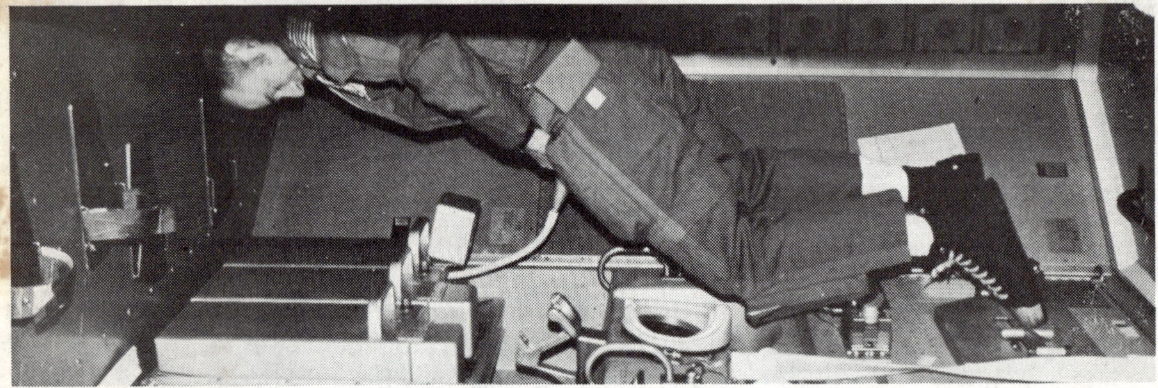

. . . said Oscar Wilde, and a lot of writers today seem to agree. They tell you about moods, feelings, general ideas, trends. But they don't tell you the details you want to know.

They don't, for example, tell you how astronauts use the bathroom in space.

The CoEvolution Quarterly does.

The CoEvolution Quarterly is a 144 page, advertising-free mass of details you can think about and use — and you can get it four times for $8. Much of what's in this book on exactly what space colonies would be like and exactly what life in space has been like so far first appeared in **The CoEvolution Quarterly.** Including the article on urination and defecation in zero g on page 116.

CQ people know it's hard to understand something, to believe it, or to do it yourself without hearing about the small hassles and the small joys that are part of anything real. So past **Quarterly**s have given our readers:

- *Details from people who were there.* Margaret Mead and Gregory Bateson arguing about using cameras in anthropology. Theodora Kroeber, who's 45 years older than her husband, on cross-generational marriage. Malcolm Wells, who has three feet of dirt on his office roof, on underground architecture. Jack Mundey, who invented them, on Australian Green Bans — strikes against projects that hurt the environment.

- *Details on how to do it.* How to buy the best tools. How to begin in Buddhism. How to start a community garden. How to make wine, gamble, buy horse gear, save energy, store food, make Zen pants, build your own computer. All by people who have actually done it.

- *Details on what it really says.* All our book reviews have <u>long</u> quotes from the book itself so you can decide if you're interested. And we have 60-100 book reviews in each **Quarterly.**

The CoEvolution Quarterly wants to help you live in a real world that is entirely made up of details. So you want lots of details that are occasionally vulgar, often useful, and always interesting, send us $8 today.

— Anne Herbert
Assistant Editor

The **CoEVOLUTION** Quarterly

BOX 428, SAUSALITO, CA 94965

The research continues in The CoEvolution Quarterly

$8 for four issues

Whole Earth Catalog – $6

Winner of the 1972 National Book Award, the **LWEC** functions as an evaluation and access device. It contains reviews of books and tools in the original Whole Earth categories: whole systems, community, shelter, craft, nomadics, communications, land use and learning.

Whole Earth Epilog – $4

The **Epilog** is, in effect, Volume II of the **Last Whole Earth Catalog**. It begins with page 449, and contains 320 pages of all new material, cross-referenced and cross-indexed to the **Catalog**.

Mailing List

I'm not ready to order anything yet, but please put me on your **Mailing List** and keep me informed of developments at **Whole Earth.**

Name

Zip

4904

Order Form

☐ $8—one year (4 issues) CoEvolution Quarterly

☐ $16—two years (8 issues) CoEvolution Quarterly

☐ $2 each—CQ back issues:

(see overleaf for availability)

☐ $6—WHOLE EARTH CATALOG

☐ $4—WHOLE EARTH EPILOG

Name

Zip

4902

Please make checks payable to: **The CoEvolution Quarterly Box 428 Sausalito, California 94965**

$ ______
Total this order

All subscriptions start with current issue. For your convenience we've included an envelope, overleaf. Optional foreign <u>airmail</u> rates (add to subscription price): Canada —$7, Mexico & Central America—$6, South America—$9, Europe—$9, all others—$12.

Order Form

☐ $8—one year (4 issues) CoEvolution Quarterly

☐ $16—two years (8 issues) CoEvolution Quarterly

☐ $2 each—CQ back issues:

(see overleaf for availability)

☐ $6—WHOLE EARTH CATALOG

☐ $4—WHOLE EARTH EPILOG

Name

Zip

4901

Please make checks payable to: **The CoEvolution Quarterly Box 428 Sausalito, California 94965**

$ ______
Total this order

All subscriptions start with current issue. For your convenience we've included an envelope, overleaf. Optional foreign <u>airmail</u> rates (add to subscription price): Canada —$7, Mexico & Central America—$6, South America—$9, Europe—$9, all others—$12.

Contents of Past Issues *(Issues not listed are Out of Print)*

2 — **Summer 1974**

Setting Food By *Doris Herrick*
Ehrlich's Guide to the Apocalypse: Food
  Anne H. Ehrlich and Paul R. Ehrlich
Tongue Fu *Paul Krassner*
Sanctity and Adaptation *Roy A. Rappaport*
Natural History Comes to Whole Earth
  Peter Warshall
Land Banking *Huey Johnson*
Gravity Engines and the Diving Engine *Steve Baer*
Sexual Honesty *Shere Hite*
New Games Tournament *Pat Farrington*
Salons and Their Keepers *Stephanie Mills*
Notes on Provisioning a Small Boat for Extended
  Cruising *Kathleen Pumphrey*
Apple Picking *Rick Fields*
Bookmaking Access *Robin Rycraft*
Making and Playing the Shakuhachi
  Monty H. Levenson
Gorf *Michael McClure*
The POINT/Penguin Books Publisher-Distributor
  Contract for the Whole Earth Epilog
  Lawrence Klein

10 — **Summer 1976**

Olbers' Paradox *Lawrence Ferlinghetti*
Jacques Cousteau at NASA Headquarters
Free Lunch Comix and Stories *Dan O'Neill*
On Observing Natural Systems *Francisco Varela in
  conversation with Donna Johnson*
Carl Ortwin Sauer 1889-1975 *James J. Parsons*
Theme of Plant and Animal Destruction in Economic
  History *Carl Sauer*
A Carl Sauer Checklist *Robert Callahan*
Trees *Jean Giono*
The Hoedads *J.D. Smith*
Environmental Impact Reports *Peter Warshall*
The B & G, The Bomb *Steve Baer*
Three Political Ideas *Paul Maag*
Theory of Game-Change *Stewart Brand*
Some Mice *Robert Horvitz*
Get Down Tonight *Deborah Haynes*
Sex Notes *Kay Weiss*
The Space Crone *Ursula Le Guin*
Memory One: Getting Started in Orygun *David Shetzline*
Winning *Ron Jones*
The Education of Joni Mitchell
Its our very effort to make reality repeatable which
  Buddhism calls suffering *Zentatsu Baker-Roshi*

11 — **Fall 1976**

From Present to Future *Herman Kahn with William Brown
  and Leon Martel*
The Shift from a Market Economy to a Household Economy
  Scott Burns
Future Primitive *Raymond Dasmann*
Biogeographical Provinces *Raymond Dasmann*
Controversy Is Rife on Mars *interviewing Carl Sagan
  and Lynn Margulis*
Odd Bodkins *Dan O'Neill*
Boone *Gurney Norman*
Mind/Body Dualism Conference Position Papers
  Invitational Paper *Gregory Bateson*
  Inside & Outside *Richard Baker-roshi*
  Position Paper *Ramon Margalef*
  Not One, Not Two *Francisco Varela*
Community Gardening *Rosemary Menninger*
The Foredads *Gary Rurkun*
Effects of Sublethal Exposure to 2,4,5-Trichlorophenoxy-
  Acetic Acid *J.R. Felix and Pierre Mortmain Sylvestre*
Underground Architecture *Malcolm Wells*
Cross-Generation Marriage *Theodora Kroeber-Quinn*
Earthshoes & Other Remarks *Ken Kesey*
May 12 *Louise Nussbaum*
A Street Man's Answer to Rape: Humiliate the Raper
  Willie "Small Banana" Williams
Animal Stories *J.D. Smith*
The American Anti-Whaling Movement Is Racist
  Michael Phillips
Lucy's Blueprints *Lucy Burrows Morley and
  Christopher Thompson*

12 — **Winter 1976**

Streaming Wisdom *Peter Warshall*
Watershed Form and Process: The Elegant Balance
  Robert R. Curry
John Wesley Powell: Conscience of the Colorado
  Peter Wild
Watershed Vignettes *Thomas Dunne and
  Luna B. Leopold*
Another Watershed Vignette: Lake Tahoe, California
  Peter Warshall and J.T. and Carin Winneberger
Diagnosing Stream Health *Thomas Dunne and
  Luna B. Leopold*
El Rancho Los Pesares *Jaime de Angulo*
Forests and the Purposes of Man *Roy A. Rappaport*
Headwaters: A Short History of a Backwoods Battle
  Randal Lee O'Toole
Whose Trees? Whose Science? *Jack Westoby*
Christo's Running Fence *Mike Alexander,
  Barbara and Paul Kayfetz, Peter Warshall*
Healing the Land: The Redwood Creek Renewal
  Project *Gerald Myers*
That Which Makes You Move: A Conversation
  With Taj Mahal *Lewis MacAdams*
Interspecies Music *Jim Nollman*
Access to Watershed Music *Peter Warshall*
From Urgencies to Essentials *speech by Canadian
  Prime Minister Pierre Elliott Trudeau*
The New Alchemists *J. Baldwin*
The Acorn People *Ron Jones*
I would like to write a silence *Anne Herbert*
Odd Bodkins *Dan O'Neill*

14 — **Summer 1977**

Voluntary Simplicity (3) *Duane Elgin and Arnold Mitchell*
Voluntary Simplicity (1) *Richard Gregg*
SRI Is Wrong About Voluntary Simplicity *Michael Phillips*
The Poverty of Power Reviewed *Kenneth Boulding*
Remarks on Recombinant DNA *James Watson*
There Ain't No Graceful Way *Astronaut Russell Schweickart
  talking to Peter Warshall*
Mushroom Hunting in Oregon *Andrew Weil*
Time Landscape *Alan Sonfist*
The Transformation of the Tract Home *Richard Nilsen*
Burning Wood *J. Baldwin*
The Ultimate City — Part 1 *J.G. Ballard*
Handspinning *Diana Sloat*
The Price of Marijuana Misinformation *Russell Falck*
Death Does Not Exist *Dr. Elisabeth Kubler-Ross*
Two by Rabindranath Tagore
Three by Anne Herbert
The Death of Ivan Ilych *Leo Tolstoy*
Hospice in America *Lynette Jordan*
Goethe in the Pea Patch *J.D. Smith*
Turkey Drop *Terrence Williams*
Dan O'Neill's Comics and Stories
Left Over in Your Heart *Will Davis*
Computer Hobbyist Publications *Marc Le Brun*
The Poem Is That Voice in Between *Acoma Poet Simon
  Ortiz talking with Lewis MacAdams*

15 — **Fall 1977**

Solar Water Heaters in California, 1891 - 1930
  Ken Butti and John Perlin
The Wandervogel *John de Graaf*
The Ideas of Greenwich Village *Malcolm Cowley*
Sanctuary in Cuba *Huey P. Newton*
Havana Province, 1977 *Conn Nugent*
Valuable Deficiencies *John McKnight*
Neighborhood Preservation is an Ecology Issue
  Stewart Brand
Living Lease *Michael Phillips*
Notes for Things to Do in the Future *Donald Burgy*
Space Day Symposium *R. Crumb*
Voluntary Simplicity Followup and Comments: Response
  to the Voluntary Simplicity Questionnaire
Who Cut Down the Sacred Tree? *Michael W. Foley*
The Hawksbill Turtle *Kevin Stevenson*
Odd Bodkins *Dan O'Neill*
New Crops *Rosemary Menninger Talking with
  Richard Felger*
Insulation Heresy *Richard Nilsen*
The Ultimate City - Part 2 *J. G. Ballard*
Wilderness Plots *Scott Sanders*
Two Dangling Animal Stories and a Heavy Bear Story
  J. D. Smith
The Custer Wolf *Patrick Holland*
Broken Circuits, Smoke and Fire *Judy Melvoin*
If the Spirits of the Dead Do Walk Among Us, Where Do
  They Spend Their Summer Vacations?
  Robert Goldman
Giraffe's Neck *Terry Lawhead*